The
Mainstream
of Physics

Department of Physics — New York University

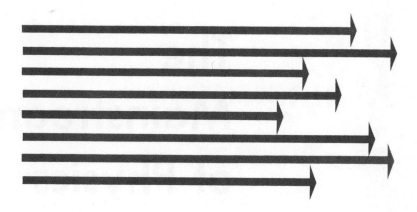

ARTHUR BEISER

The
Mainstream
of Physics

Addison-Wesley Publishing Company
Reading, Massachusetts
Menlo Park, California · London · Amsterdam · Don Mills, Ontario · Sydney

Preface

The Mainstream of Physics is an elementary introduction to the funda-
mentals of physics. As in the case of its shorter version, *Basic Concepts of
Physics*, no more than the simplest mathematics is used, a review of which
is given in an appendix, and no prior knowledge of physics or chemistry
is assumed. The progression of topics is straightforward, from the laws
of motion through classical and modern physics to a survey of elementary
particles. The treatment concentrates on those physical principles primary
to our understanding of the natural world; to make such emphasis possible,
I have omitted peripheral material and avoided a primarily historical approach, since whatever its merits the latter would have meant delving into
superseded ideas at the expense of still valid ones. Wherever possible I
have tried to present derivations of important results, thereby exhibiting,
rather than merely describing, the inductive and deductive methods of
science and encouraging in the reader a feeling of participation that I
believe essential for learning. Because modern physics has provided the
key insights that underlie our understanding of the nature and properties
of matter, approximately a third of the book is devoted specifically to it,
and references are made elsewhere to various atomic phenomena.

The problems included in each chapter serve the two ends of helping to
develop skill in analytical thought and numerical calculation, both of
which are required for the successful study of physics, and promoting the
mastery of the text proper by actually putting its ideas into practice.
The exhortation in James, i, 22, is apt: "But be ye doers of the word, and
not hearers only, deceiving your own selves." Answers to the odd-
numbered problems are given at the end of the book. A separate study
guide for *The Mainstream of Physics* has been prepared to provide supplementary assistance for those needing it.

Physics, like any other science, involves the active pursuit of knowledge,
and it contains many elements besides its basic concepts. However, while
the study of physics cannot realistically exclude its philosophical, historical,
and cultural aspects, I do not believe that a discussion of these aspects
can make much sense to a reader who does not first know just what it is
that they concern. My experience is that a text that tries to cover all the
facets of physics simultaneously is more confusing than enlightening. For
this reason I have limited *The Mainstream of Physics* largely to the formal

content of its subject. However, I also feel that its use as a course text should be accompanied by appropriate laboratory work and supplementary reading. Examples of books suitable for the latter purpose are *The Evolution of Physics*, by Einstein and Infeld (Simon and Schuster); *Foundations of Modern Physical Science*, by Holton and Roller (Addison-Wesley); and my own *The World of Physics* (McGraw-Hill); there are many others.

October, 1961   A. B.

Contents

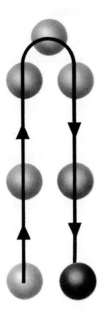

Describing Motion | 1

So much of physics is concerned with things in motion that it is appropriate to begin our study of this subject with an analysis of the simplest kinds of motion normally encountered. Besides obtaining useful and interesting information, we shall be introduced to the ways the physicist thinks, how he obtains general relationships from specific data, and how he goes about applying these relationships to other situations.

1–1 Exponential notation and significant figures

Physics is the most quantitative of the sciences, and we must become accustomed to its insistence upon accurate measurements and precise relationships if we are to appreciate its results. Since numbers are of paramount importance in physics, our very first step will be to examine how the physicist expresses them and just what they mean and do not mean.

Exponential notation is a convenient and widely used method for abbreviating large and small numbers. The method is based upon the fact that all numbers (in decimal form) may be expressed as a number between 1 and 10 multiplied by a power of 10. To see this we must first construct a table of powers of 10, such as Table 1–1. Now let us look at a few examples:

$$600 = 6 \times 100 = 6 \times 10^2,$$
$$7940 = 7.94 \times 1000 = 7.94 \times 10^3,$$
$$0.023 = 2.3 \times 0.01 = 2.3 \times 10^{-2},$$
$$93,000,000 = 9.3 \times 10,000,000 = 9.3 \times 10^7.$$

This method of writing large and small numbers has two great advantages. The first is that it makes numerical calculations less cumbersome and less prone to errors in many cases. An example will illustrate this:

$$\frac{3800 \times 0.0054 \times 0.000,001}{430,000,000 \times 73} = \frac{3.8 \times 10^3 \times 5.4 \times 10^{-3} \times 10^{-6}}{4.3 \times 10^8 \times 7.3 \times 10^1}$$

$$= \frac{3.8 \times 5.4}{4.3 \times 7.3} \times \frac{10^3 \times 10^{-3} \times 10^{-6}}{10^8 \times 10^1}$$

$$= 0.65 \times 10^{(3-3-6-8-1)}$$

$$= 0.65 \times 10^{-15}$$

$$= 6.5 \times 10^{-16}.$$

The second advantage of exponential notation is that it gives no false impressions of the degree of accuracy with which a number is known. Suppose we write the average distance d from the earth to the moon as $d = 236,000$ miles. Does this mean that d has been measured so precisely that we know it to be exactly 236,000 mi? Or even that it lies between 235,999 and 236,001 mi? Of course not. The last three zeros merely indicate the correct location of the decimal point. If we write d as

$$d = 2.36 \times 10^5 \, \text{mi},$$

TABLE 1-1

Powers of 10 from 10^{-10} to 10^{10}

10^{-10} = 0.000,000,000,1	10^{0} = 1
10^{-9} = 0.000,000,001	10^{1} = 10
10^{-8} = 0.000,000,01	10^{2} = 100
10^{-7} = 0.000,000,1	10^{3} = 1000
10^{-6} = 0.000,001	10^{4} = 10,000
10^{-5} = 0.000,01	10^{5} = 100,000
10^{-4} = 0.000,1	10^{6} = 1,000,000
10^{-3} = 0.001	10^{7} = 10,000,000
10^{-2} = 0.01	10^{8} = 100,000,000
10^{-1} = 0.1	10^{9} = 1,000,000,000
	10^{10} = 10,000,000,000

then how large the number is and how accurately we know it are both clear. The accurately known digits are called *significant figures;* in the above case d has three significant figures, **236**. Sometimes one or more zeros in a number are significant figures, and it is proper to retain them when expressing the number in exponential notation:

$$a = 3.40 \times 10^{10} \text{ ft},$$

$$b = 7.000 \times 10^{-4} \text{ sec}.$$

The *theory of error*, which treats questions regarding the accuracy of experimental measurements, prescribes ways for determining the proper number of significant figures in a specific quantity and for combining quantities with the same and different numbers of significant figures. We shall not go into this aspect of physics here, but a rather obvious rule should be kept in mind: when we combine quantities arithmetically, the result has no more significant figures than the least accurately known quantity. Suppose that a man weighing 162 lb climbs into a truck weighing 15 tons (*not* 15.000 . . . tons). The combined weight of the man and the truck must still be considered 15 tons, because all we know of the original truck weight is that it is somewhere between 14.5 and 15.5 tons, which represents an uncertainty of 2000 lb. Similarly, if we divide 1.4×10^5 by 6.70×10^3, we are not justified in writing

$$\frac{1.4 \times 10^5}{6.70 \times 10^3} = 20.89552 \dots .$$

We may properly retain only two significant figures, corresponding to the two significant figures in the least accurately known number (1.4×10^5), and so the correct answer is just **21**.

FIGURE 1-1

1-2 Constant speed

When something has changed its position with respect to its surroundings, we say that it has moved. First we shall look into situations in which a body is restricted to motion along a straight line. A good example is an automobile on a straight road. Let us imagine that we have marked off 100-ft intervals with chalk on such a road (Fig. 1–1) and are now standing next to it with a watch, pencil, and paper in hand. Along comes a car, and we note the exact time at which the front of the car passes each of the marks we have made on the road, as in the drawing. Our measurements might look like those shown in Table 1–2.

To analyze these measurements a sheet of graph paper is helpful. Along the bottom of the sheet we establish a time scale, perhaps letting each division represent 1 sec. On the left-hand side we establish a distance scale, perhaps letting each small division represent 50 ft. Next we plot each of our measurements as a point on the graph, as in Fig. 1–2, where the point is the intersection of the lines on the graph paper that pass through the time and distance measurements on the two scales. The result is shown in Fig. 1–3. Clearly, the various points form a regular pattern. In fact, if we draw lines that join adjacent points, we see that this pattern is almost a straight line. We may reasonably attribute deviations from perfect straightness to experimental errors. We have no way of knowing from the graph itself, of course, whether additional measurements between those we did take would also fall on the same line, but let us suppose that if we had made such additional measurements, they too would conform to this line.

When a graph of one quantity versus another results in a straight line, each quantity is *directly proportional* to the other. We can verify this proportionality from the graph of Fig. 1–3, where we see that doubling t means that d also doubles, tripling t means that d also triples, and so on. Thus we can write

$$d = vt,$$

(1–1)

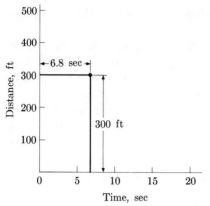

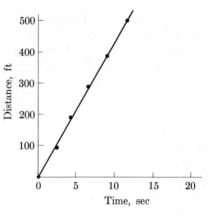

FIG. 1-2. The method by which the data of Table 1-2 are plotted as a graph. The point representing 300 ft and 6.8 sec is shown.

FIG. 1-3. A graph of the data of Table 1-2. Each point represents the results of a measurement, and the straight line indicates a nearly exact direct proportionality between distance and time. Note that the line does not go through each point.

TABLE 1-2

Measurements made with the arrangement in Fig. 1-1

Total distance, ft	Elapsed time, sec
0	0
100	2.3
200	4.5
300	6.8
400	9.1
500	11.4

TABLE 1-3

Speeds calculated from the data of Table 1-2

Total distance, ft	Elapsed time, sec	Speed ft/sec
0	0	
100	2.3	43.48
200	4.5	44.44
300	6.8	44.12
400	9.1	43.96
500	11.4	43.86

where v is the constant of proportionality. To determine the value of v, we rewrite Eq. (1-1) as

$$v = d/t \qquad (1-2)$$

and calculate d/t for each of the measurements that we made. The results are shown in Table 1-3 to four figures. However, only two are significant, and when the values of v are rounded off accordingly, we see that $v = 44$ ft/sec. This is not surprising, since the straight line on the graph indicates a direct proportionality.

The quantity v is called the *speed* of the moving car. It is the rate at which the distance covered by the car changes with time. Because here the distance the car travels is directly proportional to the elapsed time, v is a constant and the car is said to move with a *constant speed*.

The advantage of knowing that the speed of a particular car (or other body) is constant is that we can predict exactly how far it will go in a given amount of time; or, given how far it has gone, we can determine the amount of time that was required. For example, assuming that the present car continues moving at 44 ft/sec, we can compute how long it will require to travel one mile. From Eq. (1–1),

$$t = d/v \tag{1–3}$$

and, since (see Appendix III)

$$1 \text{ mi} = 5280 \text{ ft,}$$

we have

$$t = \frac{5280 \text{ ft}}{44 \text{ ft/sec}} = 120 \text{ sec.}$$

The car takes two minutes to travel one mile.

Or we might want to know how far the car will travel in one hour at this speed. We solve the problem by writing

$$d = vt = 44 \frac{\text{ft}}{\text{sec}} \times 60 \frac{\text{sec}}{\text{min}} \times 60 \frac{\text{min}}{\text{hr}} = 158{,}400 \text{ ft.}$$

Because there are 5280 ft/mi, this distance is equal to

$$d = \frac{158{,}400 \text{ ft}}{5280 \text{ ft/mi}} = 30 \text{ mi.}$$

The car travels 30 miles in an hour. (Note that the units are treated as though they are algebraic quantities, and are carried along in the calculation. We shall do this throughout the book, and it is a good habit to follow.)

1–3 Instantaneous speed

Not all automobiles that pass us on the road have constant speeds. Another car, which began moving from rest at $t = 0$, might yield the data shown in Table 1–4. When we plot these data on a graph, as in Fig. 1–4, we find that the line joining the various experimental points is not straight but shows a definite upward curve. In each successive interval of time (as marked off at the bottom of the graph) the car goes a greater distance; in other words, the car is going faster and faster. Because d is not directly proportional to t, the car's speed is not constant.

Even though v varies, at every instant it has a definite value. To find the *instantaneous speed* of the car at a particular time t, we draw a straight-line segment on the graph tangent to the v-t curve at that point. Then, from the graph, we determine v as shown in Fig. 1–4. Since the curve changes its slope (that is, the angle between a particular segment of the curve and a horizontal line) continuously, v is different at different times. Table 1–5 shows the instantaneous speeds of the car calculated for the times shown.

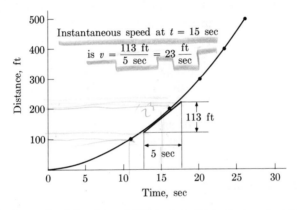

Fig. 1–4. A graph of the data of Table 1–4. Note that a straight line cannot be drawn that lies close to all the points. The procedure for determining the instantaneous speed of the car is shown for $t = 15$ sec.

TABLE 1–4

Measurements made with the arrangement in Fig 1–1 for the car which started from rest at $t = 0$

Total distance, ft	Elapsed time, sec
0	0
100	11
200	16
300	20
400	23
500	26

TABLE 1–5

Instantaneous speeds at 5-sec intervals as determined from Fig. 1–4

Elapsed time, sec	Instantaneous speed, ft/sec
0	0
5	7.5
10	15
15	23
20	30
25	38
30	45

1–4 Average speed

In some situations we are interested in the instantaneous speed of a moving object, but in others our chief concern is with its *average speed*. The average speed, whose symbol is $\bar{v}$, of a body moving along a straight path during a period of time t is the total distance d it has traveled in that period of time divided by t:

$$\bar{v} = \frac{d}{t}. \tag{1–4}$$

We note that this is the same as the definition of v for a body moving with constant speed, since in that case v and $\bar{v}$ are the same. For a body not moving with uniform speed, however, the instantaneous speed v is, in general, different from the average speed $\bar{v}$.

Applying Eq. (1–4) to the data for the second car, we substitute $d = 500$ ft and $t = 26$ sec and find that its average speed in the entire 500-ft distance was

$$\bar{v} = \frac{d}{t} = \frac{500 \text{ ft}}{26 \text{ sec}} = 19 \, \frac{\text{ft}}{\text{sec}}.$$

1–5 Speed and velocity

The speed of a body is not, by itself, a complete description of its state of motion at any time. The *direction* in which the body is moving is also important. The path of an airplane going east at 500 mi/hr is certainly very different from that of an airplane going west at 500 mi/hr. Or we might consider two boats: a boat traveling at 20 mi/hr in a constant direction will have a very different path from one traveling in a circle at 20 mi/hr. To specify motion completely, then, we must include the direction as well as the speed. Such a specification is called *velocity*. For instance, the speed of a boat might be 20 mi/hr, while its velocity is 20 mi/hr to the north. This might seem a minor distinction, but it is not. Given direction and speed, we can readily calculate how far a body will travel in a given length of time. However, without knowing its direction we cannot determine the path of the body at all.

The instantaneous velocity of some object may change in direction as well as in magnitude, but the rule for finding average velocity is essentially the same as that for finding average speed. We compute the average speed from Eq. (1–4) and determine the direction from the starting point to the terminal point. Regardless of the vagaries of the object, this procedure will yield its average velocity (Fig. 1–5).

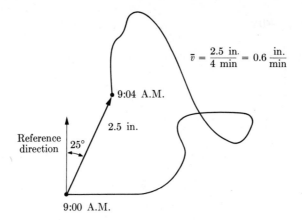

$$\bar{v} = \frac{2.5 \text{ in.}}{4 \text{ min}} = 0.6 \frac{\text{in.}}{\text{min}}$$

FIG. 1–5. The path of a caterpillar on a sheet of graph paper. The average velocity of the caterpillar is 0.6 in/min at an angle of 25° from the reference direction.

The terms *speed* and *velocity* are often used interchangeably, since the direction in which something is moving may be understood by implication. However, it is good practice to reserve "speed" for the rate at which something covers distance and "velocity" for the complete picture of its motion in space. The boat traveling in a circle has a constant speed of 20 mi/hr, but its instantaneous and average velocities are both continually changing, with the latter equaling zero periodically. Each of these terms, then, has a different meaning and each has its own area of usefulness in describing motion.

Physical quantities that require both direction and magnitude to be completely specified are called *vector quantities*. Those that are completely specified by a magnitude alone are called *scalar quantities*. Thus velocity is a vector quantity and speed a scalar quantity.

1–6 Acceleration

A body whose velocity is not constant is said to be *accelerated*. This term is applied regardless of whether v is increasing, decreasing, or changing in direction. Since we are restricting ourselves to motion in a straight line in this chapter, we shall not consider here the acceleration arising from a change in direction; this interesting topic will be discussed in Chapter 4.

Just as speed is the rate of change of distance with time, acceleration is the rate of change of speed with time in straight-line motion. (We shall continue to refer to "speed" rather than to "velocity" here because they

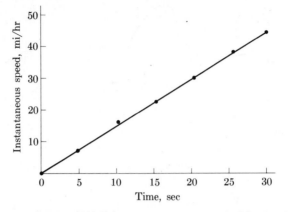

FIG. 1–6. A graph of instantaneous speed versus time for the data of Table 1–5. Although the speed is not constant, it varies in a uniform way with time. This is an example of constant acceleration.

have the same meaning in one-dimensional motion.) For the second car in our example (pages 6–7), let us plot the instantaneous speed v versus time t on a graph, as in Fig. 1–6. Here the points all lie along a straight line, which implies that v is directly proportional to t. Therefore we can write

$$v = at, \qquad (1\text{–}5)$$

where a is a constant of proportionality, called the *acceleration* of the car. Not all accelerations are constant, of course, but a great many real motions are best understood by idealizing them in terms of constant accelerations, and we shall accordingly restrict ourselves to the latter.

The unit of acceleration is, from its definition, expressed in terms of

$$\frac{\text{Speed}}{\text{Time}} = \frac{\text{distance/time}}{\text{time}} = \frac{\text{distance}}{\text{time}^2}.$$

A body whose speed increases by 10 ft/sec each second would thus have its acceleration expressed as

$$10 \, \frac{\text{ft/sec}}{\text{sec}} = 10 \, \frac{\text{ft}}{\text{sec}^2}. \qquad (1\text{–}6)$$

In order to calculate the acceleration a in general we rewrite Eq. (1–5) as

$$a = v/t,$$

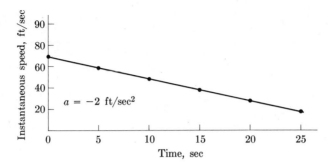

FIG. 1–7. Instantaneous speed versus time for car No. 3.

where v is the change in speed which has occurred during a time interval t. When we apply this formula to the data of Table 1–5 for the car that started from rest, we obtain the result that

$$a = 1.5 \frac{\text{ft}}{\text{sec}^2}$$

in each interval.

Let us return to the road we have calibrated and make measurements of several more automobiles. The results might yield graphs like those in Figs. 1–7, 1–8, and 1–9. Car No. 3 (Fig. 1–7) is slowing down as it passes us; v decreases with t, and we calculate that

$$a = -2 \frac{\text{ft}}{\text{sec}^2}.$$

The minus sign indicates that the acceleration is negative, which is another way of saying that v is decreasing.

The data for car No. 4 in Fig. 1–8 are more complicated, but it is not hard to make out their meaning. A horizontal line in a v versus t graph

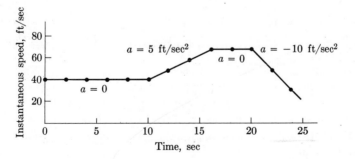

FIG. 1–8. Instantaneous speed versus time for car No. 4.

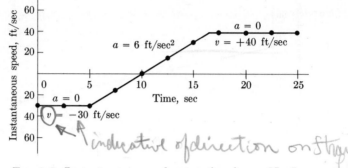

FIG. 1–9. Instantaneous speed versus time for car No. 5.

means an acceleration of zero; a line sloping upward means a positive acceleration; and a line sloping downward means a negative acceleration. Hence car No. 4 at first moves with the constant speed of 40 ft/sec, then is accelerated at 5 ft/sec² to the speed of 70 ft/sec, again travels at constant speed for 4 sec, and finally undergoes a negative acceleration of 10 ft/sec².

At first glance the graph for car No. 5 (Fig. 1–9) appears puzzling: the velocity is negative at first and then becomes positive. The answer is, of course, that car No. 5 began by moving in the opposite direction from the first four cars, from left to right in Fig. 1–1 rather than from right to left. Even when motion is restricted to a straight line, there are two directions possible. In every case we must assign one of these directions as positive and the other as negative so that we can represent the situation correctly in our equation. The car whose motion is graphed in Fig. 1–9 appeared going from left to right, slowed down to a stop after 10 sec, and then began going in the reverse direction.

1–7 Acceleration, velocity, distance, and time

The four quantities a, v, d, and t we have been discussing are not independent of one another. Let us look at a situation in which we have an automobile starting out at $t = 0$ from a point we designate as $d = 0$, with the initial speed v_0 and the constant acceleration a. What will the car's speed be at some later time, and how far will it have gone? We shall assume that the car is moving along a straight road.

The first question is the easier. After a time t the car's speed will be the sum of its original speed v_0 plus the additional speed at (see Eq. 1–5) which it has gained. Thus

$$v = v_0 + at. \tag{1-7}$$

This equation can be used even when a is negative, implying that the car is slowing down. Suppose we have a car whose initial speed is 88 ft/sec and whose acceleration is -4 ft/sec^2. The speed after 10 sec is

$$v = v_0 + at = 88 \, \frac{\text{ft}}{\text{sec}} - 4 \, \frac{\text{ft}}{\text{sec}^2} \times 10 \text{ sec}$$

$$= (88 - 40) \, \frac{\text{ft}}{\text{sec}} = 48 \, \frac{\text{ft}}{\text{sec}},$$

which is less than it was initially. After 50 sec the speed of the car, if a stays the same, is

$$v = v_0 + at = 88 \, \frac{\text{ft}}{\text{sec}} - 4 \, \frac{\text{ft}}{\text{sec}^2} \times 50 \text{ sec}$$

$$= (88 - 200) \, \frac{\text{ft}}{\text{sec}} = -112 \, \frac{\text{ft}}{\text{sec}},$$

which is in the opposite direction and greater in magnitude than the original speed v_0.

Let us now answer the second question. How far will an object of initial speed v_0 and constant acceleration a go in a time t? We know from Eq. (1–4) that

$$d = \bar{v}t,$$

so that if we can determine the average velocity $\bar{v}$ during the time interval t we can find d. Since the acceleration is constant, v is increasing at a uniform rate, and

$$\bar{v} = \frac{\text{initial velocity} + \text{final velocity}}{2}.$$

Here the initial velocity is v_0 and the final velocity is $v_0 + at$; hence

$$\bar{v} = \frac{v_0 + v_0 + at}{2} = v_0 + \tfrac{1}{2}at.$$

The distance traveled is therefore

$$d = \bar{v}t = v_0 t + \tfrac{1}{2}at^2, \tag{1–8}$$

a formula which is often useful. When the initial speed $v_0 = 0$, Eq. (1–8) reduces to

$$d = \tfrac{1}{2}at^2. \tag{1–9}$$

In the first 10 sec the car of the present example went

$$d = v_0 t + \tfrac{1}{2}at^2$$
$$= 88 \frac{ft}{sec} \times 10 \; sec - \frac{1}{2} \times 4 \frac{ft}{sec^2} \times (10 \; sec)^2$$
$$= (880 - 200) \; ft$$
$$= 680 \; ft,$$

and, after 50 sec,

$$d = v_0 t + \tfrac{1}{2}at^2$$
$$= 88 \frac{ft}{sec} \times 50 \; sec - \frac{1}{2} \times 4 \frac{ft}{sec^2} \times (50 \; sec)^2$$
$$= (4400 - 5000) \; ft$$
$$= -600 \; ft.$$

The latter result means that the car is 600 ft *behind* its starting point after 50 sec have elapsed.

We can obtain a relationship between d, v, and a without involving t directly. Equation (1–7),

$$v = v_0 + at,$$

may be rewritten as

$$t = \frac{v - v_0}{a}.$$

If we substitute this value of t into Eq. (1–8),

$$d = v_0 t + \tfrac{1}{2}at^2,$$

we obtain

$$d = v_0 \frac{(v - v_0)}{a} + \frac{1}{2} a \frac{(v - v_0)^2}{a^2}$$
$$= \frac{v_0 v}{a} - \frac{v_0^2}{a} + \frac{v^2}{2a} - \frac{v_0 v}{a} + \frac{v_0^2}{2a}$$
$$= \frac{v^2}{2a} - \frac{v_0^2}{2a},$$

so that, multiplying both sides of the equation by $2a$ and rearranging,

$$v^2 = v_0^2 + 2ad. \tag{1–10}$$

In the case of the car for which $v_0 = 88$ ft/sec and $a = -4$ ft/sec^2, we might want to know how far the car will have gone when it comes to a stop and begins to accelerate in the negative (backward) direction. When this happens, $v = 0$, and so

$$v^2 = v_0^2 + 2ad,$$

$$0 = \left(88 \, \frac{\text{ft}}{\text{sec}}\right)^2 - 2 \times 4 \, \frac{\text{ft}}{\text{sec}} \times d,$$

$$0 = 7744 \, \frac{\text{ft}^2}{\text{sec}^2} - 8 \, d \, \frac{\text{ft}}{\text{sec}^2},$$

or

$$d = 968 \text{ ft.}$$

1–8 Falling bodies

We can apply the ideas of this chapter to the motion of falling bodies because of an important observation made by Galileo (1564–1642). He found, after a number of experiments, that all bodies near the earth fall toward the earth with the *same* acceleration (Fig. 1–10). This acceleration, which is 32 ft/sec² in British units and 9.8 m/sec² (meters per second per second) in metric units, is called the acceleration of gravity, denoted by g.

Galileo's work was especially significant because, for the first time, experimental results were accepted despite the fact that they contradicted well-established ideas, in this case the notion that heavy things "should" fall faster than light ones. Modern science owes its success in understanding and utilizing natural phenomena to its reliance upon experiment. It may be hard for us today to understand the reluctance of our ancestors to shed their intuitive beliefs in the face of experimental evidence to the contrary, although if we look carefully, we will see that this unfortunate tendency still persists in the world.

FIG. 1–10. All bodies in free fall near the earth have the same acceleration.

Another aspect of Galileo's work deserves comment. His conclusion that all things fall with the same constant acceleration is only an *idealization* of reality. The actual accelerations with which objects fall depend upon many factors: the location on the earth, the size and shape of the object, and the density and state of the atmosphere. For example, an acorn reaches the ground before a leaf does that falls from the same limb at the same time because of the greater effect of buoyancy and air resistance on the leaf. Galileo perceived that the essence of the phenomenon was a constant acceleration downward, with other factors acting merely to cause deviations from the constant value.

We can apply the formulas we derived earlier for motion under constant acceleration to bodies in free fall. It must be kept in mind, of course, that the direction of g is always downward, no matter whether we are dealing with a dropped object or one which is initially thrown upward. A few examples will show the correct procedure.

A stone is dropped from the top of the Empire State Building, which is 1472 ft high. How long does it take to reach the ground, neglecting air resistance? What is its speed when it strikes the ground? We begin by writing Eq. (1–9) as

$$d = \tfrac{1}{2}gt^2.$$

Substituting for d and g,

$$1472 \text{ ft} = \frac{32}{2}\,\frac{\text{ft}}{\text{sec}^2}\,t^2,$$

$$t^2 = 92 \text{ sec}^2,$$

$$t = 9.6 \text{ sec.}$$

Knowing the duration of the fall, we find it easy to compute the final speed:

$$v = gt = 32\,\frac{\text{ft}}{\text{sec}^2} \times 9.6 \text{ sec} = 307\,\frac{\text{ft}}{\text{sec}}.$$

This is also the speed with which a stone must be thrown upward from the ground if it is to reach the top of the building. To prove this statement, we refer to Eq. (1–10), namely

$$v^2 = v_0^2 + 2ad.$$

When the stone is dropped, the acceleration a equals the acceleration of gravity g, and the initial speed v_0 of the stone is zero. Hence

$$v^2 = 0 + 2gd, \qquad v = \sqrt{2gd}.$$

When the stone is thrown upward, on the other hand, $a = -g$ (since the downward acceleration is in the opposite direction to the upward motion), and at the top of its path, $v = 0$. Hence

$$0 = v_0^2 - 2gd, \qquad v_0 = \sqrt{2gd},$$

the same numerical value as before.

Next we tackle a slightly more elaborate problem. A stone is thrown upward with an initial speed of 50 ft/sec. What will its maximum height be? When will it strike the ground? Where will it be in $1\frac{1}{8}$ sec? in 2 sec? To find the highest point the stone will reach, we again use Eq. (1–10) with $a = -g$ and $v = 0$, corresponding to the top of its path. The result is

$$0 = v_0^2 - 2gd,$$

$$d = \frac{v_0^2}{2g} = \frac{2500 \ (\text{ft/sec})^2}{64 \ \text{ft/sec}^2} = 39 \ \text{ft}.$$

When will the stone strike the ground? It is easy to show that it takes a body precisely as long to fall from a height d as it does to rise that high (provided that d is its maximum height, as it is here), just as we showed that an object's final speed when dropped from a height d is the same as the initial speed required for it to go that high. From Eq. (1–9)

$$d = \tfrac{1}{2}gt^2,$$

$$t^2 = \frac{2d}{g} = \frac{2 \times 39 \ \text{ft}}{32 \ \text{ft/sec}^2} = 2.42 \ \text{sec}^2,$$

$$t = 1.6 \ \text{sec}.$$

Because the stone takes as long to rise as to fall, the total time between when it is thrown and when it strikes the ground is twice 1.6 sec, or 3.2 sec.

To compute the position of the stone a given time after it has been thrown, Eq. (1–8), namely

$$d = v_0 t + \tfrac{1}{2}at^2,$$

rewritten as

$$d = v_0 t - \tfrac{1}{2}gt^2,$$

is required. Substituting $1\frac{1}{8}$ sec (9/8 sec) for t,

$$h = 50 \ \frac{\text{ft}}{\text{sec}} \times \frac{9}{8} \ \text{sec} - \frac{32}{2} \ \frac{\text{ft}}{\text{sec}^2} \left(\frac{9}{8} \ \text{sec} \right)^2$$

$$= (56 - 20) \ \text{ft} = 36 \ \text{ft}.$$

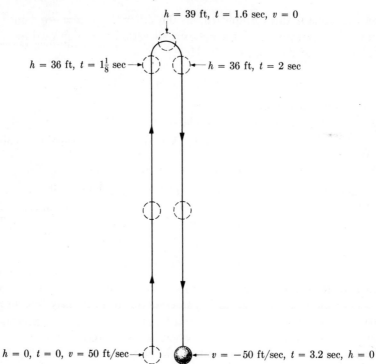

$h = 39$ ft, $t = 1.6$ sec, $v = 0$

$h = 36$ ft, $t = 1\frac{1}{8}$ sec →

← $h = 36$ ft, $t = 2$ sec

$h = 0$, $t = 0$, $v = 50$ ft/sec →

← $v = -50$ ft/sec, $t = 3.2$ sec, $h = 0$

FIG. 1–11. The path of a stone thrown upward with the initial speed of 50 ft/sec. Air resistance is neglected.

When we substitute $t = 2$ sec, we find that

$$h = 50 \,\frac{\text{ft}}{\text{sec}} \times 2 \text{ sec} - \frac{32}{2} \,\frac{\text{ft}}{\text{sec}^2} \times (2 \text{ sec})^2 = 36 \text{ ft}$$

also! All that this apparently paradoxical result means is that at $1\frac{1}{8}$ sec the stone is at a height of 36 ft on its way up, then it goes on up to its maximum height and returns to 36 ft on its way down (Fig. 1–11).

These examples have been worked out not because they are, in themselves, especially significant, but because they illustrate the power of the mathematical approach to physical phenomena. By defining certain quantities and relating them to each other and to events that actually occur in the real world, a whole theoretical structure of equations may be built up. This structure is an instrument enabling us to solve problems which otherwise would each require a separate, perhaps difficult or impossible, experiment. We must remember that the validity of the theoretical structure depends upon its experimental basis; but once this is established, we may proceed to work out its consequences with pencil and paper.

IMPORTANT TERMS

In *exponential notation* a number is expressed as a number between 1 and 10 multiplied by a power of 10. Thus $673 = 6.73 \times 10^2$ and $0.0009 = 9 \times 10^{-4}$.

The digits whose value is precisely known in a particular number are its *significant figures;* the more accurate the number, the more significant figures it has. When different numbers are combined arithmetically, the result has no more significant figures than the least accurately known of the numbers.

The *speed* of a moving body is the rate at which the distance covered by the body changes with time. When the distance is directly proportional to the elapsed time, the body moves at *constant speed.* The *instantaneous speed* of a body is its speed at a specific instant. The *average speed* of a body is the total distance through which it has moved in a time interval divided by the interval.

The *velocity* of a body is a specification of both its speed and the direction in which it is moving.

A *scalar quantity* is one that possesses magnitude only. A *vector quantity* is one that possesses both magnitude and direction. Thus speed is a scalar quantity and velocity is a vector quantity.

The *acceleration* of a body is the rate at which its velocity changes with time; the change in velocity may be in magnitude or direction or both.

The *acceleration of gravity* is the acceleration of a freely falling body near the earth's surface. The symbol of the acceleration of gravity is g, and its value is 32 ft/sec^2 in British units and 9.8 m/sec^2 in metric units.

PROBLEMS*

1. Express the following numbers in exponential notation:

6,460,000,000	1,000,000	351,600
8400	70	3.81
0.14	0.007,890	0.000,013
0.000,000,007,819		

2. Express the following numbers in exponential notation:

13,008,000,000	790,000	12,345.09
800	12.45	0.01
0.0054	0.000,000,890	

*Appendix III gives conversion factors that will be helpful in solving some of these problems, for example, Problem 8.

3. Express the following numbers in decimal notation:

4.51×10^8	5.1×10^0	1.003×10^{-8}
2.0×10^5	8×10^{-2}	10^{-10}
7.819×10^2	9.56×10^{-5}	

4. Express the following numbers in decimal notation:

10^9	9×10^2	6.350×10^{-5}
6.78×10^5	7.416×10^1	7×10^{-9}
8.000×10^4	9.4×10^{-1}	

5. A ship travels 400 mi in 1 day and 3 hr. What is its average speed in mi/hr?

6. A car travels 200 mi in 4 hr and 35 min. What is its average speed in mi/hr?

7. A boat has a cruising speed of 9 knots (1 knot = 1 nautical mi/hr). How long will it take to go from New London, Connecticut, to Nantucket, Massachusetts, a distance of 101 nautical miles?

8. The speedometer of a European car is calibrated in km/hr. What is the car's speed in mi/hr when the speedometer reads 90?

9. The speed of sound in air under ordinary conditions is 1130 ft/sec. What is this speed in (a) m/sec? (b) mi/sec? (c) mi/hr?

10. A man hears the sound of thunder exactly 4 sec after seeing a lightning flash. How many feet away was the lightning? (Assume that the man sees the flash a negligible time after it occurs.)

11. A man observes lightning strike a building 2.1 miles away from him. How long will he have to wait until he hears the thunder?

12. Echoes return in 6 sec to a man standing in front of a cliff. How far away from the cliff is the man?

13. A car travels at 60 mi/hr along a straight road for 2 hr and then at 40 mi/hr for the next 3 hr. What is the car's average speed for the entire 5 hr?

14. A car travels at 40 mi/hr along a straight road for 60 mi and then at 50 mi/hr for the next 60 mi. What is the car's average speed for the entire 120 mi?

15. (a) What is the speed with respect to the ground of an airplane whose air speed is 200 mi/hr when it is bucking a head wind of 56 mi/hr? (b) What distance can it cover relative to the ground in 5 hr?

16. What is the air speed of an airplane that requires 3 hr and 30 min to go 1200 mi from one city to another when it has a 70 mi/hr tail wind?

17. The speed of light is 3×10^8 m/sec. How many minutes does it take light to reach us from the sun, which is approximately 1.5×10^8 km away?

18. A car starts from rest and reaches a speed of 50 mi/hr in 20 sec. (a) What was its acceleration (assumed constant) in (mi/hr)/sec? (b) in ft/sec^2?

19. The brakes of a car moving at 50 mi/hr are suddenly applied and the car comes to a complete stop in 4 sec. (a) What was its acceleration (assumed constant) in (mi/hr)/sec? (b) in ft/sec^2?

20. How long would the car of Problem 18 take to reach 80 mi/hr?

21. How long would the car of Problem 19 take to come to a complete stop starting from 80 mi/hr?

22. A car whose initial speed is 20 mi/hr receives an acceleration of 3 (mi/hr)/sec. *EXM* How far will it go from the point where the acceleration is applied until its speed is 45 mi/hr?

23. The brakes of a particular car are capable of producing an acceleration of -20 ft/sec^2. How far will the car go in the course of slowing down from 90 ft/sec to 30 ft/sec?

24. A stone is dropped from a cliff 64 ft high. How long will it take to fall to the *EXM* foot of the cliff?

25. A boy throws a ball to a height of 50 ft. How long a period elapses between the time the ball left the boy's hand and the moment he catches it on the way down?

26. With what speed will the stone of Problem 24 strike the ground?

27. With what speed did the boy of Problem 25 have to throw the ball?

28. A girl throws a ball vertically upward with a speed of 20 ft/sec from the roof of a building 60 ft high. (a) How long will it take the ball to reach the ground? (b) What will its speed be when it strikes the ground?

29. A girl throws a ball vertically downward with a speed of 20 ft/sec from the roof of a building 60 ft high. (a) How long will it take the ball to reach the ground? (b) What will its speed be when it strikes the ground?

30. An orangutan throws a ball vertically upward at the foot of a cliff 200 ft high while his mate simultaneously drops another ball from the top of the cliff. The two balls collide at an altitude of 100 ft. What was the initial speed of the ball thrown upward?

31. List four vector and four scalar quantities.

32. A man in an elevator 100 ft above the ground drops an apple 6 ft above the elevator's floor. With a stop watch he determines how long the apple takes to fall to the floor. What does he find (a) when the elevator is ascending with an acceleration of 2 ft/sec^2; (b) when it is descending with an acceleration of 2 ft/sec^2; (c) when it is ascending at the constant speed of 8 ft/sec; (d) when it is descending at the constant speed of 8 ft/sec; (e) when the cable has broken and it is descending in free fall?

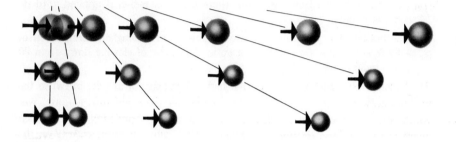

Force and Motion | 2

Thus far we have only discussed *how* motion is described mathematically. But *why* does anything move in the first place? Why do some things move faster than others? Why are some accelerated and others not? How can a body traveling in one direction be accelerated in the opposite direction? All these are reasonable questions and ones we must be able to answer if we are to understand the factors at work in the world around us that produce the physical phenomena we observe.

2–1 The laws of motion

Almost three centuries ago Isaac Newton (1642–1727) formulated three principles, based upon observations he and others had made, which are so fundamental that they have become known as the laws of motion. These laws, like most physical principles, must be approached with an awareness of the processes of selection and idealization inherent in the scientific description of nature.

The first step in analyzing a physical situation is to select those aspects of it which are essential and disregard the others. Thus if we wish to understand the motion of a ball after it has been thrown into the air, we would want to know its initial speed and direction, how high it goes and where it reaches the ground, and so on; we would dismiss as irrelevant the time of day, the color of the ball, and the manner of throwing it.

The second step is to idealize the data we obtain. If we perform careful measurements on the flight of the ball, we would find that its path is almost, but not quite, a regular curve called a parabola (Fig. 2–1). We might reasonably suppose that the deviation from a parabola arises from such factors as air resistance and wind, with which we are not at the moment concerned, and assume that the ball's trajectory in the absence of these factors would be a perfect parabola. In this way we reduce our problem to the basic one of explaining the parabolic path of an object released with a certain initial speed in a certain direction. If we are able to solve this problem, we might then be able to calculate the effects of air resistance and wind to see whether they account for the observed deviations of the ball's path from a parabola. But it would have been difficult or impossible to proceed with the problem of projectile motion unless we had first selected the essential elements of the problem and then idealized them to their simplest form. It is within this framework of selection and idealization that we must approach Newton's three laws of motion.

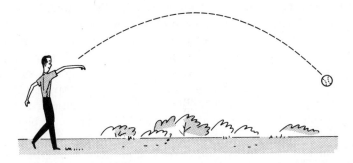

FIG. 2–1. In the absence of air resistance and wind, the path of a ball after it has been thrown is a parabola.

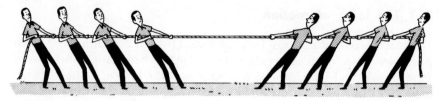

Fig. 2–2. When several forces act on a body, they may cancel one another out to leave no net force.

2–2 The first law of motion

The first law of motion states that

> *In the absence of a net force, a body at rest will remain at rest and a body in motion will continue in motion in a straight line at constant speed.*

We notice immediately that a new word has been introduced: *force.* We all have an intuitive idea of what a force is, namely a "push" or a "pull." But our own experience is quite limited and not always consistent. We usually think of a force as acting between bodies in direct contact with one another, but gravitational, electric, and magnetic forces, for example, operate without any apparent contact between the origin of the force and the body being affected by it. Furthermore, the relationship between force and motion is not always obvious.

The first law of motion clears up these ambiguities with a simple definition: a force is any influence that can cause a body to be accelerated. If a body moves along a straight path at constant speed or is stationary, no net force is acting on it. Of course, several individual forces may be present, but if this is the case their magnitudes and directions are such that they cancel one another out to leave no *net* force (Fig. 2–2). We shall return to the notion of net force later in this chapter.

We might object that, while everyday experience does indicate that bodies in motion *tend* to remain in motion along a straight line at constant speed, sooner or later they invariably come to a stop and, often, deviate from a straight path as well. Even celestial bodies such as the sun, moon, and planets are neither at rest nor pursue straight paths. But these observations do not invalidate the first law of motion, they merely emphasize how difficult it is to avoid the action of forces. A bowling ball rolling along a smooth, perfectly level alley will not continue forever owing to the resistance of air and to friction between it and the alley, but we are at liberty to imagine what would happen if the air were to be removed and the friction were to vanish.

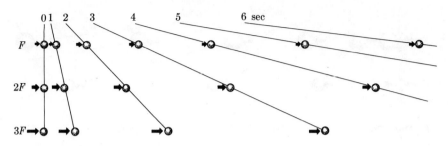

FIG. 2–3. The acceleration of a body is proportional to the net force applied to it. Successive positions of a ball are shown at 1-sec intervals while forces of F, $2F$, and $3F$ are applied.

2–3 The second law of motion

The second law of motion states that

> **When a net force acts on a body, it will be accelerated in the direction of the force with an accleration proportional to the magnitude of the force.**

The second law provides us with a method for analyzing and comparing forces in terms of the accelerations they produce. Thus a force which causes a body to have twice the acceleration another force produces must be twice as great (Fig. 2–3). A body moving to the right but going slower and slower is accelerated to the left; hence there is a force toward the left acting on it. The first law of motion is clearly a special case of the second: when the applied force is zero, the acceleration is also zero.

By itself, the second law of motion as stated above is interesting but not especially useful. However, by introducing the concept of *mass,* we can express the second law as an equation rather than as a proportionality, and in this new guise it becomes very useful indeed. Let us see how this is accomplished.

2–4 Inertia and mass

The first law of motion imputes to material bodies the property of tending to resist changes in their state of rest or uniform motion, a property that is known as *inertia*. We can specify inertia in a precise way with the help of the second law of motion. Since the response of a body to a net force F is an acceleration a proportional to F, we can write

$$F = ma, \qquad (2\text{–}1)$$

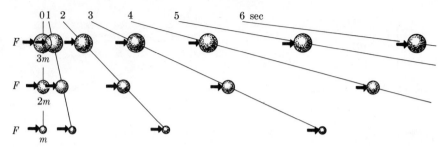

FIG. 2–4. When the same force is applied to bodies of different mass, the resulting accelerations are inversely proportional to the masses. Successive positions of balls of mass m, $2m$, and $3m$ are shown at 1-sec intervals while identical forces of F are applied.

where m is a constant of proportionality characteristic of the particular body being subjected to the force. The larger the value of m, the smaller will be the acceleration produced by a given applied force. Thus m is a measure of the inertia of a body: a large m means a great deal of inertia; small m means little inertia.

Upon what properties of a body does m depend? To find the answer we must perform experiments. What we would find is that such external characteristics of an object as color and shape are not involved in m but that what we might loosely call the *amount of matter* in the body *is* involved. The quantity m is called the *mass* of the body.

In the metric system* the unit of mass is the *kilogram*, abbreviated kg. A kilogram of anything has, by definition, the same mass as that of a certain platinum-iridium cylinder kept at Sèvres, France; measurements of mass are, in essence, comparisons with this standard. When a given force acts upon a body whose mass is 5 kg, the resulting acceleration of the body is one-fifth as great as that of a body whose mass is 1 kg; similarly, a force acting upon a mass of 0.5 kg will produce an acceleration twice as great as the same force acting upon a 1-kg mass (Fig. 2–4).

2–5 Force and motion

We are now in a position to define the unit of force in the metric system. This is the *newton* (abbreviated n), that force which when applied to a 1-kg mass gives it an acceleration of 1 m/sec^2. If we use Eq. (2–1) with metric units, F must be expressed in newtons, m in kilograms, and a in m/sec^2.

---

* Also referred to as the mks (for meter-kilogram-second) system.

Let us examine a few problems in order to become familiar with the application of the second law of motion. These problems are, of course, artificial and oversimplified, but they do illustrate the power of the second law in situations involving force and motion. We shall begin by considering a 4-kg ball on a perfectly smooth, level surface. A force of 10 newtons is applied to it, and we are to determine the speed of the ball and how far it has gone after 6 sec have passed. To begin, we write down Eq. (2–1),

$$F = ma.$$

Since we know what F and m are, we can solve for the acceleration a of the ball as follows:

$$a = \frac{F}{m} = \frac{10 \text{ n}}{4 \text{ kg}} = 2.5 \frac{\text{m}}{\text{sec}^2}.$$

The second law of motion tells us that the direction of a is the same as that of F. To find the speed of the ball after 6 sec, we return to Eq. (1–5),

$$v = at,$$

and obtain

$$v = 2.5 \frac{\text{m}}{\text{sec}^2} \times 6 \text{ sec} = 15 \frac{\text{m}}{\text{sec}}.$$

For the distance the ball has traveled in 6 sec we require Eq. (1–9),

$$d = \tfrac{1}{2}at^2,$$

with the result that

$$d = \frac{1}{2} \times 2.5 \frac{\text{m}}{\text{sec}^2} \times (6 \text{ sec})^2$$

$$= \frac{1}{2} \times 2.5 \frac{\text{m}}{\text{sec}^2} \times 36 \text{ sec}^2$$

$$= 45 \text{ m}.$$

After 6 sec a 4-kg mass acted upon by a 10-newton force will have gone 45 m and have a speed of 15 m/sec (Fig. 2–5).

Now let us consider the same ball rolling on a level but slightly rough surface without the 10-newton force acting. The ball starts at a speed of 15 m/sec, and gradually slows down and stops in a distance of 20 meters. What was the frictional resistive force that acted? To answer this question we start by determining the acceleration of the ball. Since

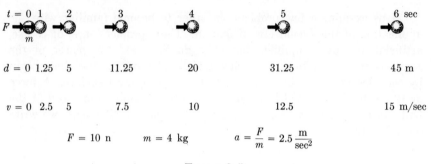

$$F = 10 \text{ n} \qquad m = 4 \text{ kg} \qquad a = \frac{F}{m} = 2.5 \frac{\text{m}}{\text{sec}^2}$$

FIGURE 2–5

we know that its speed has gone from 15 m/sec to 0 m/sec in 20 meters, we may use Eq. (1–9),

$$v^2 = v_0^2 + 2ad.$$

Here the final speed $v = 0$, the initial speed $v_0 = 15$ m/sec, and the distance $d = 20$ m; hence

$$0 = \left(15 \frac{\text{m}}{\text{sec}}\right)^2 + 2a \times 20 \text{ m}$$

$$= 225 \frac{\text{m}^2}{\text{sec}^2} + 40a \text{ m},$$

and so, solving for the acceleration a,

$$a = \frac{-225 \text{ m}^2/\text{sec}^2}{40 \text{ m}} = -5.6 \frac{\text{m}}{\text{sec}^2}.$$

The minus sign indicates a negative acceleration, that is, an acceleration directed opposite to the motion of the ball. The acceleration is given as -5.6 m/sec^2 rather than -5.625 m/sec^2 (the value of 225/40) since the initial quantities are given to only two significant figures. When we

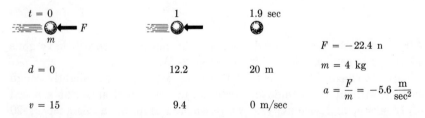

$$F = -22.4 \text{ n}$$
$$m = 4 \text{ kg}$$
$$a = \frac{F}{m} = -5.6 \frac{\text{m}}{\text{sec}^2}$$

FIG. 2–6. A mass of 4 kg with an initial speed of 15 m/sec comes to a stop in a distance of 20 m when a resistive force of 22.4 n acts on it.

know the acceleration and mass of the ball, it is easy to find the resistive force. We use Eq. (2–1) directly:

$$F = ma = 4 \text{ kg} \times (-5.6) \frac{\text{m}}{\text{sec}^2} = -22.4 \text{ n}.$$

The minus sign once more indicates that the direction of the force is opposite to that of the ball's velocity (Fig. 2–6).

2–6 Mass and weight

The second law of motion provides a method for measuring what is called the *inertial mass* of a body. All that is necessary is to apply the same force in turn to a standard of mass and to an unknown mass and then to compare their accelerations, something that can be done anywhere in the universe with the same results. No matter where we are, we can always tell a wooden ball from a lead ball of the same size by throwing them: their different inertial masses feel very different to us even though we might be isolated in interstellar space where sensations of weight would be absent (Fig. 2–7). It is worth while to discuss the relationship between inertial mass and another property of massive bodies, their mutual gravitational attraction.

Newton was the first to demonstrate convincingly that all objects in the universe attract one another in a very specific way: the force between any two objects is proportional to each of their masses and inversely proportional to the square of their separation. Gravitational forces cause

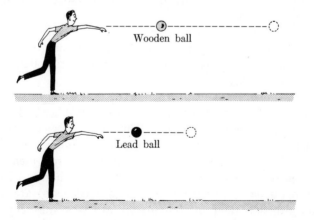

FIG. 2–7. The inertial mass of a lead ball is always greater than that of a wooden ball of the same size. When both are thrown with the same force anywhere in the universe, the lead ball accordingly has the smaller acceleration.

things to fall "down" to the earth (the earth also falls "up," but it is so huge that the motion is too small to detect), the moon to circle the earth, the earth to circle the sun, and so on. We shall discuss Newton's law of gravitation later on, but we should note that what is meant by *mass* when we speak of gravitational forces is not the same thing as inertia, but is instead related to the ability of a body to interact at a distance with others. Experience indicates to us that *gravitational mass,* like inertial mass, is related to the "quantity of matter" in an object, and therefore we might expect some kind of relationship between gravitational and inertial mass.

On the earth's surface it is simple, as a practical matter, to determine gravitational mass. At a given place we measure the acceleration g resulting from the earth's gravity. We may also measure the force of gravity w, called *weight,* on any object at that place, for example with the help of a calibrated spring. In the metric system, weight, being a force, is expressed in newtons. According to the second law of motion,

$$F = ma,$$

and so here, with

$$F = w \quad \text{and} \quad a = g$$

we have

$$w = mg,$$

$$m = \frac{w}{g}. \tag{2-2}$$

The gravitational mass of body is equal to its weight divided by the acceleration of gravity.

There is no need for any confusion between the concepts of mass and weight: the *mass* of a body is a measure of the quantity of matter it contains, while its *weight* is the force with which it is attracted toward the center of the earth.

It is a very significant empirical fact that the inertial and gravitational masses of a body are proportional to each other. (They are customarily set equal for convenience.) This is by no means "obvious"; a comparable situation is that of the electrical forces exerted by, say, protons and electrons, where electric charge plays a role identical to that of mass in gravitational forces, and the intrinsic charges of the proton and electron (as basic a property as gravitational mass) are the same in magnitude although their inertial masses are very different. The proportionality of inertial and gravitational mass is known as the principle of equivalence; it is one of the starting points of Albert Einstein's general theory of relativity.

2–7 The British system of units

In the British system of units the weight of a body, rather than its mass, is usually given. In this system the force unit is the *pound* (abbreviated lb), and so weights are specified in pounds. However, in order to apply the second law of motion, we require m, not w. We can use

$$m = \frac{w}{g}$$

to define a mass unit in the British system called the *slug*. If w is in pounds with $g = 32$ ft/sec^2, m will be expressed in slugs; thus the *weight* of a 1-slug mass is 32 lb, and the *mass* of a 1-lb weight is $\frac{1}{32}$ slug.

Several illustrative problems will make clear the procedure to be followed in using the second law in the British system of units. We shall consider first a car weighing 3000 lb which has an initial speed of 10 mi/hr. How much force is required to accelerate the car to a speed of 50 mi/hr in 9 sec?

The mass of the car is

$$m = \frac{w}{g} = \frac{3000 \text{ lb}}{32 \text{ ft/sec}^2} = 94 \text{ slugs.}$$

Before we can compute the acceleration involved in going from 10 to 50 mi/hr in 9 sec we must convert the speeds from mi/hr to ft/sec. Using Appendix III, we see that

$$1 \frac{\text{mi}}{\text{hr}} = 1.467 \frac{\text{ft}}{\text{sec}},$$

and therefore, to two significant figures,

$$10 \frac{\text{mi}}{\text{hr}} = 15 \frac{\text{ft}}{\text{sec}}$$

and

$$50 \frac{\text{mi}}{\text{hr}} = 73 \frac{\text{ft}}{\text{sec}}.$$

The car's acceleration in the proper units of ft/sec^2 is

$$a = \frac{v_{\text{final}} - v_{\text{initial}}}{t} = \frac{73 \text{ ft/sec} - 15 \text{ ft/sec}}{9 \text{ sec}} = \frac{58 \text{ ft/sec}}{9 \text{ sec}} = 6.4 \frac{\text{ft}}{\text{sec}^2}.$$

Hence the force acting upon the car during its acceleration is

$$F = ma = 94 \text{ slugs} \times 6.4 \frac{\text{ft}}{\text{sec}^2} = 600 \text{ lb.}$$

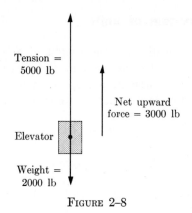

FIGURE 2–8

The next problem is somewhat different. An elevator weighing 2000 lb fully loaded is suspended by a cable whose maximum permissible tension is 5000 lb. What is the greatest upward acceleration possible for the elevator under these circumstances? What is the greatest downward acceleration?

When the elevator is at rest, or moving at constant speed, the tension in the cable is just the weight of 2000 lb. To accelerate the elevator upward, an additional tension is required in order to provide a net upward force (Fig. 2–8). Here it is stated that the maximum total tension cannot exceed 5000 lb, so that the greatest accelerating force that can be applied to the elevator is 3000 lb. The mass of the elevator is

$$ m = \frac{w}{g} = \frac{2000 \text{ lb}}{32 \text{ ft/sec}^2} = 63 \text{ slugs,} $$

and so the acceleration of this mass when a force of 3000 lb is applied to it is

$$ a = \frac{F}{m} = \frac{3000 \text{ lb}}{63 \text{ slugs}} = 48 \frac{\text{ft}}{\text{sec}^2}. \qquad a = \frac{F}{1.62} $$

For the elevator to exceed the downward acceleration of gravity, 32 ft/sec², a downward force in addition to the weight of the elevator is required. Since this cannot be provided by a cable, the maximum downward acceleration is 32 ft/sec².

The arrangement shown in Fig. 2–9 is an interesting one. We are to find the acceleration of either block; since they are joined by the string, the accelerations are the same in magnitude. We note that the force applied to the system of the two blocks is the weight of block A exclusively, since block B is resting upon the table, which we assume is

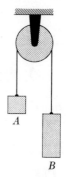

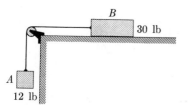

FIG. 2–9. The force on the system of the two blocks is the weight of A, while the mass being accelerated is the total mass of both. (For simplicity, the string is assumed to be massless and the pulley frictionless.)

FIG. 2–10. The force on the system of the two blocks is the difference between their weights, while the mass being accelerated is the total mass of both.

perfectly smooth. However, the *mass* being accelerated by w_A is the combined mass of both blocks, $m_A + m_B$. Hence

$$a = \frac{F}{m} = \frac{w_A}{m_A + m_B} = \frac{w_A}{(w_A/g) + (w_B/g)}.$$

Here, since w_A is 12 lb and w_B is 30 lb,

$$a = \frac{12 \text{ lb}}{[(12 \text{ lb})/(32 \text{ ft/sec}^2)] + [(30 \text{ lb})/(32 \text{ ft/sec}^2)]} = 9.1 \frac{\text{ft}}{\text{sec}^2}.$$

An elaboration of the previous example is the one shown in Fig. 2–10, where blocks A and B are suspended by a string on either side of a pulley. Here the net force acting on the system of the two blocks is $w_B - w_A$, the *difference* between their weights. A *single* force is acting upon *both* blocks since they are connected together by the string; the pulley merely changes the direction of the force. Hence the acceleration of each block can be found by determining the acceleration of the *entire* system:

$$a = \frac{F}{m} = \frac{w_B - w_A}{m_B + m_A}$$

$$= \frac{18 \text{ lb}}{(42 \text{ lb})/(32 \text{ ft/sec}^2)} = 14 \frac{\text{ft}}{\text{sec}^2}.$$

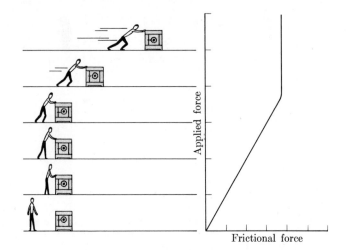

Fɪɢ. 2–11. As force is applied to a box on a level floor, the frictional resistive force increases to a certain maximum and then remains constant.

2–8 Friction

Frictional forces are of many kinds, but they all act to impede motion. The effects of friction must be distinguished from those of inertia. The term *inertia* merely refers to the fact that bodies maintain their original states of rest or of motion in the absence of net forces on them; but even a relatively minute force is sufficient to accelerate a body despite its inertia. The term *friction,* on the other hand, refers to actual forces that are exerted in opposition to applied forces.

To clearly understand the characteristic properties of frictional forces, let us consider what happens when we attempt to move a box across a level floor (Fig. 2–11). At first the box is stationary; no horizontal forces whatever act on it. As we begin to push, the box remains in place, because the floor exerts a force on the bottom of the box which opposes the force we apply. This opposition force is friction, and it arises from the nature of the contact between the floor and box. As we push harder, the frictional force also increases to match our efforts, until finally we are able to exceed the frictional force and begin to move the box. Evidently the opposing frictional force has a maximum value which it cannot go beyond, and when we apply a force greater than this maximum the box will experience a net force. As the box moves under the influence of the net force, the frictional force remains approximately constant. Because the net force on the box is the force of our push *minus* the force of

TABLE 2-1

Approximate coefficients of friction
for various substances in contact*

Materials	μ
Wood on wood	0.3
Wood on stone	0.4
Metal on metal	0.15
Metal on wood	0.5
Leather on wood	0.4
Masonry on clay	0.5
Rubber tire on dry concrete road	0.7
Rubber tire on wet concrete road	0.5

* These values hold only when there is relative motion between the substances; otherwise they are somewhat higher.

friction, it is always less than (or equal to) the force we apply (Fig. 2–11); it may even be zero, as we have seen.

It is a matter of experience that the frictional force exerted by one surface upon another surface depends upon two factors, the *normal force* holding the surfaces together, and the nature of the surfaces. To a good degree of approximation both these factors can be related to the resulting frictional force F_f by the equation

$$F_f = \mu N, \qquad (2-3)$$

where N is the normal force and μ is the *coefficient of friction*. Table 2–1 is a list of coefficients of friction for several surfaces. Note that the value of F_f computed according to Eq. (2–3) represents a maximum; when the applied force is less than F_f, the frictional force always equals the applied force. Otherwise, since F_f acts in the opposite direction to an applied force, objects would move *backward* when pushed weakly, something that does not, of course, occur.

When an object is being pushed or pulled horizontally, the normal force N holding it against the surface it is on is simply its weight mg. Thus in such cases,

$$F_f = \mu N = \mu mg.$$

We must apply a force greater than μmg in order to move a body of mass m across a level surface where the coefficient of friction is μ.

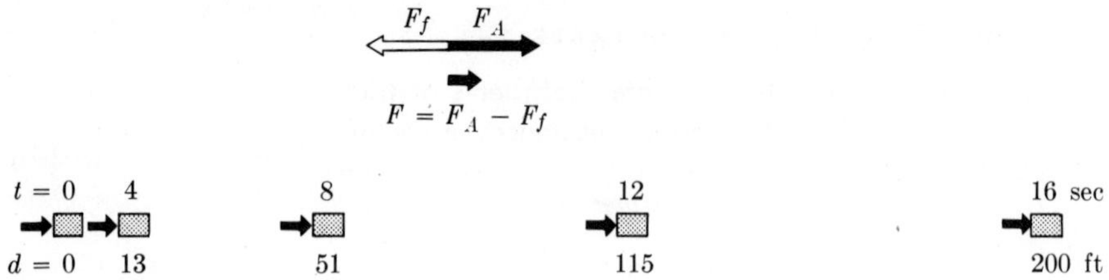

$$F = F_A - F_f$$

$t = 0$	4	8	12	16 sec
$d = 0$	13	51	115	200 ft

FIG. 2–12. A 100-lb crate travels 200 ft in 16 sec under the influence of a 45-lb applied force and a 40-lb frictional force.

As an illustration of the manner in which frictional forces are taken into account in specific problems, let us consider a 100-lb wooden crate being pushed across a stone floor with a horizontal force of 45 lb (Fig. 2–12). How long will it take for the crate, starting from rest, to travel 200 ft?

The first step is to calculate the acceleration of the crate. It is acted upon by an applied force F_A of 45 lb and, in the reverse direction, by a frictional force F_f. The latter is

$$.2 \times 49 = 9.8$$

$$F_f = \mu N = 0.4 \times 100 \text{ lb} = 40 \text{ lb},$$

since the coefficient of friction μ is, from Table 2–1, equal to 0.4 here and the normal force is the 100-lb weight of the block. The mass of the block is

$$m = \frac{w}{g} = \frac{100 \text{ lb}}{32 \text{ ft/sec}^2} = 3.1 \text{ slugs.}$$

Inserting these values in the second law of motion, we have

$$F = ma, \qquad F_A - F_f = ma,$$
$$45 \text{ lb} - 40 \text{ lb} = 3.1 \text{ slugs} \times a,$$

so that

$$a = \frac{(45 - 40) \text{ lb}}{3.1 \text{ slugs}} = \frac{5 \text{ lb}}{3.1 \text{ slugs}} = 1.6 \frac{\text{ft}}{\text{sec}^2}.$$

Now that we know the acceleration of the block, we can determine the time required for it to travel 200 ft by rewriting Eq. (1–9), $d = \frac{1}{2}at^2$, in the form $t = \sqrt{2d/a}$. Substituting 200 ft for d and 1.6 ft/sec^2 for a, we find that

$$t = \sqrt{\frac{2 \times 200 \text{ ft}}{1.6 \text{ ft/sec}^2}} = 16 \text{ sec.}$$

2–9 The third law of motion

The third of Newton's laws of motion states that

When a body exerts a force on another body, the second body exerts a force on the first body of the same magnitude but in the opposite direction.

Let us apply the third law to several different situations in order to see what it really means. A 2-lb book lies stationary on a table, pressing down on the table with a force of 2 lb (Fig. 2–13). The table pushes upward on the book with the reaction force of 2 lb. Why does the book not fly upward into the air? The answer is that the upward force of 2 lb on the book merely balances its weight of 2 lb, which acts downward. If the table were not there to cancel out the latter 2-lb force, the book would, of course, be accelerated downward.

Another illustration of the third law of motion is the operation of walking. We push backward with one foot, and the earth pushes forward on us. The forward reaction force exerted by the earth causes us to move forward, and at the same time, the backward force of our foot causes the earth to move backward (Fig. 2–14). Owing to the earth's relatively larger mass, its motion cannot be detected practically, but it is there. Why is it that there is no reaction force on us, responding to the earth's push on our foot, to keep us from moving? The explanation is that every action-reaction pair of forces acts on *different* bodies. We push on the earth, the earth pushes back on us. If there are no *additional* forces present to impede our motion (for instance, pressure by a wall directly in front of us), we proceed to undergo an acceleration.

We might conceivably find ourselves on a frozen lake with a perfectly smooth surface. Now we cannot walk because the absence of friction prevents us from exerting a backward force on the ice which would

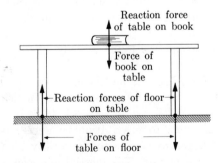

Fig. 2–13. Action-reaction forces between a book and a table and between the table and the floor.

Figure 2–14

FIGURE 2-15

produce a forward force on us. But what we can do is exert a force on some object we may have with us, say a rock. We throw the rock forward by applying a force to it; at the same time the rock is pressing back on us with the identical force but in the opposite direction, and in consequence we find ourselves moving backward (Fig. 2-15).

We shall learn in Chapter 5 how these notions are expressed in the *law of conservation of momentum,* one of the most useful formulations of the laws of motion.

IMPORTANT TERMS

Newton's *first law of motion* states that, in the absence of a net force, a body at rest will remain at rest and a body in motion will continue in motion in a straight line at constant speed.

A *force* is any influence that can cause a body to be accelerated. The unit of force in the metric system is the *newton,* in the British system the *pound.*

Newton's *second law of motion* states that when a net force acts on a body, it will be accelerated in the direction of the force with an acceleration proportional to the magnitude of the force.

The term *inertia* refers to the apparent resistance a body offers to changes in its state of motion.

The property of matter that manifests itself as inertia is called *mass.* The unit of mass in the metric system is the *kilogram,* in the British system the *slug.*

What is meant by *mass* with reference to gravitational forces has nothing to do with inertia; mass in this sense refers to the ability of a body to interact at a distance with other bodies. The inertial and gravitational masses of a body are proportional to one another and are customarily set equal for convenience.

The *weight* of a body is the force with which gravity pulls it toward the center of the earth.

The term *friction* refers to the resistive forces that arise to oppose the motion of a body past another with which it is in contact. The *coefficient of friction* is the constant of proportionality for a given pair of contacting surfaces that relates the frictional force between them to the normal force holding them together.

Newton's *third law of motion* states that when a body exerts a force on another body, the second body exerts a force on the first body of the same magnitude but in the opposite direction.

PROBLEMS

1. A force of 20 n acts upon a body whose mass is 4 kg. (a) What is the weight of the body? (b) What is its acceleration?

2. A force of 100 lb acts upon a body whose mass is 1.5 slugs. (a) What is the weight of the body? (b) What is its acceleration?

3. A force of 20 n acts upon a body whose weight is 8 n. (a) What is the mass of the body? (b) What is its acceleration?

4. A force of 100 lb acts upon a body whose weight is 96 lb. (a) What is the mass of the body? (b) What is its acceleration?

5. Show that the first law of motion can be derived from the second. Can the third law of motion also be derived from the second?

6. Can the gravitational mass of a body possessing inertial mass ever be zero? Can its weight ever be zero?

7. An elevator having a mass of 1500 kg descends with an acceleration of 1 m/sec². What is the tension in its supporting cable?

8. An unloaded truck weighing 4000 lb has a maximum acceleration of 1 ft/sec². What is its maximum acceleration when it is carrying a 2000-lb load?

9. An 80-lb kangaroo exerts a constant force on the ground in the first 2 ft of its jump, and rises 6 ft higher. When it carries a baby kangaroo in its pouch, it can rise only $5\frac{1}{2}$ ft higher. What is the weight of the baby kangaroo?

10. When a force equal to its weight is applied to a body, what is its acceleration?

11. Prove that the ratio between the acceleration of a body and the acceleration of gravity is equal to the ratio between the net force applied to the body and its weight.

12. A 200-lb man stands on a scale in an elevator. What does the scale read (a) when the elevator is ascending with an acceleration of 3 ft/sec²; (b) when it is descending with an acceleration of 3 ft/sec²; (c) when it is ascending at the constant speed of 10 ft/sec; (d) when it is descending at the constant speed of 10 ft/sec; (e) when the cable has broken and it is descending in free fall?

13. A wooden crate weighing 240 lb rests on a level wooden floor. What is the minimum force required to move it horizontally?

14. The coefficient of friction between a rubber tire and a dry concrete road is 0.7. (a) What is the maximum acceleration possible for a car on such a road? (b) What is the minimum distance in which a car on such a road can be stopped when it is moving at 60 mi/hr?

15. A 5-kg block resting on a horizontal surface is attached to a 5-kg block that hangs freely by a string passing over a pulley, an arrangement similar to that shown in Fig. 2–9. (a) If there is no friction between the first block and the surface, what is the block's acceleration? (b) If the coefficient of friction between the first block and the surface is 0.2, what is the block's acceleration?

16. A 10-kg block resting on a horizontal surface is attached to a 5-kg block that hangs freely by a string passing over a pulley, an arrangement similar to that shown in Fig. 2–9. (a) If there is no friction between the first block and the surface, what is the block's acceleration? What is the block's acceleration if the coefficient of friction between the first block and the surface is (b) 0.4? (c) 0.5? (d) 0.6?

17. A 100-lb block resting on a horizontal surface is attached to a 50-lb block that hangs freely by a string passing over a pulley, an arrangement similar to that shown in Fig. 2–9. The coefficient of friction between the first block and the surface is 0.4. (a) How long does it take the blocks to move 20 ft starting from rest? (b) What is their speed at that time?

18. A mass of 8 kg and another of 12 kg are suspended by a string on either side of a pulley, an arrangement similar to that shown in Fig. 2–10. What is the acceleration of the blocks?

19. A ladder rests against the side of a wall. (a) What forces are exerted by the foot of the ladder on the ground? (b) What are the reactions to these forces? (c) What forces are exerted by the top of the ladder on the wall? (d) What are the reactions to these forces? Use a diagram to illustrate your answers.

20. A block is pushed along a rough tabletop. (a) What forces are exerted by the block on the table? (b) What are the reaction forces exerted by the table on the block? Use a diagram to illustrate your answers.

Vector Analysis | 3

In discussing the laws of motion we have referred to the *net force* acting on a body. The reason for this is that forces seldom act singly on bodies in the real world, and we must take into account all of the various forces that may be present. By the *net force* on a body we mean that one force whose effect on its motion would be the same as that produced by the several actual forces. In this chapter we shall learn how to determine net forces, and how the procedures used for this purpose may be extended to treat problems of projectile motion and equilibrium as well.

FIG. 3–1. The net force on a body rather than the individual forces themselves is what affects the motion of the body.

3–1 Vector addition

When all the forces on a body act in the same direction, the net force on it is simply their arithmetical sum. Thus a net force of 100 lb is required to lift a 100-lb weight, regardless of whether the force is applied by one, two, or ten men (Fig. 3–1).

How do we determine the net force when the applied forces act in *different* directions? A suitable procedure is to make a scale drawing of the forces, which is called a *vector diagram* because forces are vector quantities and the manner in which they are treated mathematically is the same as for other vector quantities such as velocity and acceleration. In a vector diagram of forces, arrows, called *vectors*, are used to represent the forces that act. The length of each arrow is proportional to the magnitude of the force it represents, and its direction is that of the force. The scale to be used depends upon the problem: in one case it might be convenient to let 1 in. on the diagram represent 1 lb, while in another a better scale might be 1 in. = 1000 lb.

It is customary to print the symbols of vector quantities in boldface type, while italics are used both for scalar quantities (such as mass, time, and area) and for the magnitudes of vectors. Thus the magnitude of the vector **A** is *A*. In handwriting vectors are represented by placing arrows over their symbols, for instance $\overrightarrow{A}$.

The rule for adding vectors is illustrated in Fig. 3–2. To add **B** to **A**, shift **B** parallel to itself until its tail is at the end of **A**. In its new position **B** must have its original length and direction. The sum **A** + **B** is a vector **R** drawn from the tail of **A** to the head of **B**. The magnitude of **R** may be found by measuring its length on the diagram and making the appropriate conversion, and its direction with respect to **A** or **B** is that shown in the diagram. If **A** and **B** are forces acting on a body, their

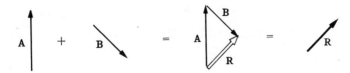

FIG. 3–2. Addition of two vectors.

sum **R** is the net force acting on it. The single vector **R** can replace the two vectors **A** and **B** without any change in the behavior of the body. This procedure may be used with any number of vectors: place the tail of each vector at the head of the previous one, keeping their lengths and original directions unchanged, and draw a vector **R** from the tail of the first vector to the head of the last. **R** is the sum required. Figure 3–3 shows how four vectors, whose initial magnitudes and directions differ, are added together. We note that the order in which the vectors are added does not affect the result; that is,

$$\mathbf{A} + \mathbf{B} = \mathbf{B} + \mathbf{A}. \tag{3-1}$$

Vector diagrams may also be used with vector quantities other than forces. For instance, let us consider a boat heading east at 10 mi/hr in a river flowing south at 3.0 mi/hr. What is the boat's velocity relative to the earth? Figure 3–4 shows how the problem is solved by using

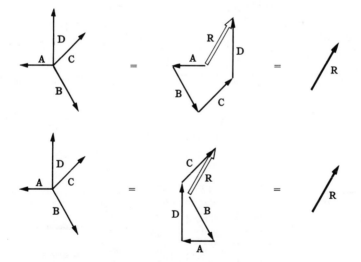

FIG. 3–3. Addition of more than two vectors. The order in which the vectors are added is immaterial.

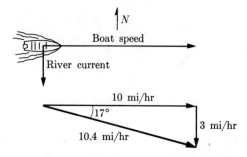

FIG. 3–4. A boat heads east at 10.0 mi/hr in a river flowing south at 3.0 mi/hr. Adding the vectors representing their velocities yields a single vector, which indicates that the boat's velocity relative to the earth is 10.4 mi/hr in a direction about 17° south of east.

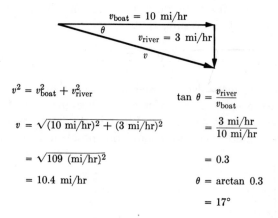

$$v^2 = v_{\text{boat}}^2 + v_{\text{river}}^2 \qquad\qquad \tan\theta = \frac{v_{\text{river}}}{v_{\text{boat}}}$$

$$v = \sqrt{(10 \text{ mi/hr})^2 + (3 \text{ mi/hr})^2} \qquad = \frac{3 \text{ mi/hr}}{10 \text{ mi/hr}}$$

$$= \sqrt{109 \ (\text{mi/hr})^2} \qquad\qquad = 0.3$$

$$= 10.4 \text{ mi/hr} \qquad\qquad \theta = \arctan 0.3$$

$$= 17°$$

FIG. 3–5. How the problem of Fig. 3–4 is solved by using trigonometry.

TABLE 3–1

Angle	Sine	Cosine	Tangent
14°	.242	.970	.249
15°	.259	.966	.268
16°	.276	.961	.287
17°	.292	.956	.306
18°	.309	.951	.325
19°	.326	.946	.344
20°	.342	.940	.364

vectors to represent the boat's velocity and the river's current. The boat is traveling at 10.4 mi/hr in a direction about 17° south of east.

For rough calculations it is sufficient to determine the magnitude and direction of **R** directly from a vector diagram with ruler and protractor. More accurate results may be obtained with the help of trigonometry. Figure 3–5 illustrates how the problem of the previous paragraph may be solved by using trigonometric methods. Because $\mathbf{v}_{\text{boat}}$ is perpendicular to $\mathbf{v}_{\text{river}}$, the velocity v of the boat relative to the earth may be obtained from the Pythagorean theorem:

$$v^2 = v_{\text{boat}}^2 + v_{\text{river}}^2,$$

$$v = \sqrt{v_{\text{boat}}^2 + v_{\text{river}}^2}$$

$$= \sqrt{100 + 9}\,\text{mi/hr}$$

$$= 10.4\,\text{mi/hr}.$$

The angle θ between $\mathbf{v}$ and $\mathbf{v}_{\text{boat}}$ is specified by

$$\tan\theta = \frac{v_{\text{river}}}{v_{\text{boat}}} = \frac{3\,\text{mi/hr}}{10\,\text{mi/hr}} = 0.3.$$

When we examine a table of trigonometric functions, we find that a portion of it reads as shown in Table 3–1. We see that the angle whose tangent is closest to 0.3 is 17°, since tan 17° = 0.306. Hence θ is slightly less than 17°. To the nearest degree, the value of θ is

$$\theta = 17°.$$

It is sometimes necessary to subtract one vector from another. If we wish to subtract **A** from **B,** for example, we first form the negative of **A,** denoted −**A,** which is a vector with the same length as **A** but which points in the opposite direction (Fig. 3–6). Then we add −**A** to **B** in the usual manner. This procedure may be summarized as

$$\mathbf{B} - \mathbf{A} = \mathbf{B} + (-\mathbf{A}). \tag{3–2}$$

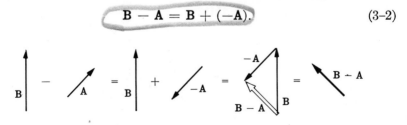

Fig. 3–6. Vector subtraction.

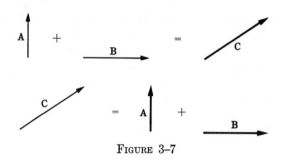

FIGURE 3–7

3–2 Resolution of vectors

Just as we can add together several vectors to give a single vector, so we can break up a single vector into two or more others. From Fig. 3–7 we see that if two vectors **A** and **B** are equivalent to the one vector **C**, it is also true that the one vector **C** is equivalent to the two vectors **A** and **B**. The process of replacing one vector by two or more others is called the *resolution* of the vector, and the new vectors are called the *components* of the original vector. We shall find it useful later in this chapter and elsewhere in our study of physics to resolve vectors into components in certain definite directions that are suggested by the problem at hand.

Almost invariably the components of a vector are chosen to be perpendicular to one another. Figure 3–8 shows a boy pulling a wagon with a rope at the angle θ above the ground. Only part of the force he exerts is reflected in the motion of the wagon, since it moves horizontally while the force **F** is not a horizontal one. Let us resolve **F** into two components, F_x and F_y, where

$$F_x = \text{horizontal component of } \mathbf{F},$$

$$F_y = \text{vertical component of } \mathbf{F}.$$

The horizontal component F_x is responsible for moving the wagon, while the vertical component F_y merely pulls upward on it. The components of

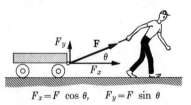

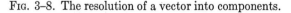

$F_x = F \cos \theta, \qquad F_y = F \sin \theta$

FIG. 3–8. The resolution of a vector into components.

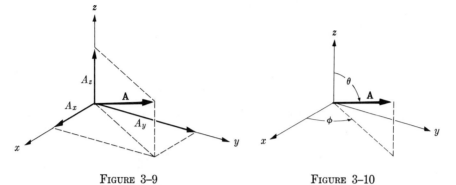

FIGURE 3–9   FIGURE 3–10

F may be calculated from its magnitude F and the angle θ, using the trigonometric formulas

$$F_x = F \cos \theta, \qquad (3\text{–}3)$$

$$F_y = F \sin \theta. \qquad (3\text{–}4)$$

We can think of F_x as being the projection of **F** in the horizontal direction and of F_y as being its projection in the vertical direction.

In the above example the boy and the wagon are both in a vertical plane, and so only two quantities are needed to specify the force **F**. These quantities can be either the force's magnitude F and the angle θ it makes with the horizontal, or equally well its horizontal and vertical components F_x and F_y. In general, however, vectors represent quantities that need not lie in a plane but may have any orientation in space whatever. Under such circumstances *three* quantities are needed to completely describe a vector. Figures 3–9 and 3–10 show how a vector **A** may be specified in either of two ways: by its magnitude A and the angles θ and ϕ, or by its components A_x, A_y, and A_z. The latter method is usually the more convenient.

3–3 Projectile motion

An excellent example of the value of vector resolution is the case of a body moving through space but not restricted to a straight path. As we all know from personal observation, such a body usually has a curved path, while the methods we employed in the previous two chapters for analyzing motion apply only to perfectly straight paths. Our strategy in attacking problems of this kind is to resolve the body's velocity v into two components, v_x in a horizontal direction and v_y in a vertical direction, and then examine separately how the body behaves in each of these directions.

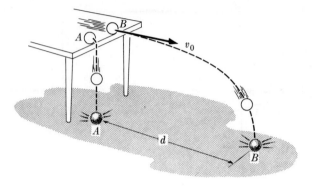

Figure 3–11

Suppose we drop a ball A from the edge of a table while rolling an identical ball B off to the side (Fig. 3–11). At the moment the balls leave the table, A has zero velocity while B has the horizontal velocity v_0. The velocity components of the balls are therefore

$$v_{Ax} = 0, \qquad v_{Bx} = v_0,$$

$$v_{Ay} = 0, \qquad v_{By} = 0.$$

We observe that both balls reach the floor at the same time, even though B has traveled some distance d away from the table. The reason for this behavior is that the acceleration of gravity is the same for all bodies near the earth regardless of their state of motion; both A and B started out with no vertical velocity, both underwent the same downward acceleration, and so both took the same period of time to fall.

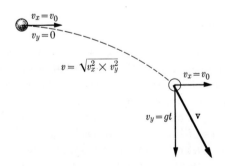

Figure 3–12

is zero. At this time t,

$$0 = v_0 \sin \theta - gt \qquad \text{and} \qquad t = \frac{v_0 \sin \theta}{g}.$$

The rocket requires the same period of time to return to the ground, and so its total time of flight T is

$$T = 2t = \frac{2v_0 \sin \theta}{g}. \tag{3-10}$$

We can now find the range R of the rocket; that is, we can see how far from its launching point it will strike the ground. Since the horizontal component v_x of the rocket's velocity is constant,

$$
\begin{aligned}
R &= v_x T \\
&= v_0 \cos \theta \times \frac{2v_0 \sin \theta}{g} \\
&= \frac{2v_0^2}{g} \sin \theta \cos \theta. \tag{3-11}
\end{aligned}
$$

Equation (3–11) can be simplified by making use of the trigonometric formula

$$\sin \theta \cos \theta = \tfrac{1}{2} \sin 2\theta. \tag{3-12}$$

The rocket's range may therefore be written

$$R = \frac{v_0^2}{g} \sin 2\theta. \tag{3-13}$$

This formula gives the range of a rocket (or any other projectile, for that matter, provided it obeys the restrictions given earlier) in terms of its initial speed v_0 and the angle θ at which it is launched. We note that R is a maximum when $\sin 2\theta = 1$, since 1 is the highest value the sine function can have. Since $\sin 90° = 1$, the maximum range occurs when the initial angle θ is 45°. Any other angle, greater or smaller, will result in a shorter range (Fig. 3–14).

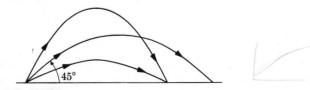

FIG. 3–14. In the absence of air resistance, the maximum range of a projectile occurs when it is fired at an angle of 45°.

What is the minimum initial velocity a rocket must have if it is to be able to reach a target 1000 km (621 miles) away? From Eq. (3–13) we obtain

$$v_0 = \sqrt{Rg/\sin 2\theta}.$$

Setting $\theta = 45°$ for maximum range,

$$v_0 = \sqrt{Rg/\sin 90°} = \sqrt{Rg}$$

$$= \sqrt{10^3 \text{ km} \times 10^3 \text{ m/km} \times 9.8 \text{ m/sec}^2}$$

$$= 3100 \text{ m/sec}.$$

This is a little less than 2 mi/sec.

3–4 Equilibrium of a particle

Another branch of mechanics in which vector resolution is a necessary tool is *statics*. In contrast with *dynamics*, which is the study of bodies in motion, statics is concerned exclusively with bodies at rest. While dynamics is naturally of greater interest to the physicist, an introduction to the basic notions of statics is valuable both as practice in the use of vector methods and because these notions have important applications elsewhere.

A particle which has no net force acting on it is said to be in *equilibrium*. According to the first law of motion, such a particle need not be at rest, but may instead be moving along a straight path at constant speed. The condition for particle equilibrium may be expressed in the form

$$\Sigma \mathbf{F} = 0, \tag{3–14}$$

where the symbol Σ (Greek capital letter *sigma*) means "sum of" and $\mathbf{F}$ refers to the various forces acting on a specific particle. Since every force $\mathbf{F}$ may be resolved into the three perpendicular components F_x, F_y, and F_z, we can replace Eq. (3–14), which is a vector equation, with the three scalar equations

$$\Sigma F_x = 0, \tag{3–15}$$

$$\Sigma F_y = 0, \tag{3–16}$$

$$\Sigma F_z = 0. \tag{3–17}$$

In considering the equilibrium of a particle, it is usually easier to calculate the components of each force present and to make use of Eqs. (3–15) through (3–17) than it is to work with the forces themselves directly in Eq. (3–14).

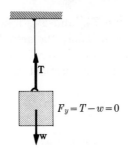

$$F_y = T - w = 0$$

FIGURE 3–15

In Fig. 3–15 we have the simple situation of a weight w being supported by a single rope. The weight will be in equilibrium when all the forces acting on it cancel one another out, which here means that

$$\sum F_y = 0,$$

since there are no forces in the x- or z-directions. The tension T in the rope acts upward (the $+y$-direction) and the weight w acts downward (the $-y$-direction); hence

$$\sum F_y = T - w = 0 \qquad \text{and} \qquad T = w.$$

The tension in the rope must equal the weight being supported.

In Fig. 3–16 the same weight is suspended from two ropes, A and B, which are at the angles θ and ϕ, respectively, with the horizontal. We begin by resolving the tension in each rope into components in the x- and

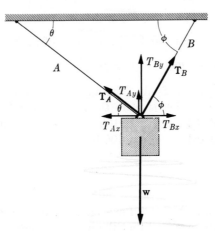

FIG. 3–16. At equilibrium the sums of the horizontal and vertical force components equal zero.

y-directions, so that we have T_{Ay} and T_{By} upward, T_{Ax} to the left, and T_{Bx} to the right. We may now ignore the actual tensions T_A and T_B and treat their components as individual forces acting at the same point as the weight w. For the forces in the vertical and horizontal directions to separately cancel,

$$\sum F_x = T_{Bx} - T_{Ax} = 0 \tag{3–18}$$

$$\sum F_y = T_{Ay} + T_{By} - w = 0. \tag{3–19}$$

Let us suppose that in the above situation $w = 100$ lb, $\theta = 37°$, and $\phi = 60°$, and that we are to find the tension in each rope. From Eq. (3–18) we find that

$$T_{Bx} - T_{Ax} = 0,$$

$$T_B \cos \phi - T_A \cos \theta = 0,$$

$$T_B = T_A \frac{\cos \theta}{\cos \phi} = T_A \frac{\cos 37°}{\cos 60°} = 1.6 T_A.$$

From Eq. (3–19) we find that

$$T_{Ay} + T_{By} - w = 0,$$

$$T_A \sin \theta + T_B \sin \phi - 100 \text{ lb} = 0.$$

Substituting $1.6 T_A$ for T_B in the last equation,

$$T_A \sin \theta + 1.6 T_A \sin \phi - 100 \text{ lb} = 0,$$

$$T_A(\sin 37° + 1.6 \sin 60°) = 100 \text{ lb},$$

$$T_A = 50 \text{ lb}.$$

The tension in rope A is 50 lb. Since $T_B = 1.6 T_A$ here,

$$T_B = 1.6 \times 50 \text{ lb} = 80 \text{ lb}.$$

The tension in rope B is 80 lb. The algebraic sum of the tensions in the two ropes is 130 lb, which is more than the weight being supported, but the vector sum of $\mathbf{T}_A$ and $\mathbf{T}_B$ is 100 lb acting vertically upward, which cancels the downward force $\mathbf{w} = 100$ lb.

3–5 Torque

The condition that $\sum \mathbf{F} = 0$ for a particle to be in equilibrium applies also to a large body provided that the lines of action of the various forces acting on it intersect at a common point. If the lines of action of the forces do *not* intersect, the body may be set into rotation even though

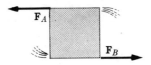

FIG. 3–17. A body will not be in equilibrium when equal and opposite forces are applied unless the forces have a common line of action.

the vector sum of the forces may equal zero. In Fig. 3–17 force $\mathbf{F}_A$ acts to the left and force $\mathbf{F}_B$ of the same magnitude acts to the right. Their combined effect is to start the object spinning. If we want the term *equilibrium* to imply the absence of a rotational acceleration as well as the absence of a linear acceleration, we must supplement $\sum \mathbf{F} = 0$ with another condition which the forces on a body must obey if it is to be in equilibrium.

A hint as to the nature of this additional condition may be obtained by watching a see-saw in operation at a playground (Fig. 3–18). A small child can balance a large child merely by sitting farther from the pivot. Two children of the same weight will not balance unless they sit the same distance from the pivot, though the exact distance does not matter. Evidently both the magnitudes of the forces (here the weights of the children) and their lines of action determine whether or not the object is in equilibrium. If we were to try various combinations of weights and distances from the pivot, we would find that the see-saw is balanced when the product of the weight w and distance from the pivot L of one child is equal to the product wL of the other child's weight and distance from the pivot.

To make our discussion perfectly general, let us consider a force $\mathbf{F}$ acting upon a body free to rotate about some pivot point O (Fig. 3–19). The perpendicular distance L from O to the line of action of the force is called the *moment arm* of the force about O. The product of the mag-

FIG. 3–18. A see-saw is balanced when $w_1 L_1 = w_2 L_2$.

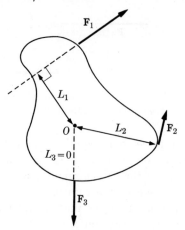

Fig. 3–19. Forces and their moment arms about the pivot point O. Lines of action are shown as dashed lines. The line of action of $\mathbf{F}_3$ passes through O, and its moment arm is therefore zero. The torque $\tau_1 = F_1 L_1$ tends to produce clockwise rotation, and is therefore considered negative; the torque $\tau_2 = F_2 L_2$ tends to produce counterclockwise rotation, and is therefore considered positive.

nitude F of the force and its moment arm L is known as the *torque* of the force about O. The symbol for torque is τ (Greek letter *tau*), so that

$$\tau = FL. \tag{3–20}$$

In the metric system torque is expressed in newton meters; in the British system, in lb·ft. By convention a torque that tends to produce a counterclockwise rotation is considered positive and a torque that tends to produce a clockwise rotation is considered negative. Thus the condition for a body to be in rotational equilibrium is that the sum of the torques acting upon it about any point, using the above convention for plus and minus signs, be zero:

$$\Sigma\tau = 0. \tag{3–21}$$

Of course, if the various forces that act do not all lie in the same plane, it is necessary that the sum of the torques in each of three mutually perpendicular planes be zero.

If the sum of the torques on a body is zero at any point, it is also zero about all other points. Hence the location of the point about which torques are calculated in an equilibrium problem is completely arbitrary; *any* point will do. Let us verify this statement with an example. Figure 3–20 is a diagram of a rod 6 ft long having weights of 10 lb at one end and 30 lb at the other. We assume that the rod is weightless. Obviously a force of 40 lb is necessary to lift the rod. A more difficult question might

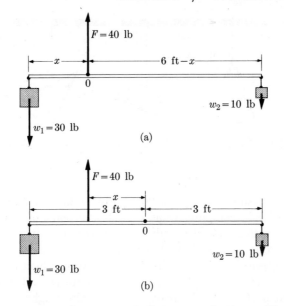

$w_1 = 30$ lb

(a)

(b)

Fig. 3-20. Torques may be computed about any point in equilibrium problems.

be asked. At what point should the rod be picked up if it is to have no tendency to rotate? In other words, where is the balance point of the rod? To solve the problem we apply Eq. (3-21). Let us choose the unknown balance point to compute torques about, and let x be the distance of the 30-lb weight from this point. The 10-lb weight is therefore $(6 - x)$ ft from the balance point. The torques these weights exert are therefore

$$\tau_1 = w_1 L_1 = +30x,$$

$$\tau_2 = w_2 L_2 = -10(6 - x),$$

and equilibrium will result when

$$\sum \tau = 30x - 10(6 - x) = 0,$$

$$x = 1.5 \text{ ft}.$$

When the rod is picked up 1.5 ft from the 30-lb weight, the two weights exert opposite torques of the same magnitude (45 lb·ft) about this point, so the rod is in balance. If we choose instead to calculate torques about the middle of the rod (Fig. 3-20b), we have

$$\tau_1 = w_1 L_1 = +30 \text{ lb} \times 3 \text{ ft} = +90 \text{ lb·ft},$$

$$\tau_2 = F L_2 = -40x,$$

$$\tau_3 = w_2 L_3 = -10 \text{ lb} \times 3 \text{ ft} = -30 \text{ lb·ft}.$$

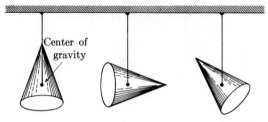

FIG. 3–21. A body suspended from its center of gravity is in equilibrium in any orientation.

Equilibrium will result when $\sum\tau = 90 - 40x - 30 = 0$, and $x = 1.5$ ft. It is usually wise to compute torques about the point of application of one of the forces that act on a body, since this makes it unnecessary to consider the torque produced by that force.

The *center of gravity* of a body is that point from which it can be suspended in any orientation without tending to rotate (Fig. 3–21). Each of the constituent particles of the body has a certain weight, and therefore exerts a torque about whatever point the body is suspended from. There is only a single point in a body about which all of these torques cancel out no matter how the body is oriented; this is its center of gravity. We can therefore regard the entire weight of a body as concentrated at its center of gravity. If the rod in the previous example had the weight w_r instead of being weightless, we could take into account its effect on the location of the balance point by including the torque due to a force of magnitude w_r acting downward at the center of the rod. The center of gravity of a uniform body of regular shape is located at its geometrical center. An experimental method for determining the center of gravity of a flat body of any shape is illustrated in Fig. 3–22.

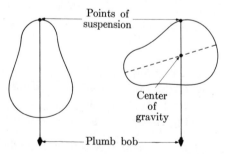

FIG. 3–22. To find the center of gravity of a flat body, suspend it and a plumb bob successively from two different points on its edge. The center of gravity is located at the intersection of the two lines of action of the plumb bob.

IMPORTANT TERMS

A *vector* is an arrow whose length is proportional to the magnitude of some vector quantity and whose direction is that of the quantity. A *vector diagram* is a scale drawing of the various forces, velocities, or other vector quantities involved in the motion of a body. In *vector addition*, the tail of each successive vector is placed at the head of the previous one, with their lengths and original directions kept unchanged. The sum is a vector drawn from the tail of the first vector to the head of the last.

A vector can be *resolved* into two or more other vectors whose sum is equal to the original vector. The new vectors are called the *components* of the original vector, and are normally chosen to be perpendicular to one another.

A particle upon which no net force is acting is said to be in *equilibrium*. A particle in equilibrium may be at rest or may have a constant velocity.

The *torque* of a force about a particular pivot point is the product of the magnitude of the force and the perpendicular distance from the line of action of the force to the pivot point. The latter distance is called the *moment arm* of the force. For a body to be in equilibrium there must be neither net force nor net torque acting upon it.

The *center of gravity* of a body is that point from which it can be suspended in any orientation without tending to rotate.

PROBLEMS

1. In going from one city to another, a car travels 120 mi west, 50 mi north, and 40 mi southeast. How far apart are the two cities?

2. The *Queen Mary* is heading due west at 20 mi/hr in the presence of a southeast wind of 15 mi/hr. In what horizontal direction will smoke from the liner's funnel go?

3. A boat heads directly across a river with a speed through the water of 8 mi/hr. The river current is 2 mi/hr. (a) What is the magnitude and direction of the boat's velocity relative to the earth? (b) If the river is 1.5 mi wide, how long does it take the boat to cross to the other bank? (c) When the boat reaches the other bank, how far downstream will it be?

4. The sum of two perpendicular forces is 300 lb. If one of the forces has a magnitude of 120 lb, what is the magnitude of the other?

5. Find the magnitude of the sum of the following displacements: 6 m north and 10 m west, both in a horizontal plane, and 12 m vertically upward.

6. Three forces act in a horizontal plane as follows: **A** has a magnitude of 6 n and points north; **B** has a magnitude of 10 n and points west; and **C** has a magnitude of 8 n and points southeast. Find the magnitude and direction of $\mathbf{A} + \mathbf{B} - \mathbf{C}$.

7. The following forces act on a body resting on a level, frictionless surface: 10 n at an angle of 10° clockwise from a reference direction; 40 n at an angle of 60° clockwise from the reference direction; 20 n at an angle of 180° clockwise from the reference direction; and 20 n at an angle of 300° clockwise from the reference direction. What is the magnitude and direction of the net force on the body?

8. A force of 20 lb is applied to the lower end of a window pole that is 70° above the horizontal. How much vertical force is being applied to the window?

9. Find the vertical and horizontal components of a 100-lb force that is directed 40° above the horizontal.

10. A body is moving in a plane at 12 m/sec in a direction 40° clockwise from the $+x$ axis. What are the x- and y-components of its velocity?

11. An airplane is heading southwest when it takes off at an angle of 25° from the horizontal. Its speed is 220 mi/hr. (a) What is the upward component of its speed? (b) What is the component of its speed toward the south?

12. Two particles are moving in a plane; one has the velocity components $v_x = 1$ m/sec, $v_y = 5$ m/sec, and the other has the velocity components $v_x = 4$ m/sec, $v_y = 3$ m/sec. If both started from the same point, what is the angle between their paths?

13. A ball is thrown horizontally from the roof of a building 60 ft high at a speed of 100 ft/sec. What will the magnitude and direction of the ball's velocity be when it strikes the ground?

14. A bomb is dropped by an airplane in level flight whose speed is 400 mi/hr when it is 12,000 ft above the ground. Where will the bomb strike the ground relative to the location above which it was dropped?

15. A blunderbuss can fire a slug 300 ft vertically upward. (a) What is its maximum horizontal range? (b) With what speeds will the slug strike the ground when fired upward and when fired so as to have maximum range?

16. A football leaves the toe of a punter at an angle of 50° above the horizontal. What was its initial speed if it travels 40 yards?

17. A shell is fired at a speed of 300 m/sec at an angle of 30° above the horizontal. (a) Neglecting friction, how far does it go? (b) How long a time does its flight take?

18. A rifle located at $x = 0$, $y = 0$ is to fire at a target located at the coordinates x, y. What is the minimum muzzle velocity required in terms of x and y?

19. A ball is thrown horizontally toward the north from a rooftop at a speed of 8 m/sec. A 10-m/sec wind is blowing from the east. (a) What is the speed of the ball relative to the ground after 2 sec? (b) What angle does its velocity make relative to the vertical at this time? (c) What angle does its velocity make relative to due north at this time?

20. A 50-kg body is suspended from two ropes which each make an angle of 30° with the vertical. What is the tension in each rope?

21. A horizontal force of 18 lb acts upon an unsupported body whose weight is 10 lb. What is the direction and magnitude of the force needed to keep the body at rest?

22. A crate weighing 500 lb is being slid down a ramp which makes an angle of 20° with the horizontal. The coefficient of friction is 0.3. How much force parallel to the plane must be applied to the crate if it is to slide down at constant speed? In which direction must the force be applied?

23. The coefficient of friction between a plane and an object is μ. Prove that the object will remain stationary provided the plane makes an angle of less than θ with the horizontal, where $\tan \theta = \mu$.

24. A car is stuck in the mud. To get it out, the driver ties one end of a rope to the car and the other to a tree 100 ft away. He then pulls sideways on the rope at its midpoint. If he exerts a force of 120 lb, how much force is applied to the car when he has pulled the rope 5 ft to one side?

25. A 1-kg bird sits on a telephone wire midway between two poles 50 m apart. The wire, assumed weightless, sags by 10 cm. What is the tension in the wire?

26. A fishing rod 3 m long is attached to a pivoting holder at its lower end. The rod is 50° above the horizontal. A fisherman grasps the rod 1 m from its lower end. What horizontal force must the fisherman exert to keep the rod at this angle when reeling in a 20-lb fish from directly below the rod's upper end?

27. A horizontal beam 6 m long projects from the wall of a building. A guy wire that makes a 40° angle with the horizontal is attached to the outer end of the beam. When a weight of 100 n is attached to the end of the beam, what is the tension in the guy wire? (Neglect the weight of the beam.)

28. The front wheels of a certain car are found to support 1600 lb and the rear wheels 1400 lb. The car's wheelbase (distance between axles) is 8 ft. (a) What is the total weight of the car? (b) How far from the forward axle is the center of gravity of the car?

29. A 30-lb child and a 50-lb child sit at opposite ends of a 12-ft seesaw pivoted at its center. Where should a third child weighing 35 lb sit in order to balance the seesaw?

30. A 25-n bag of cement is placed on a 4-m-long plank 1.8 m from one end. The plank itself weighs 8 n. Two men pick up the plank, one at each end. How much weight must each support?

31. A weightless ladder 20 ft long rests against a frictionless wall at an angle of 60° from the horizontal. A 150-lb man is 4 ft from the top of the ladder. What horizontal force at the bottom of the ladder is needed to keep the ladder from slipping?

32. A ladder 12 ft long rests against a vertical frictionless wall with its lower end 4 ft from the wall. The ladder weighs 40 lb and its center of gravity is at its center. If the ladder is to stay in place, what must be the minimum coefficient of friction between the bottom of the ladder and the ground?

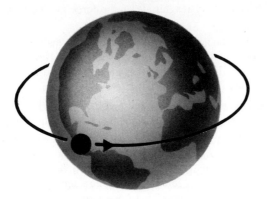

Circular Motion and Gravitation | 4

Bodies in the natural world exhibit a remarkable tendency to travel in curved paths. The orbits of the earth and planets around the sun and of the moon around the earth are very nearly circular, a fact employed by Newton in discovering his law of gravitation. On a microcosmic scale, a useful visualization of the atom invokes electrons circling about a central nucleus, an idea employed by Bohr in devising his theory of the atom. And, of course, circular motion is no novelty in our own experience. We shall find it both interesting and essential for our later work to consider circular motion in some detail. In the latter part of this chapter we shall go on to study gravitation, the most obvious of the natural forces whose action has shaped our universe.

4–1 Uniform circular motion

We shall restrict ourselves to *uniform circular motion,* which is that motion executed by a body traveling in a circle at a constant speed. We can see from Fig. 4–1 that even though the speed of the body has the same numerical value all along its path, the *direction* of the body's motion changes constantly. According to the first law of motion, a body which is neither stationary nor moving along a straight line at constant speed is being acted upon by a force. Since the body's path, being circular, is constantly deflected inward, there must be an *inward* force on it to keep it moving in a circle. This force is called *centripetal force,* literally, "force moving toward the center." Without it, circular motion cannot occur.

To verify directly the crucial role of centripetal force in circular motion, we can whirl a ball at the end of a string (Fig. 4–2). As the ball

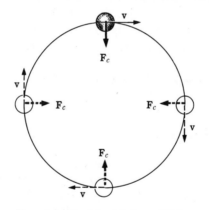

Fig. 4–1. Centripetal force ($\mathbf{F}_c$).

Figure 4–2

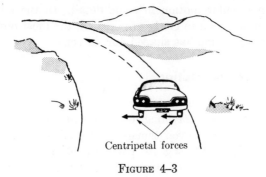

Centripetal forces

FIGURE 4-3

swings around, we must continually exert a centripetal force inward by means of the string. If we let go of the string, the ball flies off tangent to its original circular path. With no centripetal force on it, the ball then proceeds along a straight path at constant speed as the first law of motion predicts. (For clarity, the effect of gravitation on the ball's motion is omitted from Fig. 4–2; actually, of course, it would fall to the ground in a parabolic path when released.)

A car rounding a curve is also an example of the necessity of a centripetal force if circular motion is to occur (Fig. 4–3). In the case of a car, the centripetal force is provided by the friction between the road and its tires. If the tires are worn and the road wet or icy, the frictional force is small and may not be enough to permit the car to turn.

4–2 Centripetal acceleration

To compute the centripetal force on a particular body undergoing uniform circular motion, we shall begin by determining the acceleration of such a body. Figure 4–4 shows the position and velocity of a particle moving along a circle of radius r at two different times. In this drawing we have represented the velocities by vectors; as usual, the length of each vector is proportional to the magnitude of the velocity, and the vector points in the direction of the motion. When the particle is at point A, its velocity is $\mathbf{v}_A$, and when it is at B, its velocity is $\mathbf{v}_B$. The acceleration of the particle is, by definition,

$$\mathbf{a} = \frac{\Delta \mathbf{v}}{\Delta t},$$

where

$$\Delta \mathbf{v} = \mathbf{v}_B - \mathbf{v}_A$$

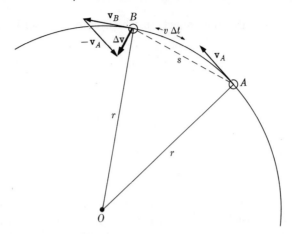

FIG. 4–4. *A* and *B* are two successive positions Δt sec apart of a particle undergoing uniform circular motion at the speed v in a circle of radius r. The velocity of the particle at A is $\mathbf{v}$ and at B is $\mathbf{v}_B$; the change in its velocity in going from A to B is $\Delta \mathbf{v} = \mathbf{v}_B - \mathbf{v}_A$. The chord joining A and B is s, while the actual distance the particle traverses is $v\,\Delta t$. In calculating the instantaneous acceleration of the particle we are restricted to having A and B an infinitesimal distance apart, in which case the chord and arc have the same length and $\Delta \mathbf{v}$ points radially inward to the center of the circle at O.

is the change in the particle's velocity in the time Δt it takes to go from A to B.

We recall from the previous chapter that to subtract a vector $\mathbf{v}_A$ from another vector $\mathbf{v}_B$, we first construct $-\mathbf{v}_A$ by redrawing $\mathbf{v}_A$ so that it points in the opposite direction and then add $-\mathbf{v}_A$ to $\mathbf{v}_B$ by placing what is now its tail at the head of $\mathbf{v}_B$. As shown in Fig. 4–4, the magnitude Δv of the vector difference $\Delta \mathbf{v}$ is related to v (the magnitude of both $\mathbf{v}_A$ and $\mathbf{v}_B$), r (the radius of the circle in which the particle revolves), and s (the distance between A and B) by the proportionality

$$\frac{\Delta v}{v} = \frac{s}{r}, \qquad (4\text{–}1)$$

since the velocity vector triangle whose sides are $\mathbf{v}_A$, $\mathbf{v}_B$, and $\Delta \mathbf{v}$ is similar to the space triangle whose sides are the radii OA and OB and the chord s.

The distance the particle actually covers in going from A to B is the arc joining these points, the length of which is $v\,\Delta t$. The distance s that appears in Eq. (4–1), however, is the chord joining A and B. Now the closer A and B are, the more nearly equal the chord joining them is to the arc. What we are calculating is the *instantaneous* acceleration of the particle, and we are therefore concerned with the case where A and B

are an infinitesimal distance apart, in which case the chord and arc are equal. Hence

$$s = v \, \Delta t,$$

and Eq. (4–1) becomes

$$\frac{\Delta v}{v} = \frac{v \, \Delta t}{r},$$

or, rearranging,

$$\frac{\Delta v}{\Delta t} = \frac{v^2}{r}. \tag{4–2}$$

The quantity $\Delta v/\Delta t$ on the left side of Eq. (4–2) is the magnitude of the instantaneous acceleration of a particle undergoing uniform circular motion with the speed v in a path of radius r. This acceleration is called the *centripetal acceleration* of the particle, the symbol for which is $\mathbf{a}_c$, and its magnitude is thus

$$a_c = \frac{v^2}{r} \, ; \tag{4–3}$$

the centripetal acceleration of a particle in uniform circular motion is proportional to the square of its speed and inversely proportional to the radius of its path.

In Fig. 4–4 the points A and B are a finite distance apart in order to make the geometry of the problem clear, and under these circumstances $\Delta\mathbf{v}$ does not point toward O, the center of the particle's circular path. When A and B are an infinitesimal distance apart, however, $\Delta\mathbf{v}$ does point toward O. Because the centripetal acceleration $\mathbf{a}_c$ is an instantaneous acceleration, being defined as the particle's acceleration when Δt and hence s are vanishingly small, we are solely concerned with the latter case, and the direction of $\mathbf{a}_c$ is accordingly radially inward.

To illustrate the application of Eq. (4–3) we shall compute the centripetal acceleration of a point on the earth's equator. The radius of the earth is 6.4×10^6 m, and it rotates once each day. Since the circumference of a circle of radius r is $2\pi r$, a point on the equator travels a distance of

$$s = 2\pi r_{\text{earth}} = 2\pi \times 6.4 \times 10^6 \text{ m} = 4.0 \times 10^7 \text{ m}$$

per day. Since there are 24 hr/day, 60 min/hr, and 60 sec/min, the number of seconds in a day is

$$1 \text{ day} = 24 \text{ hr} \times 60 \, \frac{\text{min}}{\text{hr}} \times 60 \, \frac{\text{sec}}{\text{min}}$$

$$= 86{,}400 \text{ sec} = 8.64 \times 10^4 \text{ sec}.$$

The speed of a point on the equator is therefore

$$v = \frac{s}{t} = \frac{4.0 \times 10^7 \text{ m}}{8.64 \times 10^4 \text{ sec}} = 4.6 \times 10^2 \frac{\text{m}}{\text{sec}},$$

and its centripetal acceleration is

$$a_c = \frac{v^2}{r} = \frac{(4.6 \times 10^2 \text{ m/sec})^2}{6.4 \times 10^6 \text{ m}} = 3.3 \times 10^{-2} \frac{\text{m}}{\text{sec}^2}.$$

4–3 Centripetal force

From the second law of motion,

$$\mathbf{F} = m\mathbf{a},$$

we see that the centripetal force $\mathbf{F}_c$ which must be acting upon a body of mass m that is in uniform circular motion is

$$\mathbf{F}_c = m\mathbf{a}_c,$$

and its magnitude is

$$F_c = \frac{mv^2}{r}. \tag{4–4}$$

Evidently the centripetal force that must be exerted to maintain a body in uniform circular motion increases with increasing mass and with increasing speed, with the force more sensitive to a change in speed since it is the square of the speed that is involved. Doubling the mass doubles the required centripetal force, but doubling the speed quadruples it. Increasing the radius of the path, however, reduces the force required; doubling the radius reduces the force to half its previous value. Figure 4–5 illustrates the dependence of centripetal force $\mathbf{F}_c$ on these factors.

Examples of centripetal forces. Let us calculate the centripetal force required by a car making a turn (Fig. 4–6). The car might weigh 3000 lb and be traveling at 15 mi/hr around a turn of radius 100 ft. The mass of the car is

$$\text{Mass} = \frac{\text{weight}}{\text{acceleration of gravity}},$$

$$m = \frac{3000 \text{ lb}}{32 \text{ ft/sec}^2} = 94 \text{ slugs}.$$

Since 15 mi/hr = 22 ft/sec (Appendix III), the centripetal force is

$$F_c = \frac{94 \text{ slugs} \times (22 \text{ ft/sec})^2}{100 \text{ ft}} = 455 \text{ lb}.$$

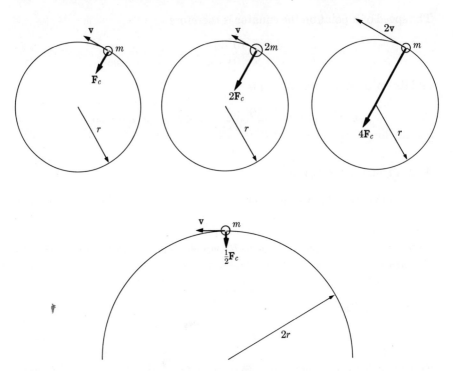

FIG. 4–5. The centripetal force $\mathbf{F}_c$ on a body in uniform circular motion is directly proportional to m, its mass, and to v^2, the square of its velocity, and inversely proportional to r, the radius of its path.

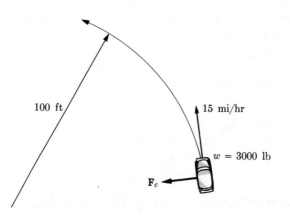

FIGURE 4–6

The centripetal force of 455 lb must be provided by the pavement acting on the car's tires through the agency of friction. We can readily compute the minimum coefficient of friction μ that must be present if the car is to make the turn without skidding. The frictional force is

$$F_f = \mu N$$

in general. To find μ, we substitute 455 lb for F_f and 3000 lb for the normal force N, which here is the car's weight w. Thus

$$\mu = \frac{F_f}{N} = \frac{F_c}{w} = \frac{455 \text{ lb}}{3000 \text{ lb}} = 0.15$$

to two significant figures, which is a reasonable value under good driving conditions (Table 2–1).

Another calculation we can make is even more interesting. Let us determine the minimum velocity an earth satellite must have in order to pursue a stable orbit without falling to the ground (Fig. 4–7). (We neglect the resistance of the atmosphere, which, even at the great altitudes of actual satellites, will ultimately bring them down.) Near the earth the gravitational force on a body of mass m is its weight, so that

$$F = mg, \qquad (4\text{–}5)$$

where g is the acceleration of gravity. For uniform circular motion about the earth this force must provide the body with the centripetal force

$$F_c = \frac{mv^2}{r}.$$

Hence, equating the gravitational force mg with the centripetal force F_c,

$$mg = \frac{mv^2}{r},$$

$$v^2 = rg,$$

$$v = \sqrt{rg}.$$

FIG. 4–7. The minimum speed an earth satellite must have in order to pursue a stable orbit is 7.9×10^3 m/sec, which is about 18,000 mi/hr.

Substituting $r = 6.4 \times 10^6$ m (the earth's radius) and $g = 9.8$ m/sec², we find that

$$v = \sqrt{6.4 \times 10^6 \text{ m} \times 9.8 \text{ m/sec}^2} = 7.9 \times 10^3 \text{ m/sec},$$

which is about 5 mi/sec or

$$5\,\frac{\text{mi}}{\text{sec}} \times 60\,\frac{\text{sec}}{\text{min}} \times 60\,\frac{\text{min}}{\text{hr}} = 18{,}000\,\frac{\text{mi}}{\text{hr}}\,!$$

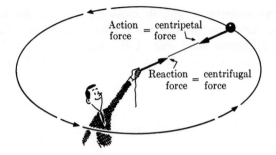

Fɪɢ. 4–8. Centrifugal force.

4–4 Centrifugal force

Let us return for a moment to the case of a ball being whirled at the end of a string. By means of the string we transmit the centripetal force

$$F_c = \frac{mv^2}{r}$$

to the ball, pulling it inward in a circular path. At the same time that we pull inward on the ball, however, by Newton's third law of motion the ball pulls *outward* on us, as in Fig. 4–8. This outward reaction force on us is called *centrifugal force*, literally, "force fleeing from the center." Centrifugal force may properly be thought of as a manifestation of inertia in the sense that the tendency of the ball is to move in a straight path rather than be accelerated in a curved path. If the string breaks, the ball does not fly off because "centrifugal force pushes it out," as we might be tempted to think; when the string breaks there is no longer any centripetal force on the ball, and so it simply proceeds tangentially to its former path.

The distinction between centripetal and centrifugal forces is illustrated by a person in a car which is rounding a curve (Fig. 4–9). As the car begins turning, the upper part of his body continues along a straight

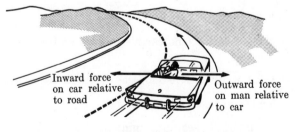

Fɪɢ. 4–9. Centripetal and centrifugal forces.

path while the lower part moves with the car. To a stationary observer at the side of the road, the lower part of the passenger's body is acted upon by a centripetal force pulling it inward toward the center of the curve. The person in the car, on the other hand, sees himself tending to move outward *with respect to the car,* and attributes this acceleration to a centrifugal force.

We shall not refer to centrifugal forces further in this book. Centripetal forces, however, the forces that cause circular motion from the point of view of a stationary observer, will figure in a number of important discussions in what follows.

4–5 Gravitation

As we have noted, the earth and planets pursue approximately circular orbits around the sun. (The orbits are actually ellipses, but the difference is not important for our analysis.) Isaac Newton proposed that the inward force exerted by the sun that makes these orbits possible is merely one example of a universal force, called *gravitation,* that acts between all bodies in the universe. According to Newton, the centripetal acceleration of the moon and the downward acceleration g of the objects dropped at the earth's surface have an identical cause, namely the gravitational attraction of the earth.

An important empirical observation enables us to determine in detail the properties of the gravitational forces that tend to pull all bodies toward one another. This is the discovery by Johannes Kepler (1571–1630) of the fact that the ratio between the square of the time T required by a planet to make a complete revolution about the sun and the cube of the radius r of the planet's orbit is a constant throughout the solar system. In symbols,

$$\frac{T^2}{r^3} = K, \tag{4-6}$$

where K has the same value for every planet.

The distance s traveled by a body at the constant speed v in a time interval t is

$$s = vt.$$

Here, in uniform circular motion, the distance the body travels in each revolution is the circumference of the circle, $2\pi r$, and so

$$2\pi r = vT,$$

$$T = \frac{2\pi r}{v}. \tag{4-7}$$

Thus we may express Eq. (4–6) as

$$\frac{T^2}{r^3} = \left(\frac{2\pi r}{v}\right)^2 r^{-3} = \frac{4\pi^2}{v^2 r}.$$

Since Kepler found that the ratio T^2/r^3 equals the constant K,

$$K = \frac{4\pi^2}{v^2 r}$$

and so

$$v^2 = \frac{4\pi^2}{Kr}. \tag{4–8}$$

Equation (4–8) states that the square of the speed of a planet is inversely proportional to its distance r from the sun.

According to Newton's hypothesis, the gravitational force F_g exerted by the sun on each planet is the same as the centripetal force

$$F_c = \frac{mv^2}{r}$$

acting upon it, where m is the planet's mass. Substituting F_g for F_c and $4\pi^2/Kr$ for v^2, we have

$$F_g = \frac{4\pi^2}{K}\frac{m}{r^2}; \tag{4–9}$$

the gravitational force on a planet varies inversely with the square of its distance from the sun and directly with its mass.

To complete our analysis we shall go back to the third law of motion:

When one body exerts a force on another body, the second body exerts a force on the first body of the same magnitude but in the opposite direction.

For this law to hold in the solar system, the gravitational force F_g on each planet must be proportional to the sun's mass M as well as to the mass m of the planet. If Eq. (4–9) is correct at all, we should be able to apply it to *either* a planet or the sun and get the same value of F_g. Since r is the same either way, F_g must be proportional to *both* m and M. Because the sun's mass M is a constant, we may express the constant quantity $4\pi^2/K$ as GM in Eq. (4–9). That is, we shall let

$$\frac{4\pi^2}{K} = GM,$$

where G is a universal constant. Our final expression for the gravitational

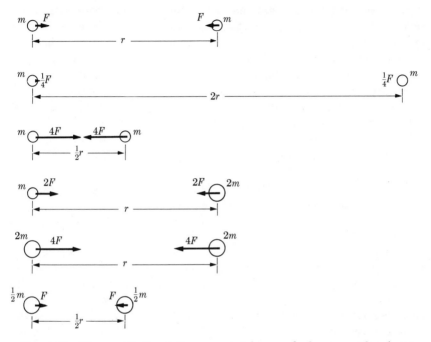

FIG. 4–10. The gravitational force exerted by one body on another is proportional to each of their masses and inversely proportional to the square of the distance between them.

force between the sun and any planet is therefore

$$F_g = \frac{GmM}{r^2}.$$ (4–10)

To emphasize the general applicability of this formula, we may express it as

$$F_g = \frac{Gm_1m_2}{r^2},$$ (4–11)

where m_1 and m_2 are the masses of *any* two bodies in the universe and r the distance between them (Fig. 4–10). Equation (4–11) is Newton's law of universal gravitation:

Every body in the universe attracts every other body with a force directly proportional to both of their masses and inversely proportional to the square of the distance separating them.

What is our justification for extending Eq. (4–10), valid for the sun and planets, to Eq. (4–11), which we assert is valid for the entire universe, describing the gravitational attraction of bodies both larger and smaller than the members of the solar system? There is no simple answer to this legitimate query; instead we must invoke a broad body of knowledge that bears upon the extrapolation of Eq. (4–10). For example, we observe that all the matter on the earth's surface experiences the same acceleration in free fall, suggesting identical gravitational behavior; careful analysis of the light reaching us from the stars and nebulae throughout the visible universe indicates that the matter of which these bodies is composed behaves identically with the matter found on earth; and so on. Nowhere do we find reason to suspect that there should be any masses in the universe that do not obey Newton's law of gravitation, and it is unreasonable to postulate the existence of such masses with no evidence for the necessity of doing so.

Another problem arises regarding Eq. (4–11): between what points in the respective bodies concerned should r be measured? The planets are sufficiently far from the sun for their dimensions to be negligible compared with the appropriate values of r, but this is certainly not true for, say, the moon and the earth. In computing the gravitational force the earth exerts on the moon, should r be the distance between their facing surfaces, between their centers, or between some other points? Newton himself succeeded in solving this problem with the help of the calculus he had invented, and showed that in the case of spherically symmetric bodies, r is to be taken as the distance between their centers. That is, a sphere behaves gravitationally as if its entire mass were concentrated at its center (Fig. 4–11). For bodies of other shapes the problem is more difficult, but it is always possible to establish a *center of mass* from which r is to be taken for the purpose of gravitational calculations. In the event that the two interacting bodies are so close together that r is small compared with their dimensions, for instance an apple at the earth's surface, local irregularities become significant. In fact, a common tool of the mineral prospector is an instrument for determining variations in the gravitational force on a standard object at different locations.

FIG. 4–11. A spherical body that is either uniform or consists of uniform spherical shells behaves gravitationally as though its mass is concentrated at its center.

4–6 The constant of gravitation

The constant of gravitation G cannot be determined from astronomical data alone, as Newton realized. A direct measurement of the gravitational force between two known masses a known distance apart is required. The difficulty here is that gravitational forces are minute between objects of laboratory size: two 1-ton spheres whose centers are 10 ft apart attract each other with a force of only 0.0000013 lb! The measurement of such forces had to wait for over a century after Newton's work. It was finally accomplished by Henry Cavendish in 1798. The principle of the instrument he used is shown in Fig. 4–12; called a *torsion balance*, it measures the forces exerted on the small spheres when the large ones are brought close to them in terms of the resulting twist in the fine suspending thread. (The torsion balance may be thought of as the rotational analog of an ordinary spring balance which measures force in terms of the extension of a spring.) The currently accepted value of G is

$$G = 6.67 \times 10^{-11} \frac{\text{n·m}^2}{\text{kg}^2}.$$

4–7 The mass of the earth

We are now able to determine the mass of the earth itself, something we could not hope to do by direct experiment. Let us consider an object of mass m near the earth's surface. According to Eq. (4–11), it is attracted to the earth with a force

$$F_g = \frac{Gmm_e}{r_e^2},$$

where m_e is the earth's mass and r_e the earth's radius. Hence the body

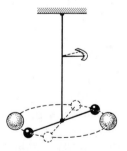

FIG. 4–12. A method of determining the constant of gravitation G.

will experience an acceleration a given by

$$F_g = ma,$$

or, substituting for the gravitational force F_g,

$$\frac{Gmm_e}{r_e^2} = ma.$$

Canceling the mass m on both sides of this equation and solving for a, we have

$$a = \frac{Gm_e}{r_e^2}.$$

But this acceleration is the known acceleration of gravity g! If we substitute g for a, we find that

$$g = \frac{Gm_e}{r_e^2}$$

and

$$m_e = \frac{gr_e^2}{G}.$$

Inserting the numerical values

$$g = 9.8 \, \frac{m}{sec^2}, \qquad r_e = 6.38 \times 10^6 \, m, \qquad G = 6.67 \times 10^{-11} \, \frac{n \cdot m^2}{kg^2},$$

we find that

$$m_e = \frac{9.8 \, m/sec^2 \times (6.38 \times 10^6 \, m)^2}{6.67 \times 10^{-11} \, n \cdot m^2/kg^2}$$

$$= 5.98 \times 10^{24} \, kg.$$

The earth's mass is 5.98×10^{24} kg, or 6.6×10^{21} tons.

4–8 The gravitational field

In everyday life, forces seem to be transmitted by "direct contact": something pushes or pulls something else. Gravitational forces, however, act in the absence of direct contact and are able to produce their effects through millions of miles of empty space. There are two ways of regarding this situation. The first is merely to think of gravitational phenomena as representing the mutual interaction of separated centers of force, as

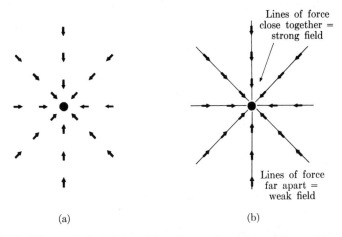

FIG. 4–13. The method of determining the configuration of lines of force near a body.

"action at a distance." In the other view the interaction is considered as involving the region in which the centers of force are located: the presence of a single mass alters the properties of the region around it, setting up a *gravitational field* which stands ready to interact with any other masses brought into it. Thus the sun is surrounded by a gravitational field, and it is the force exerted *by this field* on the planets which provides the centripetal force that holds them in their orbits. A number of different factors have led to the adoption of the field concept to replace the older one of action at a distance, and we shall consider several of them in later chapters.

To help visualize what is meant by a gravitational field we may use *lines of force*. These are constructed as shown in Fig. 4–13: at several points in space we draw arrows in the direction in which a small test mass would go if allowed to move freely there. Then we connect these arrows so as to form smooth curves, called lines of force, arranging their concentration in the vicinity of any point to be proportional to the magnitude of the force on the test mass at that point. The resulting picture may be thought of as a portrait of the gravitational field. If we were to place a mass in the field, it would experience a force along whatever line of force it was on, with the direction of the arrow and with a magnitude depending upon how close together the lines of force are in its vicinity. Figure 4–14 shows the lines of force near two spheres that are close together. A body in the center will never reach either, since the net force on it is zero. At distances large relative to the diameters of the spheres and their separation, the field reduces to that of a single point mass.

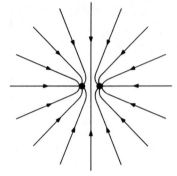

FIG. 4–14. Lines of force near two identical adjacent spheres.

It is important, of course, to keep in mind that lines of force do not actually exist threading space; they are no more than a device for giving intuitive form to our thinking about gravitational fields. We shall apply the motion of lines of force to other fields of force in addition to gravitational fields. Electric and magnetic fields, for instance, are often described in this manner.

IMPORTANT TERMS

A body traveling in a circle at constant speed is said to be undergoing *uniform circular motion.*

The velocity of a body in uniform circular motion continually changes in direction although its magnitude remains constant. The acceleration that causes the body's velocity to change is called *centripetal accelera-tion,* and it points toward the center of the body's circular path. Cen-tripetal acceleration is proportional to the square of the body's speed and inversely proportional to the radius of its path.

The inward force that provides a body in uniform circular motion with its centripetal acceleration is called *centripetal force.* From the second law of motion, the centripetal force on such a body equals the product of its mass and its acceleration and is in the same direction as the latter.

Centrifugal force is the reaction force exerted by a body in uniform circular motion on the agency that is providing the centripetal force holding it in its circular path. Centrifugal force may be thought of as a manifestation of inertia.

Newton's *law of universal gravitation* states that every body in the universe attracts every other body with a force directly proportional to both their masses and inversely proportional to the square of the distance separating them.

The *gravitational field* of a body is the alteration in the properties of the region around it caused by its presence. The gravitational field of a body is what interacts with another body brought into its vicinity to produce what we recognize as the gravitational force between them.

Lines of force are means for visualizing a gravitational (or other) field. Their direction at any point is that in which a test body would move if released there, and their concentration in the neighborhood of a point is proportional to the magnitude of the force on a test particle at that point.

PROBLEMS

1. A phonograph record 12 in. in diameter rotates $33\frac{1}{3}$ times per minute. (a) What is the linear speed of a point on its rim in ft/sec? (b) What is the centripetal acceleration of a point on its rim?

2. The minute hand of a large clock is 0.5 m long. (a) What is the linear speed of its tip in m/sec? (b) What is the centripetal acceleration of its tip?

3. What is the centripetal force needed to keep a 3-kg mass moving in a circle of radius 0.5 m at a speed of 8 m/sec?

4. What is the centripetal force needed to keep a 12-lb weight moving in a circle of radius 2 ft at a speed of 10 ft/sec?

5. A certain string 4 ft long breaks when its tension is 40 lb. (a) What is the greatest speed in ft/sec at which it can be used to whirl a stone weighing 3 lb? (b) Express this speed in revolutions/sec. (Neglect the gravitational pull of the earth on the stone.)

6. A string 1 m long is used to whirl a $\frac{1}{2}$-kg stone in a vertical circle at a speed of 5 m/sec. (a) What is the tension in the string when the stone is at the top of the circle? (b) at the bottom?

7. A string 2 ft long is used to whirl a 5-lb stone in a vertical circle. What must the minimum speed of the stone be if the string is to be just taut when the stone is at the top of the circle?

8. A string 1.5 m long is used to whirl a 1-kg stone in a vertical circle at a speed of 5 m/sec. What is the tension in the string when it is horizontal?

9. Why is it harder for a car to make a given turn at high speed than at low speed?

10. A man weighs 160 lb at the earth's surface. (a) What would he weigh at an altitude of one earth radius? (b) What would his mass be at that altitude?

11. The radius of the earth is 6.4×10^6 m. What is the acceleration of a meteor when it is 8×10^6 m from the center of the earth?

12. Even if the earth were a perfect sphere, the experimental value of g, the acceleration of gravity, would be least at the equator. Why?

Centripetal force

$F_c = ma$

$F_c = \frac{mv^2}{R}$ mg

$\frac{F_c}{R}$

For

Centripetal acceleration

$a_c = \frac{v^2}{R}$

Centrif

$F_c = \frac{mv^2}{R}$

13. Is the gravitational force between the earth and the sun the same at all times of the year? Explain.

14. The moon's mass is 1.2% of the earth's mass. Is the gravitational force exerted by the moon on the earth less than, the same as, or more than the gravitational force exerted by the earth on the moon? Explain.

15. The earth's equatorial radius is about 21 km (1 km = 10^3 m) greater than its polar radius, a phenomenon known as the "equatorial bulge." (a) How is this fact related to the daily rotation of the earth on its axis? (b) If the earth were to rotate twice as fast as it now does, would you expect the equatorial bulge to be more than, the same as, or less than it is now? Explain.

16. The mass of the planet Jupiter is 1.9×10^{27} kg and that of the sun is 2.0×10^{30} kg. The average distance between them is 7.8×10^{11} m. (a) What is the gravitational force the sun exerts on Jupiter? (b) Assuming that Jupiter has a circular orbit, what must its speed be for the orbit to be a stable one?

17. The moon's radius is 27% of the earth's radius and its mass is 1.2% of the earth's mass. (a) What is the acceleration of gravity (in ft/sec²) on the surface of the moon? (b) How much would a boy weigh there whose weight on the earth is 100 lb?

18. The mass of the planet Jupiter is 1.9×10^{27} kg and its radius is 7.0×10^7 m. What is the acceleration of gravity (in m/sec²) on the surface of Jupiter?

19. A 2-kg mass is 1 m away from a 5-kg mass. What is the gravitational force (a) that the 5-kg mass exerts upon the 2-kg mass; (b) that the 2-kg mass exerts upon the 5-kg mass? (c) If both masses are free to move, what are their respective accelerations in the absence of other forces?

20. (a) Calculate the force the sun exerts on 1 kg of water at a point nearest the sun and also at the earth's center. (b) Calculate the force the moon exerts on 1 kg of water at the same locations. (c) On the basis of these results, explain why the moon is more effective than the sun at causing tides.

21. Sketch the pattern of gravitational lines of force near (a) an egg-shaped body of uniform composition; (b) three identical point masses at the corners of an equilateral triangle; (c) a flat, square mass (like a tabletop).

22. Can lines of force ever intersect in space? Explain.

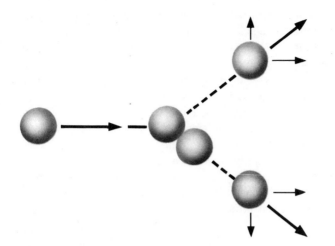

Energy and Momentum | 5

5–1 Introduction

One of the chief distinctions between the physical sciences and nearly all other scientific disciplines is that in the former certain very general conservation theorems have been found valid. A conservation theorem states that no matter what changes a system (in a general sense) undergoes, a certain quantity keeps the same value it had originally. For example, the law of conservation of mass revolutionized chemistry by holding that the total mass of the products of a chemical reaction is the same as the total mass of the original substances. The increase in mass of a piece of iron when it rusts therefore indicates that the iron has combined with some other material, rather than having decomposed, as the early chemists believed. In fact, the gas oxygen was discovered in the course of seeking this other material.

Given one or more conservation theorems applicable to a given system, we can immediately determine which classes of events can take place in the system and which cannot. Thus when iron rusts, the gain in mass means that it has combined chemically with something else. In physics it is often possible to draw some conclusions about the behavior of the particles that make up a system without a detailed investigation, basing our analysis simply upon the conservation of some particular quantities. The power of this method of approach is exemplified by the great success of physics, a field in which a large variety of different quantities may, under appropriate circumstances, be conserved. In this chapter we shall consider conservation of energy and conservation of momentum.

The laws of motion we have discussed in the past several chapters permit us, in principle, to solve all problems involving forces and moving bodies. However, these laws in their original form are seldom very useful, because in order to apply them we must take into detailed account all the various forces acting in a given situation at every point in the path of a moving body, usually a difficult and complicated procedure. The real utility of the laws of motion is that they provide the basis for conservation theorems which permit us to draw definite conclusions about the relationship between the initial and final states of motion of some mechanical system without having to investigate the process at every stage.

5–2 Work

When we push against a brick wall, nothing happens. We have applied a force, but the wall has not yielded and shows no effects. On the other hand, when we apply exactly the same force to a stone, the stone flies through the air for some distance. In the latter case something has been accomplished because of our push, while in the former there has been no result (Fig. 5–1). What is the essential difference between the two situations? In the first case, where we pushed against a wall, our hand did not move; the force remained in place. But in the second case, where we threw the stone, our hand *did* move while the force was being applied and before the stone left our hand. The *motion* of the force was what made the difference.

Fig. 5–1. For a force to cause a body to move, the force itself must undergo a displacement.

If we think carefully along these lines, we will see that, whenever a force acts so as to produce motion in a body, the force itself undergoes a displacement. In order to make this notion definite, a physical quantity called *work* is defined as follows: *work = force times the distance through which the force acts*. In symbols,

$$W = Fs. \tag{5-1}$$

This definition is a great help in clarifying the effects of forces. Unless a force acts through a distance, no work is done, no matter how great the force. And even if a body moves through a distance, no work is done unless a force is acting upon it or it exerts a force on something else.

Our intuitive concept of work is in accord with the precise definition above: when something happens because a person applies a force of some kind, we say that he has done work. Here we have simply broadened the concept to include inanimate forces. (We still must be careful, though; while we may become tired after pushing against a brick wall

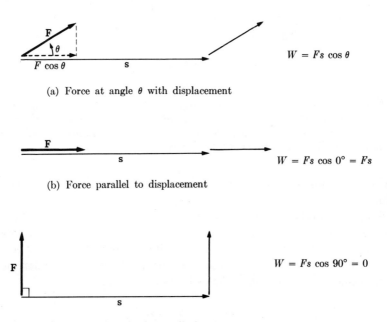

(a) Force at angle θ with displacement

(b) Force parallel to displacement

(c) Force perpendicular to displacement

FIG. 5-2. (a) The work W done by a force $\mathbf{F}$ when it acts through the distance $\mathbf{s}$ is given by $W = Fs \cos \theta$, where θ is the angle between the directions of $\mathbf{F}$ and $\mathbf{s}$. (b) When $\mathbf{F}$ is parallel to $\mathbf{s}$, $\theta = 0°$ and, since $\cos 0° = 1$, $W = Fs$. (c) When $\mathbf{F}$ is perpendicular to $\mathbf{s}$, $\theta = 90°$ and, since $\cos 90° = 0$, $W = 0$.

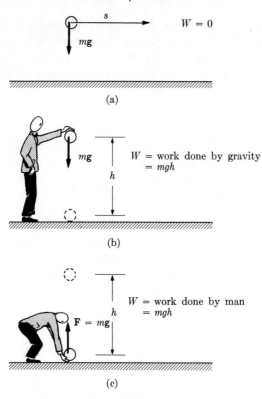

FIG. 5–3. (a) The force of gravity does no work on bodies that move parallel to the earth's surface. (b) When a body of mass m falls from a height h, the force of gravity does the work mgh on it. (c) When a man lifts a body of mass m to a height h, he does the work mgh on it.

for a long time, we still have done no work on the wall if it remains in place.)

Equation (5–1), which defines work, has an important qualification: the vector displacement **s** must be parallel to the direction of the force **F**. If **F** and **s** are not parallel, we must take into account the angle θ between them (Fig. 5–2). Since the projection of **F** along the direction of the displacement **s** is $F \cos \theta$, in this situation we have

$$W = Fs \cos \theta. \qquad (5\text{–}2)$$

This equation is always correct, since when the force and its motion are parallel, the angle $\theta = 0$ and $\cos 0° = 1$. When the force and its displacement are perpendicular to each other, $\theta = 90°$ and $\cos 90° = 0$; no work is done, even though the force has moved. An example of this is

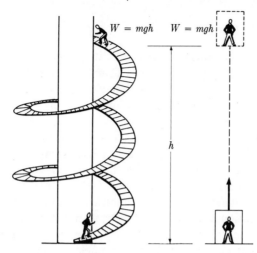

Fig. 5–4. In the absence of friction, the work done in lifting a mass m to a height h is mgh regardless of the exact path taken.

motion parallel to the earth's surface, where the force of gravity, which pulls downward, is perpendicular to all horizontal displacements (Fig. 5–3). However, when we lift an object, work *is* done, since the motion and force are now parallel.

It is easy to compute the work done in lifting a body against gravity. The force of gravity on a body of mass m is the same as the weight of the body mg. Hence if the body is raised to a height h, we have

$$F = mg, \qquad s = h,$$

and so the work done is

$$W = Fs = mgh. \tag{5–3}$$

To lift a body of mass m to a height h above its starting point requires that we perform the amount of work mgh. We should note that only the height h is involved, and not the particular route taken by the body in being lifted (Fig. 5–4); excluding any frictional effects, just as much work must be expended to climb a flight of stairs as to go up in an elevator to the same floor (though not by the same agent).

In the British system, the unit of work is the *foot-pound*, abbreviated ft·lb. One ft·lb is equal to the work done by a force of 1 lb acting through a distance of 1 ft. In the metric system, work is given a special unit, the *joule*, abbreviated j. One joule is equal to the work done by a force of 1 newton acting through a distance of 1 meter. That is,

$$1 \, \text{j} = 1 \, \text{n·m}.$$

5–3 Energy

Work is a simple enough notion, but it brings us to the complicated and many-sided concept of *energy*. Energy may be thought of as the ability to do work. When we say that something has energy, we mean that it is capable of exerting a force on something else and doing work on it. On the other hand, when we do work on something, we have added to it an amount of energy equal to the work done. The units of energy are the same as those of work, the foot-pound and the joule.

What properties can a body have that may be converted into work? In other words, what forms does energy take? We shall consider in this chapter two broad categories of mechanical energy, *kinetic energy*, which is energy of motion, and *potential energy*, which is energy of position. Later we shall find that there are other varieties of energy such as heat and electrical, magnetic, chemical, and nuclear energy, as well as mass energy, which is that energy possessed by a body by virtue of its mass alone.

5–4 Kinetic energy

When we perform work on a stone by throwing it, what becomes of this work? Let us suppose that we apply the uniform force of magnitude F to the stone for a distance s before it leaves our hand. If the stone's mass is m, the second law of motion,

$$F = ma,$$

tells us that the acceleration of the stone while it is in our hand is

$$a = \frac{F}{m}. \tag{5-4}$$

We know from Eq. (1–9), that when a body starting from rest ($v_o = 0$) undergoes an acceleration a through a distance s, its final velocity v is related to a and s by the formula

$$v^2 = 2as. \tag{5-5}$$

Inserting for a the value of Eq. (5–4),

$$v^2 = 2\frac{F}{m}s,$$

or rewriting this equation,

$$Fs = \tfrac{1}{2}mv^2. \tag{5-6}$$

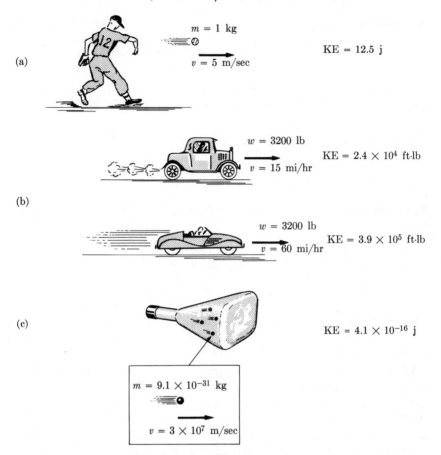

(a)

$m = 1$ kg

KE = 12.5 j

$v = 5$ m/sec

$w = 3200$ lb

KE = 2.4×10^4 ft·lb

$v = 15$ mi/hr

(b)

$w = 3200$ lb

KE = 3.9×10^5 ft·lb

$v = 60$ mi/hr

(c)

KE = 4.1×10^{-16} j

$m = 9.1 \times 10^{-31}$ kg

$v = 3 \times 10^7$ m/sec

FIG. 5–5. Some kinetic energies.

The quantity on the left-hand side of Eq. (5–6), Fs, is the work our hand has done in throwing the stone. The quantity on the right-hand side, $\frac{1}{2}mv^2$, must therefore be the energy acquired by the stone as a result of the work we did on it. Accordingly we define the *kinetic energy* of a moving body as

$$\text{Kinetic energy} = \text{KE} = \tfrac{1}{2}mv^2, \qquad (5\text{–}7)$$

one-half the product of its mass and the square of its velocity. (We use KE as the abbreviation for kinetic energy.) A moving body is therefore able to perform an amount of work equal to $\frac{1}{2}mv^2$ in the course of being stopped.

To obtain a feeling for the magnitudes of the kinetic energies characteristic of various common motions, we shall compute a few such values. A ball of mass 1 kg (Fig. 5–5a) is thrown with a speed of 5 m/sec.

Its kinetic energy is

$$\mathrm{KE} = \frac{1}{2}\, mv^2 = \frac{1}{2} \times 1\ \mathrm{kg} \times \left(5\ \frac{\mathrm{m}}{\mathrm{sec}}\right)^2 = 12.5\ \mathrm{j}.$$

An automobile which weighs 3200 lb (Fig. 5–5b) is traveling at a speed of 15 mi/hr (22 ft/sec). Its kinetic energy is

$$\mathrm{KE} = \frac{1}{2}\, mv^2 = \frac{1}{2}\, \frac{w}{g}\, v^2 = \frac{1}{2} \times \frac{3200\ \mathrm{lb}}{32\ \mathrm{ft/sec^2}} \times \left(22\ \frac{\mathrm{ft}}{\mathrm{sec}}\right)^2$$

$$= 24{,}200\ \mathrm{ft\cdot lb} = 2.4 \times 10^4\ \mathrm{ft\cdot lb}.$$

The same automobile at a speed of 60 mi/hr (88 ft/sec) has a kinetic energy of

$$\mathrm{KE} = \frac{1}{2}\, \frac{w}{g}\, v^2 = \frac{1}{2} \times \frac{3200\ \mathrm{lb}}{32\ \mathrm{ft/sec^2}} \times \left(88\ \frac{\mathrm{ft}}{\mathrm{sec}}\right)^2$$

$$= 387{,}200\ \mathrm{ft\cdot lb} = 3.9 \times 10^5\ \mathrm{ft\cdot lb},$$

which is 16 times greater than at a velocity of 15 mi/hr. The fact that kinetic energy is proportional to the *square* of the velocity is responsible for the severity of automobile accidents at high speeds.

Electrons are the smallest particles present in ordinary matter. The electron mass (provided that the electron speed is not too close to the speed of light, a requirement we shall discuss in Chapter 18) is 9.1×10^{-31} kg. Electrons constitute the beam in a television picture tube (Fig. 5–5c), whose impacts on the tube screen produce tiny flashes of light. Such electrons might have speeds of 3×10^7 m/sec; hence their kinetic energies are

$$\mathrm{KE} = \frac{1}{2}\, mv^2 = \frac{1}{2} \times 9.1 \times 10^{-31}\ \mathrm{kg} \times \left(3 \times 10^7\ \frac{\mathrm{m}}{\mathrm{sec}}\right)^2$$

$$= 4.1 \times 10^{-16}\ \mathrm{j}.$$

5–5 Potential energy

When we drop a stone from a height h, it falls faster and faster and finally strikes the ground. In striking the ground the stone does work; if it is sufficiently heavy and has fallen from a great enough height, the work done by the stone is manifest as a hole. Obviously the stone at its original position h above the ground had the capacity to do work, even though it was stationary at the time. We have already calculated that the work we must do to raise the stone to the height h is

$$W = mgh,$$

where m is the stone's mass. This amount of work can also be done *by* the stone after dropping from the height h. Since the raised stone has the potentiality of doing the amount of work mgh, we define its *potential energy as*

$$\text{Potential energy} = \text{PE} = mgh, \tag{5-8}$$

the product of its mass, the acceleration of gravity, and its height. (We use PE as the abbreviation for potential energy.) In the British system of units, weights rather than masses are usually specified. Since $w = mg$, we may also write Eq. (5–8) as

$$\text{PE} = wh, \tag{5-9}$$

which is more convenient in treating problems in this system of units.

The kinetic energy of a body dropped from a height h is

$$\tfrac{1}{2}mv^2 = mgh$$

just before it strikes the ground, since at this point all its potential energy has become kinetic energy (Fig. 5–6). Its speed at this point is therefore

$$v = \sqrt{2gh},$$

the same as the result we found in Chapter 1. The advantage of working out problems in terms of energy is that we compute kinetic-energy and potential-energy values at the beginning and end of the motion only.

Fig. 5–6. When a body of mass m falls from a height h, all its initial potential energy mgh has been converted to kinetic energy just before it strikes the ground. Its final speed is therefore $2gh$ (in the absence of friction) regardless of the precise path it takes.

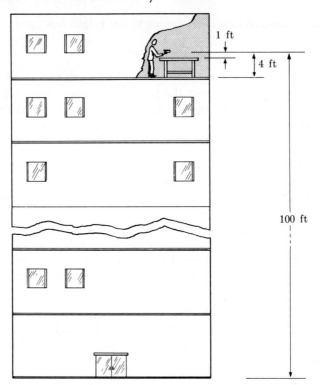

Fig. 5–7. The potential energy of a body depends upon the reference level from which its height h is measured.

Even if the body were not dropped vertically downward, but were, perhaps, allowed to slide down a smooth, curved track, we can still equate kinetic and potential energies to find its final velocity. If we used the methods of Chapter 1, we would instead have to consider the details of the stone's path and find ourselves with a difficult problem indeed.

It is important to note that the potential energy of a body depends upon where we choose the base height $h = 0$ to be. The potential energy of a 1-lb book (Fig. 5–7) held 1 ft above a desk is 1 ft·lb *with respect to the desk;* but it is 4 ft·lb with respect to the floor of the room, and conceivably 100 ft·lb or more with respect to the earth's surface. Hence we should be careful to interpret the h of Eq. (5–8) properly in the problem at hand.

We shall now compute a few potential-energy values. An apple of mass 0.5 kg is on a branch 5 meters from the ground. Its potential energy relative to the ground is

$$\text{PE} = mgh = 0.5 \,\text{kg} \times 9.8 \,\frac{\text{m}}{\text{sec}^2} \times 5 \,\text{m} = 24.5 \,\text{j}.$$

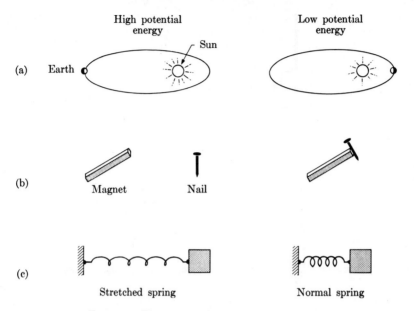

Fɪɢ. 5–8. Three examples of potential energy.

A 3200-lb automobile is at the top of a 100-ft hill. Its potential energy relative to the foot of the hill is

$$PE = mgh = wh = 3200 \text{ lb} \times 100 \text{ ft} = 320,000 \text{ ft·lb},$$

which is less than the kinetic energy of the same car when its velocity is 60 mi/hr. In other words, a crash at 60 mi/hr into a stationary obstacle will do more damage than dropping the car 100 ft.

An electron 10 cm (0.1 meter) from the bottom of a television picture tube has a potential energy relative to the bottom of

$$PE = mgh = 9.1 \times 10^{-31} \text{ kg} \times 9.8 \frac{m}{sec^2} \times 0.1 \text{ m} = 8.8 \times 10^{-31} \text{ j}.$$

This is almost 15 powers of 10 less than its kinetic energy!

Until now we have been considering only one type of potential energy, namely that possesed by a body by virtue of being raised above some reference level in the earth's gravitational field. The concept of potential energy is a much more general one, however, for it refers to the energy something has as a consequence of its position anywhere. The earth itself, for instance, has potential energy with respect to the sun, since if its orbital motion were to cease it would fall toward the sun (Fig. 5–8a). An iron nail has potential energy with respect to a nearby magnet, since it will fly to the magnet if released (Fig. 5–8b). An object at the end

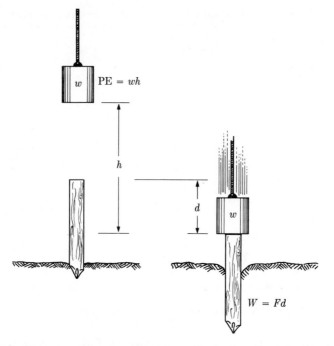

Fɪɢ. 5–9. The operation of a pile driver, illustrating the interchanges of potential energy, kinetic energy, and work.

of a stretched spring has potential energy with respect to its position when the spring has its normal extension, since if let go the object will move as the spring collapses (Fig. 5–8c). In each of these cases the body in question has the potentiality of doing work in its original position.

Most mechanical processes represent interchanges of energy among kinetic energy, potential energy, and work. A simple illustration is the operation of a pile driver (Fig. 5–9). First an engine performs an amount of work wh in raising a hammer of weight w to a height h above the top of the pile. Then the hammer, with a potential energy of wh, is released, and as it drops, its potential energy becomes kinetic energy. When the pile is struck by the hammer, the kinetic energy of the latter is converted to work as the pile is driven into the ground. If it requires a force F to drive the pile downward, the depth d to which the pile will go is

$$d = \frac{w}{F} h,$$

a formula we obtain simply by equating the initial potential energy of the hammer with the work it does.

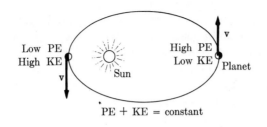

FIGURE 5-10

Planets in their orbital motions about the sun exhibit a constant inter-change of kinetic and potential energy. Planetary orbits are elliptical, so that at different points in its orbit a planet is at different distances from the sun (Fig. 5–10). When the planet is close to the sun, it has a low potential and a high kinetic energy; when it is far from the sun, it has a high potential and a low kinetic energy. The sum of KE + PE remains the same at all times.

5–6 Conservation of energy

As we have seen, it is possible to do work on something and thereby give it kinetic or potential energy which, in turn, is capable of doing work. When we throw a stone upward into the air, the work we do appears first as kinetic energy; as the stone rises, the kinetic energy gradually becomes potential energy; at its highest point, the stone has potential energy exclusively, which, as the stone begins to fall, is con-verted back into kinetic energy; and, when the stone strikes the ground, the kinetic energy does work. But what happens to the work done in opposing frictional forces?

To make this question more definite, let us consider the work done in moving a box whose weight is 100 lb a distance of 30 ft on a level floor where the coefficient of friction μ between the box and floor is 0.40. This is

$$W = F_f s = \mu w s = 0.4 \times 100 \text{ lb} \times 30 \text{ ft} = 1200 \text{ ft·lb.}$$

When the box is in its new position, it has no kinetic energy and no more potential energy than it had initially, even though we have done 1200 ft·lb of work in moving it. The work is dissipated; that is, it has not gone into a form capable of being directly converted back again into work, such as kinetic or potential energy.

Where has the work gone? To discover the answer, all we need do is rub one thing against another: two blocks of wood, perhaps, or even our hands. When this is done, we find that the objects are warmer than before. This is a quite general conclusion. We might do work against *viscosity*, which is friction in a fluid, by stirring some water vigorously, and again find that it becomes warmer. Evidently there is some connection between energy and heat, and this connection is a most significant one. While we shall discuss experiments involving heat energy in detail somewhat later, we can state here the general conclusion that whenever mechanical energy disappears in a system of bodies isolated from its surroundings, all the energy is converted into either work or heat. (This conclusion holds even when some other form of energy intervenes between the extremes of mechanical and heat energy.) The above statement is one way of expressing the law of conservation of energy.

Einstein's discovery of the interchangeability of mass and energy (Chapter 18) has compelled the extension of the law of conservation of energy to include situations involving the conservation of mass to energy and vice versa. A modern statement of this law might therefore be

The total amount of energy in a system isolated from its surroundings always remains constant although energy transformations from one form to another, reckoning mass as a form of energy, may occur within the system.

5-7 Momentum

The second law of motion states that the force $\mathbf{F}$ applied to a body of mass m that undergoes an acceleration $\mathbf{a}$ is given by

$$\mathbf{F} = m\mathbf{a}.$$

We can express the acceleration $\mathbf{a}$ of a body in terms of its velocities $\mathbf{v}_1$ and $\mathbf{v}_2$ at the two different times t_1 and t_2 as

$$\mathbf{a} = \frac{\mathbf{v}_2 - \mathbf{v}_1}{t_2 - t_1}.$$

Hence the second law may be rewritten

$$\mathbf{F} = \frac{m(\mathbf{v}_2 - \mathbf{v}_1)}{t_2 - t_1} = \frac{m\mathbf{v}_2 - m\mathbf{v}_1}{t_2 - t_1},$$

or

$$\mathbf{F}(t_2 - t_1) = m\mathbf{v}_2 - m\mathbf{v}_1. \tag{5-10}$$

In rewriting $\mathbf{F} = m\mathbf{a}$ as above we have merely made use of the definition that acceleration is the rate of change of velocity; when the force $\mathbf{F}$

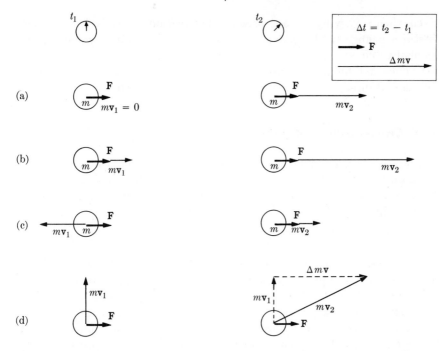

Fɪɢ. 5–11. Applying a constant force $\mathbf{F}$ to a mass m for a time Δt changes its momentum by $\Delta m\mathbf{v} = \mathbf{F}\,\Delta t$. At (a) the mass is initially at rest, at (b) its initial momentum is in the same direction as $\mathbf{F}$, at (c) its initial momentum is in the opposite direction to $\mathbf{F}$, and at (d) its initial momentum is perpendicular to $\mathbf{F}$. Since momentum is a vector quantity, the momentum change $\Delta m\mathbf{v}$ must be added to the initial momentum $m\mathbf{v}$ by the process of vector addition.

is applied to a mass m at the time t_1, the velocity of the mass changes from its initial value $\mathbf{v}_1$ to the value $\mathbf{v}_2$ at the later time t_2. We may consider that the effect of applying a force for the time $t_2 - t_1$ to the mass m is to change the quantity $m\mathbf{v}$ from its original value to a different value, where $(m\mathbf{v}_2 - m\mathbf{v}_1)$ equals the force multiplied by the time during which it acts (Fig. 5–11). Another way of writing Eq. (5–10) is

$$\mathbf{F}\,\Delta t = \Delta m\mathbf{v}, \qquad (5\text{–}11)$$

where Δt is the time interval and $\Delta m\mathbf{v}$ the change in $m\mathbf{v}$.

The quantity $m\mathbf{v}$ is known as the *linear momentum* of a particle of mass m and velocity $\mathbf{v}$. The product of a constant force $\mathbf{F}$ and the time during which it acts $(t_2 - t_1)$ is called the *impulse* of the force. Thus

Impulse = change in momentum.

Even when a varying force is present, its effect on the body it acts upon is still a change in the body's momentum from its initial value $m\mathbf{v}_1$ to its final value $m\mathbf{v}_2$.

Impulse and momentum are vector quantities which require both magnitude and direction in order to be completely specified.

5–8 Conservation of momentum

The motivation behind introducing the idea of momentum appears when we consider a *system* of many particles rather than a single particle. If no forces from outside the system act upon its component particles, the total linear momentum of the system, which is the sum

$$MV = m_1\mathbf{v}_1 + m_2\mathbf{v}_2 + m_3\mathbf{v}_3 + \cdots \qquad (5\text{–}12)$$

of the individual momenta of its particles, cannot change; with no force there is no impulse, hence no change in momentum. However, the *distribution* of the total momentum $M\mathbf{V}$ among the various particles in the system may change without $M\mathbf{V}$ changing in the absence of an *external* force. Since we almost always can include the sources of all forces relevant to a particular process within what we choose to be our "system," in such cases we have the condition that no matter what interactions take place within the system, its total momentum never varies. This theorem is known as *conservation of linear momentum*. More formally, it states that

When the vector sum of the external forces acting upon a system of particles equals zero, the total linear momentum of the system remains constant.

Let us consider a specific example readily treated with the help of conservation of momentum. Suppose we have an isolated particle of mass m, initially at rest, that suddenly explodes into two particles of masses m_1 and m_2 that fly apart. The forces acting on the original particle that caused it to break up were internal ones, and no external force was present. Since m has the initial momentum of zero, the final momentum of m_1 and m_2, when added together, must also be zero. Hence

$$m\mathbf{v} = 0 = m_1\mathbf{v}_1 + m_2\mathbf{v}_2$$

and

$$\mathbf{v}_2 = -\frac{m_1}{m_2}\,\mathbf{v}_1, \qquad (5\text{–}13)$$

where $\mathbf{v}_1$ and $\mathbf{v}_2$ are the final velocities of the two fragments. We note

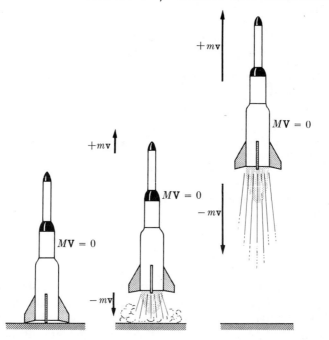

Fɪɢ. 5–12. Conservation of momentum in rocket flight. The downward momentum of the exhaust gases is exactly balanced by the upward momentum of the rocket itself.

immediately that these velocities must be in opposite directions along the same line. This problem could *not* be solved starting from $\mathbf{F} = m\mathbf{a}$, since we do not know explicitly what forces were acting during the explosion.

The principle underlying rocket flight is conservation of momentum. The total momentum of a rocket on its launching pad is zero. When it is fired, exhaust gases shoot downward at high speed, and the rocket moves upward to balance the momentum of the gases (Fig. 5–12). The total momentum of the system of rocket plus exhaust remains its initial value of zero. Rockets do not operate by "pushing" against their launching pads, the air, or anything else; in fact, they perform best in space, where there is no atmosphere to impede their motion. The energy of the rocket and its exhaust comes from chemical energy stored in its fuel.

As a numerical example, we shall compute the recoil velocity of a rifle when it fires a bullet. The rifle might weigh 5 lb and the bullet 0.03 lb, and the muzzle velocity of the bullet might be 2000 ft/sec (Fig. 5–13). Thus

$$m_1 = \frac{0.03 \text{ lb}}{g}, \qquad m_2 = \frac{5 \text{ lb}}{g}, \qquad v_1 = 2000 \frac{\text{ft}}{\text{sec}},$$

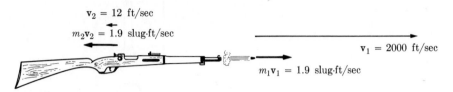

$v_2 = 12$ ft/sec

$m_2v_2 = 1.9$ slug·ft/sec

$v_1 = 2000$ ft/sec

$m_1v_1 = 1.9$ slug·ft/sec

FIG. 5–13. When a 5-lb rifle fires a 0.03-lb bullet with a muzzle velocity of 2000 ft/sec, it recoils with a velocity of 12 ft/sec. The total momentum of the *system* of rifle plus bullet is zero before and after the rifle is fired, although the momenta of both rifle and bullet have changed.

and, inserting these values in Eq. (5–13), we find that the recoil velocity of the rifle is

$$v_2 = -\frac{0.03 \text{ lb}/g}{5 \text{ lb}/g} \times 2000 \frac{\text{ft}}{\text{sec}} = -12 \frac{\text{ft}}{\text{sec}}.$$

Note that the g's have canceled out in the calculation, making it unnecessary for us to find the mass values numerically; the ratio of two masses is always the same as the ratio of the corresponding weights.

It is essential to keep in mind the directional character of momentum. In adding up the individual momenta $m\mathbf{v}$ we must consider the directions of the velocities and not merely add them algebraically. Sometimes the problem under consideration involves motion along a straight line, as in the example above, but in general it will be in two or three dimensions, and we must be sure to take this into account by a vector calculation.

5–9 Collisions

The law of conservation of momentum is indispensable in dealing with collisions between particles. In such cases no external forces act on the participants, and therefore their total momentum before they collide equals their total momentum afterward. The essential effect of the collision is to redistribute the total momentum of the particles.

An example of a collision problem might be helpful. Let us suppose that a 5-kg lump of clay moving with a velocity of 10 m/sec to the left strikes a 6-kg lump of clay moving with a velocity of 12 m/sec to the right, and that the two lumps stick together after the collision (Fig. 5–14). If we call the mass of the final body M and its velocity V, conservation of momentum requires that

$$MV = m_1v_1 + m_2v_2. \tag{5–14}$$

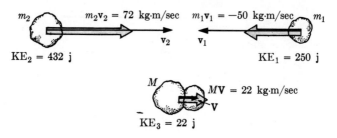

FIG. 5-14. A collision between two lumps of clay that stick together upon impact. Momentum is conserved in this collision, but kinetic energy is not.

Here we have

$$m_1 = 5 \text{ kg}, \qquad m_2 = 6 \text{ kg}, \qquad M = m_1 + m_2 = 11 \text{ kg},$$

$$v_1 = -10 \frac{\text{m}}{\text{sec}}, \qquad v_2 = 12 \frac{\text{m}}{\text{sec}},$$

where we have adopted the convention that motion to the right is + and motion to the left is −. Using these values in Eq. (5–14), we have

$$11V \text{ kg} = -50 \frac{\text{kg·m}}{\text{sec}} + 72 \frac{\text{kg·m}}{\text{sec}},$$

$$V = \frac{22 \text{ kg·m/sec}}{11 \text{ kg}} = 2 \frac{\text{m}}{\text{sec}};$$

the final body moves to the right with a velocity of 2 m/sec.

Energy and momentum are entirely independent concepts. In the above problem the lumps of clay before the collision had the kinetic energies

$$KE_1 = \frac{1}{2} m_1 v_1^2 = \frac{1}{2} \times 5 \text{ kg} \times \left(-10 \frac{\text{m}}{\text{sec}}\right)^2 = 250 \text{ j},$$

$$KE_2 = \frac{1}{2} m_2 v_2^2 = \frac{1}{2} \times 6 \text{ kg} \times \left(12 \frac{\text{m}}{\text{sec}}\right)^2 = 432 \text{ j}.$$

After the collision the new lump of clay has the kinetic energy

$$KE_3 = \frac{1}{2} M V^2 = \frac{1}{2} \times 11 \text{ kg} \times \left(2 \frac{\text{m}}{\text{sec}}\right)^2 = 22 \text{ j}.$$

The total kinetic energy prior to the collision was $432 + 250$ or 682 j, while afterward it is only 22 j. The difference of 660 j was dissipated largely into heat energy in the collision, with some probably being lost to sound energy as well.

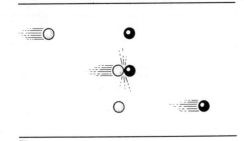

FIG. 5-15. When a ball rolling on a table makes a head-on collision with an identical stationary ball, the first ball stops and the second begins moving with the same velocity the first had initially.

While conservation of energy is an excellent guide in situations which do not involve the dissipation of mechanical energy (kinetic and potential energy) into other forms, conservation of momentum must be employed whenever there are situations involving several bodies interacting with one another. Both laws must be used together in order to understand certain processes. For instance, many of us are familiar with what happens when a ball rolling on a table strikes head-on another identical ball which is stationary; the first ball stops, while the second one begins moving with the same speed as the first and in the same direction (Fig. 5-15). Calling the initial and final speeds of the balls $\mathbf{v}_1$, $\mathbf{v}_2$ and $\mathbf{v}_1'$,$\mathbf{v}_2'$, conservation of momentum and of energy require that

$$\begin{array}{ccc} & \text{Before} & \text{After} \end{array}$$

$$\text{Momentum:} \quad m_1\mathbf{v}_1 + m_2\mathbf{v}_2 = m_1\mathbf{v}_1' + m_2\mathbf{v}_2',$$

$$\text{Energy:} \quad \tfrac{1}{2}m_1v_1^2 + \tfrac{1}{2}m_2v_2^2 = \tfrac{1}{2}m_1v_1'^2 + \tfrac{1}{2}m_2v_2'^2.$$

Since we have said that the balls are identical, $m_1 = m_2$, and since the second ball was originally at rest, $v_2 = 0$; hence

$$\mathbf{v}_1 = \mathbf{v}_1' + \mathbf{v}_2', \qquad v_1^2 = v_1'^2 + v_2'^2.$$

The *only* way of solving these equations is to have either $\mathbf{v}_1'$ or $\mathbf{v}_2'$ equal zero. If $\mathbf{v}_2'$ were zero, it would mean that the first ball traveled completely *through* the second ball. Because this is impossible, we must have as the solution

$$\mathbf{v}_1' = 0, \qquad \mathbf{v}_2' = \mathbf{v}_1;$$

the first ball stops, and the second begins to move with the original speed of the first ball.

IMPORTANT TERMS

Whenever a force acts to produce motion in a body, the force itself undergoes a displacement. The product of the force and its displacement is called *work*. When the force and its displacement are not parallel but have the angle θ between them, the work done is the product of the projection $F \cos \theta$ of the force in the direction of the displacement and the displacement s, that is, $W = Fs \cos \theta$. In the metric system the unit of work is the *joule*, and in the British system it is the foot-pound.

Energy is that which may be converted into work. When something possesses energy, it is capable of performing work. The units of energy are the same as those of work.

Kinetic energy is the energy a moving body possesses by virtue of its motion. If the body has the mass m and speed v, its kinetic energy is mv^2.

Potential energy is the energy a body has by virtue of its position. The gravitational potential energy of a body of mass m at a height h above a particular reference point is mgh, or, if its weight w is specified, its potential energy is wh. Other examples of potential energy are that of a planet with respect to the sun, that of a piece of iron with respect to a magnet, and that of a body at the end of a stretched spring.

Other forms of energy are heat and electrical, magnetic, chemical, and nuclear energy, as well as mass energy.

The *law of conservation* of energy states that the total amount of energy in a system isolated from its surroundings always remains constant although energy transformation from one form to another, reckoning mass as a form of energy, may occur within the system.

The *linear momentum* of a body is the product of its mass and velocity. Momentum is a vector quantity having the direction of the body's velocity.

The *impulse* of a force is the product of the force and the time during which it acts. Impulse is a vector quantity having the direction of the force. When a force acts on a body that is free to move, its change in momentum equals the impulse given it by the force.

The law of *conservation of momentum* states that when the vector sum of the external force acting upon a system of particles equals zero, the total linear momentum of the system remains constant.

PROBLEMS

1. (a) A force of 130 n is used to lift a 12-kg mass to a height of 8 m. How much work is done by the force? (b) A force of 130 n is used to push a 12-kg mass on a horizontal, frictionless surface for a distance of 8 m. How much work is done by the force?

2. A 20-kg wooden box is pushed a distance of 15 m on a horizontal stone floor by a force just sufficient to overcome the friction between box and floor. The coefficient of friction is 0.4. (a) What is the required force? (b) How much work does the force do?

3. Four thousand joules are used to lift a 30-kg mass. If the mass is at rest before and after its elevation, how high does it go?

4. A boy pulls a sled with a force of 10 lb for 100 ft. The rope attached to the sled is at an angle of 30° above the horizontal. How much work is done?

5. A centripetal force of 18 n is used to keep a 2-kg ball in uniform circular motion at the end of a string 1 m long. How much work does the force do in each revolution of the ball?

6. A horizontal force of 5 n is used to push a box up a ramp 5 m long that is at an angle of 15° above the horizontal. How much work is done?

7. A horse is towing a barge with a rope that makes an angle of 20° with the canal. If the horse exerts a force of 80 n, how much work does it do in moving the barge 1 km (1 km = 10^3m)?

8. A 160-lb man carrying a 32-lb knapsack climbs to the summit of a 14,000-ft mountain. The average slope of the mountain is 40°. If he starts from a base camp at an altitude of 8000 ft, how much work does he perform in making the ascent?

9. A 3200-lb car is moving at 40 mi/hr. What is its kinetic energy?

10. A 0.02-kg bullet has a speed of 500 m/sec. What is its kinetic energy?

11. A 2-kg ball is at rest when a horizontal force of 5 n is applied. In the absence of friction, what is the speed of the ball after it has gone to 10 m?

12. A 3-kg stone is lifted to a height of 100 m and then dropped. What is its kinetic energy when it is 50 m from the ground?

13. A man uses a rope and system of pulleys to lift a 160-lb weight to a height of 5 ft. He exerts a force of 45 lb on the rope, and pulls a total of 20 ft of rope through the pulleys in the course of lifting the weight. (a) How much work does the man do? (b) What is the change in the potential energy of the weight? (c) If the answers to (a) and (b) are different, explain.

14. A boy slides down a sliding pond from a starting point 3 m above the ground. His speed at the bottom is 4 m/sec. What percentage of his initial potential energy was dissipated?

15. A force of 100 lb is used to lift an 80-lb weight to a height of 20 ft. There is no friction present. (a) How much work is done by the force? (b) What is

the change in the potential energy of the weight? (c) What is the change in the kinetic energy of the weight?

16. (a) A force of 8 n is used to push a 0.5-kg ball over a horizontal, frictionless table a distance of 3 m. If the ball starts from rest, what is its final kinetic energy? (b) The same force is used to lift the same ball a height of 3 m. If the ball starts from rest what is its final kinetic energy?

17. At what point in its motion is the kinetic energy of a pendulum bob a maximum? At what point is its potential energy a maximum?

18. At her highest point, a girl on a swing is 8 ft from the ground while at her lowest point she is 3 ft from the ground. What is her maximum speed?

19. In the operation of a certain pile driver, a hammer weighing 1000 lb is dropped from a height of 19 ft above the head of a pile. If the pile is driven 0.5 ft into the ground with each impact of the hammer, what is the average force on the pile when struck?

20. A sledge hammer whose head has a mass of 5 kg is used to drive a spike into a wooden beam. The workman is tired and merely allows the hammer to drop on the spike from a height 0.4 m above it. If the spike is driven 1 cm at each blow, what is the average force on it when struck?

21. What is the momentum of the car of Problem 9?

22. What is the momentum of the bullet of Problem 10?

23. Two boys, one weighing 60 lb and the other 90 lb, are at rest on frictionless roller skates. The larger boy pushes the other so that the latter rolls away at a speed of 5 mi/hr. What is the effect of this action on the larger boy himself?

24. Give an example to show that a body may simultaneously have more kinetic energy but less momentum than another body.

25. A driverless car weighing 4000 lb is moving along a road at a speed of 50 mi/hr. In order to stop the car, a tank weighing 16,000 lb makes a head-on collision with it. (a) What should the tank's speed be in order that both tank and car come to a stop as a result of the collision? (b) How much kinetic energy is dissipated in the collision if the tank has the speed of part (a)?

26. A freight car weighing 12 tons rolls at 5 ft/sec along a horizontal railroad track. It collides with another freight car weighing 16 tons that is standing at rest on the track, and the two cars couple together. What is the speed of the cars after the collision?

27. A hunter has a rifle that can fire 0.06-kg bullets with a muzzle velocity of 900 m/sec. A 40-kg leopard springs at him at a speed of 10 m/sec. How many bullets must the hunter fire into the leopard in order to stop him in his tracks?

28. A 0.5-kg stone moving at 4 m/sec overtakes a 4-kg lump of clay moving at 1 m/sec. The stone becomes embedded in the clay. (a) What is the speed of the composite body after the collision? (b) How much kinetic energy is lost?

29. A 0.5-kg stone moving north at 4 m/sec collides with a 4-kg lump of clay moving west at 1 m/sec. The stone becomes embedded in the clay. (a) What is the velocity (speed and direction) of the composite body after the collision? (b) How much kinetic energy is lost?

30. A neutron of mass 1.67×10^{-27} kg and speed 10^5 km/sec collides with a stationary deuteron of mass 3.34×10^{-27} kg. The two particles stick together. What is the speed of the composite particle (called a *triton*)?

31. A neutron of mass 1.67×10^{-27} kg and speed 10^5 m/sec collides with a stationary deuteron of mass 3.34×10^{-27} kg. The particles do not stick together, and no kinetic energy is lost in the collision. What is the subsequent speed of each particle?

32. A 160-lb man is sliding on the frictionless surface of a frozen pond at a speed of 0.8 ft/sec. He is struck by a snowball weighing 0.5 lb whose velocity is 60 ft/sec in a direction perpendicular to his motion. What is the man's final speed?

33. A rocket motor consumes 100 kg of fuel per second, exhausting it with a speed of 5×10^3 m/sec. What force is exerted on the rocket?

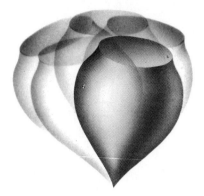

Rigid-Body Motion | 6

Until now we have been discussing only *translational motion,* motion in which the position of a body changes from one moment to the next. But it is possible for a body to remain in one place while it is rotating, as in Fig. 6–1, so that almost all its constituent particles are constantly moving even though the body as a whole does not go anywhere. In our ordinary experience, *rotational motion* is encountered nearly as often as translational motion. Wheels, driveshafts, pulleys, propellers, fan blades, phonograph records, drills, and revolving doors all rotate in carrying out their functions. In fact, the most general kind of motion a body may undergo is a combination of translation and rotation. In this chapter our chief concern will be the rotational motion of a rigid body about a fixed axis. As we shall see, all the formulas that describe such motion are exact analogs of the formulas we have already used to describe translational motion; this is a convenient circumstance indeed.

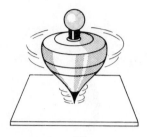

FIGURE 6-1

6-1 Angular measure

While we are accustomed to measuring angles in degrees, where 1° is defined as 1/360 of a full rotation (that is, a complete turn represents 360°), this unit is not very convenient for our present work. A more appropriate unit is the *radian,* abbreviated rad. If we draw a circle of radius r with its center at the vertex of an angle θ, as in Fig. 6–2, the angle contains

$$\theta = \frac{s}{r} \tag{6-1}$$

radians by definition, where s is the arc of the circle cut by the angle. That is, the angle θ between two radii of a circle is the ratio of the arc s to the radius r. Evidently an angle of 1 rad has an arc that is the same as the radius.

It is easy to find the conversion factor between degrees and radians and vice versa. We observe that there are 360° in a complete circle, while the number of radians in a complete circle is

$$\theta = \frac{s}{r} = \frac{2\pi r}{r} = 2\pi,$$

because the circumference of a circle of radius r is $2\pi r$. Hence

$$360° = 2\pi \text{ rad,}$$

from which we find that

$$1° = 0.01745 \text{ rad,}$$

$$1 \text{ rad} = 57.30°.$$

Often it is useful to express angles in radian measure in terms of π itself. For example, an angle of 90° is $\frac{1}{4}$ of a complete circle, and $\frac{1}{4} \times 2\pi = \pi/2$ rad. Of course, this has the same numerical value as $90 \times 0.01745 = 1.571$ rad, since $\pi/2 = 1.571$.

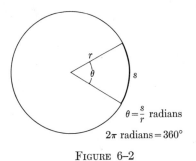

$$\theta = \frac{s}{r} \text{ radians}$$

$$2\pi \text{ radians} = 360°$$

FIGURE 6–2

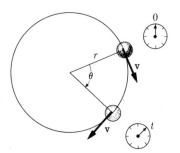

FIG. 6–3. Uniform circular motion.

6–2 Angular velocity

If a rotating body turns through the angle θ in the time t, its *angular velocity* ω (Greek letter *omega*) is

$$\omega = \theta/t. \tag{6-2}$$

If θ is in radians and t in seconds, which are the usual units for these quantities, the unit of ω is the rad/sec. Another fairly common unit of angular velocity is the revolution per minute (or rpm); evidently

$$1 \frac{\text{rev}}{\text{min}} = 1 \frac{\text{rev}}{\text{min}} \times 2\pi \frac{\text{rad}}{\text{rev}} \times \frac{1}{60 \text{ sec/min}} = 0.105 \frac{\text{rad}}{\text{sec}}.$$

Let us consider a particle moving with the uniform speed v in a circle of radius r, as in Fig. 6–3. This particle travels the distance

$$s = vt$$

in the time t. The angle through which it moves in that time is

$$\theta = \frac{s}{r} = \frac{vt}{r},$$

so that its angular velocity is

$$\omega = \frac{\theta}{t} = \frac{v}{r}. \tag{6-3}$$

The angular velocity of a particle in uniform circular motion is the ratio between its linear speed and the radius of its path. Conversely,

$$v = \omega r; \tag{6-4}$$

the linear speed of a particle in uniform circular motion is the product of its angular velocity and the radius of its path. Equations (6–3) and (6–4) are valid only when ω is expressed in radian measure.

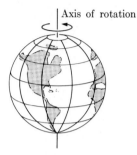

Axis of rotation

FIGURE 6–4

The *axis of rotation* of a rigid body turning in place is that line of particles which does not move (Fig. 6–4). Sometimes the axis of rotation is a line in space. All other particles of the body move in circles about the axis. Since $v = \omega r$, the farther a particle is from the axis, the greater its linear speed, although all the particles of the body (except those on the axis) have the same angular velocity. A phonograph record 12 in. in diameter rotates at $33\frac{1}{3}$ rpm, which is

$$33\tfrac{1}{3}\text{ rpm} \times 0.105\ \frac{\text{rad/sec}}{\text{rpm}} = 3.50\ \frac{\text{rad}}{\text{sec}}.$$

Hence a point 1 in. from the axis has a linear speed of

$$v = \omega r = 3.5\ \frac{\text{rad}}{\text{sec}} \times 1\text{ in} = 3.5\ \frac{\text{in}}{\text{sec}},$$

while a point on the record's rim, where $r = 6$ in., has a speed of

$$v = 3.5\ \frac{\text{rad}}{\text{sec}} \times 6\text{ in} = 21\ \frac{\text{in}}{\text{sec}}.$$

6–3 Angular acceleration

A rotating body need not have a uniform angular velocity ω, just as a moving particle need not have a uniform velocity v. If the angular velocity of a body changes from some original value ω_0 to the new value ω in a time interval t, the *angular acceleration* α (Greek letter *alpha*) of the body is

$$\alpha = \frac{\omega - \omega_0}{t}. \tag{6–5}$$

The unit of angular acceleration we shall use is the rad/sec².

Let us return to the particle moving in a circle of radius r and suppose that its angular velocity changes from ω_0 to ω in the time t. Since the linear speeds corresponding to ω_0 and ω are

$$v_0 = \omega_0 r, \qquad v = \omega r,$$

we can express the particle's angular acceleration as

$$\alpha = \frac{v/r - v_0/r}{t} = \frac{v - v_0}{tr}.$$

But, by definition, the linear acceleration a_T of the particle tangent to its path is

$$a_T = \frac{v - v_0}{t},$$

the ratio between its change in speed $v - v_0$ in a time t and the value of t. Hence the relationship between the angular and linear accelerations of the particle is

$$\alpha = \frac{a_T}{r}. \tag{6-6}$$

The angular acceleration of a particle in circular motion about a fixed axis is the ratio between its linear acceleration and the radius of its path. Conversely,

$$a_T = \alpha r; \tag{6-7}$$

the linear acceleration of a particle in circular motion about a fixed axis is the product of its angular acceleration and the radius of its path.

We must be careful to distinguish between the *tangential acceleration* a_T of a particle, which represents a change in its linear speed, and the *centripetal acceleration* a_C, which represents a change in its direction of motion. A particle in circular motion with the speed v has, as we know, the centripetal acceleration

$$a_C = \frac{v^2}{r} \tag{6-8}$$

directed toward the center of its circular path. The accelerations a_T and a_C are therefore always perpendicular. *All* particles in circular motion have centripetal accelerations. Only those particles whose speed changes in magnitude, however, have tangential accelerations. The centripetal acceleration of a particle moving in a circle of radius r can be expressed in terms of its angular velocity ω as

$$a_C = \omega^2 r, \tag{6-9}$$

because $v = \omega r$.

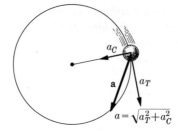

Fɪɢ. 6–5. Tangential and centripetal accelerations on a particle in circular motion.

As an example, we shall compute the total linear acceleration of a particle moving in a circle of radius 0.4 meter at the instant that its angular velocity is 2 rad/sec and its angular acceleration is 5 rad/sec². From Eqs. (6–7) and (6–9) we have for the particle's tangential and centripetal accelerations

$$a_T = \alpha r = 5\,\frac{\text{rad}}{\text{sec}^2} \times 0.4\,\text{m} = 2\,\frac{\text{m}}{\text{sec}^2},$$

$$a_C = \omega^2 r = \left(2\,\frac{\text{rad}}{\text{sec}}\right)^2 \times 0.4\,\text{m} = 1.6\,\frac{\text{m}}{\text{sec}^2}.$$

We must add a_T and a_C vectorially, since they are vector quantities which are in different directions. As shown in Fig. 6–5, the magnitude a of the vector sum $\mathbf{a}$ of $\mathbf{a_T}$ and $\mathbf{a_C}$ is

$$a = \sqrt{a_T^2 + a_C^2}$$

$$= \sqrt{2^2 + 1.6^2}\,\frac{\text{m}}{\text{sec}^2}$$

$$= 2.6\,\frac{\text{m}}{\text{sec}^2}.$$

From the definitions of angular velocity ω and angular acceleration α, we can obtain the useful formulas

$$\theta = \omega_0 t + \tfrac{1}{2}\alpha t^2, \tag{6–10}$$

$$\omega^2 = \omega_0^2 + 2\alpha\theta \tag{6–11}$$

by means of the same reasoning that enabled us to obtain the formulas

$$d = v_0 t + \tfrac{1}{2}at^2, \tag{1–8}$$

$$v^2 = v_0^2 + 2ad \tag{1–10}$$

in Chapter 1. The analogy between the angular quantities θ, ω, and α and the linear quantities d, v, and a is evidently a close one.

6–4 Rotational kinetic energy

A rotating body possesses kinetic energy because its constituent particles are in motion, even though the body as a whole remains in place. The speed of a particle that is the distance r from the axis of a rigid body rotating with the angular velocity ω is, as we know,

$$v = \omega r.$$

If the particle's mass is m, its kinetic energy is

$$\text{KE} = \tfrac{1}{2}mv^2 = \tfrac{1}{2}m\omega^2 r^2. \tag{6–12}$$

The body consists of numerous particles which need not have the same mass or be the same distance from the axis. However, all the particles have the common angular velocity ω. Hence the total kinetic energy of all the particles may be written

$$\text{KE} = \sum \tfrac{1}{2}mv^2 = \tfrac{1}{2}(\sum mr^2)\omega^2, \tag{6–13}$$

where the symbol $\sum$ means, as mentioned before, "sum of." Equation (6–13) states that the kinetic energy of a rotating rigid body is equal to one-half the sum of the mr^2 values of its constituent particles multiplied by the square of its angular velocity ω.

The quantity

$$I = \sum mr^2 \tag{6–14}$$

is known as the *moment of inertia* of the body. It has the same value regardless of the body's state of motion. The farther a given particle is from the axis of rotation, the faster it moves and the greater its contribution to the kinetic energy of the body. The moment of inertia of a body depends upon the way in which its mass is distributed relative to its axis of rotation; it is perfectly possible for one body to have a greater moment of inertia than another even though its mass may be much the smaller of the two.

The kinetic energy of a body of moment of inertia I rotating with the angular velocity ω is therefore

$$\text{KE} = \tfrac{1}{2}I\omega^2. \tag{6–15}$$

Evidently the rotational analog of mass is moment of inertia, just as the rotational analog of velocity is angular velocity. We shall find further support for the correspondence of mass and moment of inertia later in this chapter.

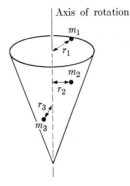

FIGURE 6–6

6–5 Moment of inertia

A rigid body may be considered to be made up of a large number of separate particles whose masses are m_1, m_2, m_3, and so on (Fig. 6–6). To compute the moment of inertia of such a body about a specified axis, we multiply the mass of each of these particles by the square of its distance from the axis — r_1^2, r_2^2, r_3^2, and so on — and add up all the mr^2 values. That is,

$$I = \sum mr^2 = m_1 r_1^2 + m_2 r_2^2 + m_3 r_3^2 + \ldots . \qquad (6\text{–}16)$$

The more particles we imagine the body to contain, the more accurate will be our value of its moment of inertia. While I can be calculated for a few simple bodies without difficulty by Eq. (6–16), in general either considerable labor or the use of integral calculus is required. Figure 6–7 gives the moments of inertia of several regularly shaped bodies in terms of the total mass M and dimensions of each. The unit of I is the kg·m² in the metric system and the slug·ft² in the British system.

The earth's mass is 6×10^{24} kg and its radius is 6.4×10^6 meters. Considering it to be a uniform sphere, its moment of inertia is

$$I = \tfrac{2}{5}MR^2 = \tfrac{2}{5} \times 6 \times 10^{24}\,\text{kg} \times (6.4 \times 10^6\,\text{m})^2 = 9.8 \times 10^{37}\,\text{kg·m}^2.$$

The angular velocity of the earth is 7.3×10^{-5} rad/sec, corresponding to one rotation per day. Hence its rotational kinetic energy is

$$\text{KE} = \tfrac{1}{2}I\omega^2$$

$$= \tfrac{1}{2} \times 9.8 \times 10^{37}\,\text{kg·m}^2 \times \left(7.3 \times 10^{-5}\,\frac{\text{rad}}{\text{sec}}\right)^2$$

$$= 2.6 \times 10^{29}\,\text{j}.$$

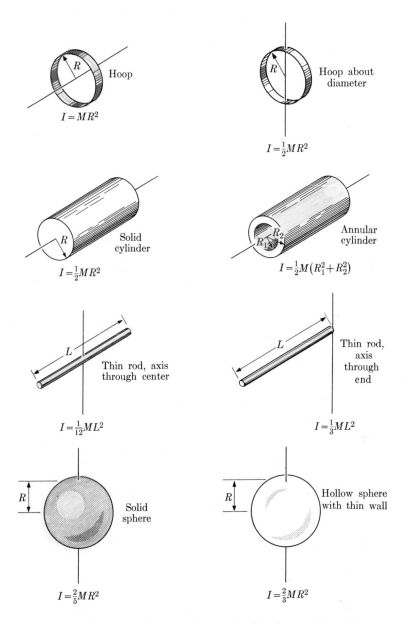

FIG. 6–7. Moments of inertia of various bodies about indicated axes.

It is interesting to compare this figure with the translational kinetic energy of the earth due to its annual revolution around the sun. Since the average orbital speed of the earth is 3×10^4 m/sec, its translational kinetic energy is

$$\text{KE} = \tfrac{1}{2}mv^2$$

$$= \tfrac{1}{2} \times 6 \times 10^{24}\,\text{kg} \times \left(3 \times 10^4\,\frac{\text{m}}{\text{sec}}\right)^2$$

$$= 2.7 \times 10^{33}\,\text{j}.$$

The kinetic energy of the earth's orbital motion is about 10,000 times greater than that of its rotational motion.

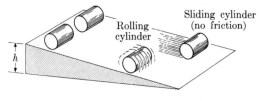

FIGURE 6–8

When a rigid body is both moving through space and undergoing rotation, its total kinetic energy is

$$\text{KE} = \tfrac{1}{2}mv^2 + \tfrac{1}{2}I\omega^2.$$

In this formula v is taken to be the speed of the body's center of gravity; I and ω are determined on the basis of an axis passing through the center of gravity. Suppose we have a cylinder of radius R and mass m that is poised at the top of an inclined plane (Fig. 6–8). Will it have a greater speed at the bottom if it slides down without friction or if it rolls down? In the first case, we set the cylinder's initial potential energy of mgh equal to its final kinetic energy of $\tfrac{1}{2}mv^2$, and find that

$$\text{PE} = \text{KE}, \qquad mgh = \tfrac{1}{2}mv^2, \qquad v = \sqrt{2gh}.$$

In the second case the cylinder has both translational and rotational kinetic energy at the bottom, so that

$$\text{PE} = \text{KE}, \qquad mgh = \tfrac{1}{2}mv^2 + \tfrac{1}{2}I\omega^2.$$

The moment of inertia of a cylinder is $I = \tfrac{1}{2}mR^2$; if it rolls without slipping, its linear and angular velocities are related by the formula $\omega = v/R$. Hence

$$mgh = \tfrac{1}{2}mv^2 + \tfrac{1}{2}(\tfrac{1}{2}mR^2)\left(\frac{v^2}{R^2}\right)$$

$$= \tfrac{1}{2}mv^2 + \tfrac{1}{4}mv^2 = \tfrac{3}{4}mv^2,$$

$$v = \sqrt{\tfrac{4}{3}gh}.$$

The cylinder moves more slowly when it rolls down the plane than when it slides without friction, because some of the available energy is absorbed by its rotation.

6–6 Torque and angular acceleration

According to Newton's second law of motion, a net force applied to a body causes it to be accelerated. What can cause a body capable of rotation to experience an angular acceleration? To fix our ideas, let us look once more at a single particle of mass m restricted to motion in a circle of radius r (Fig. 6–9). A force F that acts upon the particle tangent to its path gives it the acceleration a, according to the formula

$$F = ma. \tag{6–17}$$

Because the particle is restricted to the circular path, forces in any other direction cannot affect it. The linear acceleration a corresponds here to the angular acceleration

$$\alpha = \frac{a}{r},$$

and so Eq. (6–17) can be rewritten

$$F = mr\alpha.$$

Multiplying both sides of this equation by r,

$$Fr = mr^2\alpha. \tag{6–18}$$

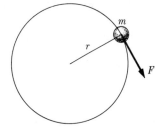

FIGURE 6–9

We recognize Fr as the torque τ of the force F about the axis of the particle's rotation and mr^2 as the particle's moment of inertia I. Equation (6–18) therefore states that

$$\tau = I\alpha. \tag{6–19}$$

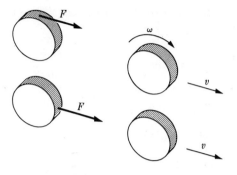

FIG. 6–10. A force whose line of action passes through the center of gravity of a body cannot cause it to rotate.

While we have derived Eq. (6–19) for the case of a single particle, it is also valid for any rotating body, provided that the torque and moment of inertia are both calculated about the same axis.

The formula $\tau = I\alpha$ is the fundamental law of motion for rotating bodies in the same sense that $F = ma$ is the fundamental law of motion for bodies moving through space. In rotational motion, torque plays the same role that force does in translational motion. The angular acceleration experienced by a body when the torque τ acts upon it is

$$\alpha = \frac{\tau}{I} \, ;$$

α is directly proportional to the torque and inversely proportional to the body's moment of inertia.

As an exercise, let us determine the tangential force acting at the equator that would be needed to stop the earth's rotation in a period of, say, one year. Here $\omega_0 = 7.3 \times 10^{-5}$ rad/sec, $\omega = 0$, and $t = 3.2 \times 10^7$ sec. The angular acceleration of the earth would therefore be

$$\alpha = \frac{\omega - \omega_0}{t} = -\frac{7.3 \times 10^{-5} \, \text{rad/sec}}{3.2 \times 10^7 \, \text{sec}} = -2.3 \times 10^{-12} \, \text{rad/sec}^2.$$

The minus sign implies that the acceleration is such as to reduce the angular velocity. We found earlier that the moment of inertia of the earth is 9.8×10^{37} kg·m², so that the torque required to stop the earth from rotating in one year is

$$\tau = I\alpha$$

$$= -9.8 \times 10^{37} \, \text{kg·m}^2 \times 2.3 \times 10^{-12} \, \text{rad/sec}^2$$

$$= -2.3 \times 10^{26} \, \text{n·m}.$$

Since the earth's radius is 6.4×10^6 meters, the tangential force that must be applied is

$$F = \frac{\tau}{r} = -\frac{2.3 \times 10^{26}\,\text{n·m}}{6.4 \times 10^6\,\text{m}} = -3.6 \times 10^{19}\,\text{n}.$$

This force is about one percent of the gravitational force that holds the earth in its orbit around the sun.

When a force acts on a body that is able to move freely, the body will experience both linear and angular accelerations unless the line of action of the force passes through the body's center of gravity (Fig. 6–10). In the latter case the body will be in rotational equilibrium, and if it was not rotating initially, it will keep its original orientation during its motion.

6–7 Angular momentum

The rotational analog of linear momentum is *angular momentum*. The angular momentum L of a body depends upon its moment of inertia I and angular velocity ω in the same way that its linear momentum depends upon its mass m and linear velocity v:

$$L = I\omega. \tag{6–20}$$

When there is no net torque on a rigid body, both its angular velocity ω and its angular momentum are constant. A deeper analysis shows that the angular momentum of a body does not change in the absence of a net torque on it even if it is *not* a rigid body, but is so altered during its motion that its moment of inertia changes. In a situation of this kind the angular velocity of the body also changes so that L stays the same. Thus we have the useful theorem of *conservation of angular momentum:*

When the sum of the external torques acting upon a system of particles equals zero, the total angular momentum of the system remains constant.

A skater or ballet dancer doing a spin capitalizes upon conservation of angular momentum. In Fig. 6–11 a skater is shown starting her spin with her arms and one leg outstretched. By bringing her arms and extended leg inward, she reduces her moment of inertia considerably and consequently spins faster. The greater orbital speed of the earth when it is closest to the sun follows from the same principle: the reduction in moment of inertia about the sun leads to an increase in angular velocity in this part of its orbit.

Angular momentum, like linear momentum, is a vector quantity that possesses both magnitude and direction. To change the orientation of

I large, ω small   *I* small, ω large

FIGURE 6–11

the axis of rotation of a spinning body requires a torque, just as a torque is required to change the magnitude of its angular velocity. The greater the value of **L**, the more torque is needed to deviate it from its original direction. This is the principle behind the spin stabilization of projectiles, such as footballs and rockets. Such projectiles are set spinning about axes in their directions of motion so that they do not tumble and thereby offer excessive air resistance. A top is another illustration of the vector nature of angular momentum. A stationary top set on its tip falls over at once, but a rotating top stays upright until its angular momentum is dissipated by friction between its tip and the ground.

IMPORTANT TERMS

The *radian* is a unit of angular measure equal to 57.30°. If a circle is drawn whose center is at the vertex of an angle, the angle in radian measure is equal to the ratio between the arc of the circle cut by the angle and the radius of the circle. A full circle contains 2π radians.

The *angular velocity* ω of a rotating body is the angle through which it turns per unit time. The *angular acceleration* α of a rotating body is the rate of change of its angular velocity with respect to time.

The *axis of rotation* of a rigid body turning in place is that line of particles which does not move.

All particles in circular motion experience centripetal accelerations, but only those particles whose angular velocity changes have *tangential accelerations*.

The *moment of inertia I* of a body about a given axis is the rotational analog of mass in linear motion. Its value depends upon the way in which the mass of the body is distributed about the axis.

The *angular momentum L* of a rotating body is the product $I\omega$ of its moment of inertia and angular velocity. The principle of *conservation of angular momentum* states that the total angular momentum of a system of particles remains constant when no net external torque acts upon it.

PROBLEMS

1. (a) Express 93 radians in degrees. (b) Express 20° in radians.

2. (a) Express 1.3 radians in degrees. (b) Express 198° in radians.

3. An apple pie 12 in. in diameter is cut into nine equal pieces. What angle (in radians) is included between the sides of each piece?

4. What is the angular velocity in rad/sec of the hour, minute, and second hands of a clock?

5. A grindstone 10 cm in radius is rotating at 1725 rpm. (a) What is its angular velocity? (b) What is the linear speed of a point on its rim?

6. A drill bit $\frac{1}{4}$ in. in diameter is rotating at 1200 rpm. (a) What is its angular velocity? (b) What is the linear speed of a point on its circumference?

7. A barrel 2 ft in diameter is rolling with an angular velocity of 5 rad/sec. What is the instantaneous linear velocity of (a) its top, (b) its center, and (c) its bottom with respect to the ground?

8. The propeller of a boat rotates at 1000 rpm when the speed of the boat is 20 ft/sec. The diameter of the propeller is 18 in. What is the speed of the tip of the propeller?

9. The shaft of a motor rotates at the constant angular velocity of 3000 rpm. How many radians will it have turned through in $\frac{1}{2}$ min?

10. A phonograph turntable slows down to a stop from an initial angular velocity of 45 rpm in 20 sec. (a) What is its acceleration? (b) How many turns does it make while slowing down?

11. The speed of a motor increases from 1200 rpm to 1800 rpm in 20 sec. What is its angular acceleration? How many revolutions does it make during these 20 sec?

12. A truck undergoes an acceleration of 0.25 m/sec². If its wheels are 1 meter in diameter, what is their angular acceleration?

13. A wheel starts from rest under the influence of a constant torque and turns through 500 radians in 10 sec. (a) What is its angular acceleration? (b) What is its angular velocity after 10 sec?

14. Will a car coast down a hill faster when it has heavy tires or light tires?

15. A hollow cylinder and a solid cylinder having the same mass and diameter are released from rest simultaneously at the top of an inclined plane. Which reaches the bottom first?

16. A solid sphere 10 cm in radius starts from rest at the top of an inclined plane 10 meters long and reaches the bottom in 7 sec. What angle does the plane make with the horizontal?

17. A rope 1 meter long is wound around the rim of a drum of radius 12 cm and moment of inertia 0.02 kg·m². The rope is pulled with a force of 2.5 newtons. (a) Assuming that the drum is free to rotate without friction, what is its final angular velocity? (b) What is its final kinetic energy? (c) How much work is done by the force?

18. A 3-kg hoop 1 meter in diameter rolls down an inclined plane 10 meters long that is at an angle of 20° with the horizontal. (a) What is the angular velocity of the hoop at the bottom of the plane? (b) What is its linear velocity? (c) What is its rotational kinetic energy? (d) What is its total kinetic energy?

19. The baton of a drum majorette is a 0.7-lb uniform rod 2 ft long. (a) What is its kinetic energy when it is twirled at an angular velocity of 10 rad/sec? (b) What is its angular momentum?

20. A 200-kg cylindrical flywheel 0.3 m in radius is acted upon by a torque of 20 n·m. (a) If it starts from rest, how much time is required to accelerate it to an angular velocity of 10 rad/sec? (b) What is its kinetic energy at this angular velocity?

21. A 16-lb bowling ball 1 ft in diameter rolls at a speed of 18 ft/sec. What is its total kinetic energy?

22. A 1-horsepower (hp) motor does work at the rate of 550 ft·lb per second by definition. If such a motor rotates at 1200 rpm, how much torque can it exert?

23. Use the vector nature of angular momentum to explain the advantage of rifling gun barrels so that their bullets emerge spinning about an axis in their direction of motion.

24. Two 0.4-kg balls are joined by a 1 meter string and set whirling through the air at 5 rev/sec about a vertical axis through the center of the string. After a while the string stretches to 1.2 meters. (a) What is the new angular velocity? (b) What was the change in kinetic energy?

25. If the polar ice caps melt, how will the length of the day be affected?

26. All helicopters have two propellers; some have both propellers on vertical axes but rotating in opposite directions, and others have one on a vertical axis and one on a horizontal axis perpendicular to the helicopter body at the tail. Why is a single propeller never used?

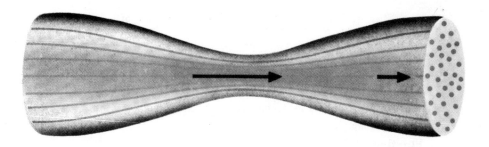

Continuous Matter | 7

The standard operating procedure of the physicist in approaching a complex situation is to first construct an abstract model that represents the essential features of the situation. Then, if this model proves successful in the sense that predictions obtained from it agree at least approximately with the results of observation and experiment, it is further refined until the agreement is even better. The study of mechanics illustrates this procedure. We began by treating objects as though they are minute particles, and went on to extend our analysis by considering them as rigid bodies. Actually, there is no such thing as a rigid body; the strongest block of steel can be stretched, compressed, or twisted by applying a suitable force or forces. In this chapter we shall pursue reality further by examining the properties of deformable bodies, a subject that includes the mechanical behavior of liquids and gases.

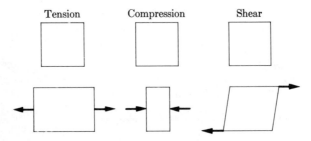

FIG. 7–1. The three types of stress.

7–1 Elasticity

While solid bodies generally seem perfectly rigid and unyielding, it is nevertheless possible to deform them either temporarily or permanently by applying stresses. Stress forces fall into three categories, *tensions*, *compressions*, and *shears*. These are illustrated in Fig. 7–1. A *tensile* stress is applied to a body when equal and opposite forces that act away from each other are exerted on its ends along the same line of action, thereby tending to elongate the body. A *compressive* stress is applied to a body when equal and opposite forces that act toward each other are exerted on its ends along the same line of action, thereby tending to decrease its length. A *shearing* stress is applied to a body when equal and opposite forces are exerted on its ends along different lines of action, thereby tending to change the shape of the body without changing its volume. In simpler terms, tensions stretch bodies on which they act, compressions shrink bodies on which they act, and shears twist bodies on which they act.

The response of a body to any of the above stresses depends upon its composition, shape, temperature, and so on, but as a general rule the amount of deformation is directly proportional to the applied stress provided that the force does not exceed a certain limit. In the case of tension, for example, we might find that supporting a 20-lb weight with a certain thin wire causes the wire to stretch 0.1 in. (Fig. 7–2). Doubling the weight to 40 lb therefore will produce a total elongation of 0.2 in., tripling the weight to 60 lb will produce a total elongation of 0.3 in., and so on. When the force is removed, the wire returns to its original length. This proportionality is called *Hooke's law*, and may be written

$$F = ks,$$

(7–1)

where F is the applied tension force, s the resulting elongation, and k a constant whose value depends upon the nature and dimensions of the

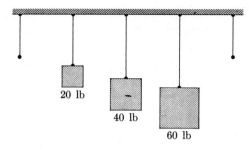

FIG. 7-2. The elongation of a wire is proportional to the stress applied to it, provided that the elastic limit is not exceeded. When the stress is removed, the wire returns to its original length.

object under stress. The force constant k is higher for materials such as steel than it is for materials such as rubber; it is directly proportional to the cross-sectional area of the object, so that a thick wire of a given material has a higher value of k than a thin wire of the same material. Relationships similar to Eq. (7-1) are found to apply to the behavior of solids under shear stresses and to all states of matter under compression stresses.

The term *elastic limit* refers to the maximum deformation an object can undergo as the result of stress forces without being permanently altered. When its elastic limit is exceeded, the object may or may not be far from rupture. Brittle substances like glass or cast iron break at or near their elastic limits. For example, a glass rod whose cross-sectional area is 1 in² obeys Hooke's law as long as the tension or compression forces on it are less than about 10,000 lb, but it will break if this figure is exceeded. Most metals may be deformed considerably beyond their elastic limits, a property known as *ductility*. Copper is a very ductile metal, and while a copper bar whose cross-sectional area is 1 in² reaches its elastic limit when a force of about 6000 lb is applied, it will not rupture

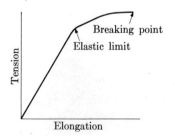

FIG. 7-3. Graph of the elongation of an iron rod as a function of the tension applied to it.

until the force has reached perhaps 33,000 lb. A graph of the elongation of an iron rod as a function of the tension applied to it is shown in Fig. 7–3. At first the graph is a straight line, corresponding to Hooke's law, but at the elastic limit it flattens out, corresponding to the shift from a reversible to an irreversible change in length. A moderate increase in tension beyond the elastic limit causes the rod to break.

7–2 Elastic moduli

Hooke's law for each kind of stress can be expressed in such a way that only a single constant need be known for a particular material in order to relate the stress applied to *any* object of this material to the resulting deformation. Experimentally it is found that the relative change in size or shape of a body under stress is proportional to the ratio between the applied force and the cross-sectional area of the object. Thus with a given force a thick rod will stretch less than a thin rod. In the case of a rod of initial length L and cross-sectional area A, a tension or compression force F (Fig. 7–4) produces a change ΔL in the length of the rod according to the relationship

$$\frac{\Delta L}{L} = \frac{1}{Y}\frac{F}{A}. \tag{7–2}$$

The rod increases in length by ΔL if it is in tension and decreases by that amount if it is in compression. The value of the quantity Y, which is called *Young's modulus*, depends upon the composition of the rod. Young's moduli for a number of common substances are given in Table 7–1; it is customary to express Y in units of lb/in² in the British system. Equation (7–2) is valid only when the elastic limit is not exceeded, of course.

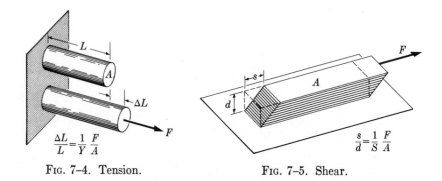

$$\frac{\Delta L}{L} = \frac{1}{Y}\frac{F}{A}$$

FIG. 7–4. Tension.

$$\frac{s}{d} = \frac{1}{S}\frac{F}{A}$$

FIG. 7–5. Shear.

Shear stresses change the shape of an object upon which they act. Let us consider a block of thickness d whose lower face is fixed in place and upon whose upper face the force F acts (Fig. 7–5). A measure of the relative distortion of the block caused by the shear stress is the ratio s/d between the displacement s of the block's faces and the distance d between them. The greater the area A of these faces, the less they will be displaced by the shear force F. The analog of Eq. (7–2) for shear stresses is therefore

$$\frac{s}{d} = \frac{1}{S}\frac{F}{A}. \tag{7-3}$$

Here the applied forces are *parallel* to the faces upon which they act and not perpendicular as in the case of tension and compression. The quantity S is called the *shear modulus*, and values of S for various substances are given in Table 7–1. It is interesting to note that the shear modulus of any material is usually a good deal less than its Young's modulus. This means that it is easier to slide the atoms of a solid past one another than it is to pull them apart or squeeze them together.

When inward forces act over the entire surface of a body, its volume decreases by some amount ΔV from its original volume of V. Only those force components that are perpendicular to the body's surface where they act are effective in compressing it, since the parallel components lead only to shear stresses. If the compression force per unit area F/A is the same over the entire surface of the body, as in Fig. 7–6, the relationship

$$\frac{\Delta V}{V} = -\frac{1}{B}\frac{F}{A} \tag{7-4}$$

holds. The minus sign corresponds to the fact that an increase in force

$B = BULK \ MODULUS$ TABLE 7.1   leads to decrease in volume

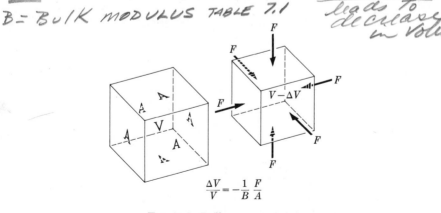

$$\frac{\Delta V}{V} = -\frac{1}{B}\frac{F}{A}$$

FIG. 7–6. Bulk compression.

TABLE 7–1

Elastic moduli

Material	Young's modulus, Y		Shear modulus, S		Bulk modulus, B	
	$\dfrac{10^{10}\,n}{m^2}$	$\dfrac{10^6\,lb}{in^2}$	$\dfrac{10^{10}\,n}{m^2}$	$\dfrac{10^6\,lb}{in^2}$	$\dfrac{10^{10}\,n}{m^2}$	$\dfrac{10^6\,lb}{in^2}$
Aluminum	7.0	10	2.4	3.4	7.0	10
Brass	9.1	13	3.6	5.1	6.1	8.5
Copper	11	16	4.2	6.0	14	20
Glass	5.5	7.8	2.3	3.3	3.7	5.2
Iron	9.1	13	7.0	10	10	14
Lead	1.6	2.3	0.56	0.8	0.77	1.1
Nickel	21	30	7.7	11	26	34
Steel	20	29	8.4	12	16	23

leads to a decrease in volume. The quantity B is called the *bulk modulus;* typical values of B are given in Table 7–1.

Liquids can support neither tensions nor shears, but they do tend to resist compression. Bulk moduli for several liquids are given in Table 7–2. Interatomic forces within a liquid are smaller than within a solid, which is reflected in the considerably smaller bulk moduli of liquids. Substantial forces are nevertheless needed to compress a liquid by more than a slight amount; to compress a volume of water by one percent requires an inward force per unit area of 3300 lb/in².

TABLE 7–2

Bulk moduli of liquids at 20°C

Liquid	Bulk modulus, B	
	$\dfrac{10^9\,n}{m^2}$	$\dfrac{10^5\,lb}{in^2}$
Alcohol, ethyl	0.90	1.3
Benzene	1.05	1.5
Kerosene	1.3	1.9
Mercury	26	38
Oil, lubricating	1.7	2.5
Sulfuric acid	3.0	4.3
Water	2.3	3.3

7–3 Pressure

When a force **F** acts perpendicular to a surface whose area is A, the pressure p being exerted on the surface is defined as the ratio between the magnitude of the force and the area (Fig. 7–7):

$$p = \frac{F}{A}. \tag{7-5}$$

Pressure is a useful quantity because fluids (gases and liquids) flow when forces are exerted upon them. The lack of rigidity exhibited by fluids has three significant consequences:

(1) The forces a fluid exerts on the walls of its container, and vice versa, always act perpendicular to the walls. If this were not so, any sideways force by a fluid on a wall would be met, according to the third law of motion, by a sideways force back on the fluid, which would cause the fluid to move constantly parallel to the wall. But fluids may be at rest in containers of any shape, and so the sole forces they can exert on their containers must be perpendicular to the walls of the latter.

(2) An external pressure exerted on a fluid is transmitted uniformly throughout the volume of the fluid. If this were not so, the fluid would flow from a region of high pressure to one of low pressure, thereby equalizing the pressure. We must keep in mind, however, that the above refers to a pressure imposed from outside the fluid. The fluid at the bottom of a container is always under greater pressure than that at the top owing to the weight of the overlying fluid. Such pressure differences are ordinarily significant only for liquids.

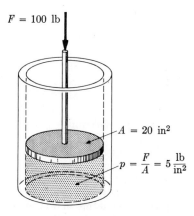

$F = 100$ lb

$A = 20$ in^2

$p = \dfrac{F}{A} = 5 \dfrac{\text{lb}}{\text{in}^2}$

FIG. 7–7. The pressure p exerted by a force acting perpendicular to a surface is equal to the magnitude of the force F divided by the area A of the surface. Pressure exerted on a fluid is transmitted throughout the volume of the fluid.

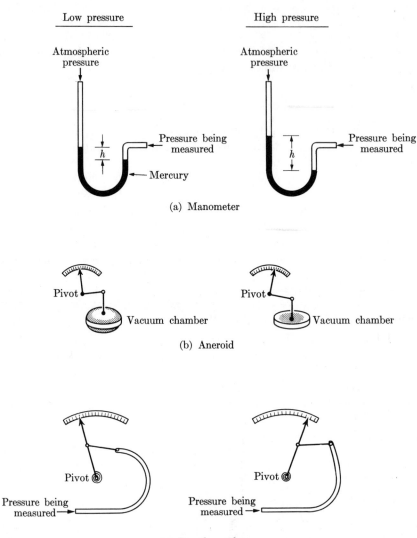

(a) Manometer

(b) Aneroid

(c) Bourdon tube

FIG. 7–8. Three types of pressure gauge. (a) A manometer measures pressure in terms of the differences in height h of two mercury columns, one open to the atmosphere and the other connected to the source of the unknown pressure. (b) An aneroid measures pressure in terms of the amount by which the thin, flexible ends of an evacuated metal chamber are pushed in or out by the external pressure. (c) A Bourdon tube straightens out when the internal pressure exceeds the external pressure.

(3) At any depth in a fluid the pressure is the same in all directions. If this were not so, again, the fluid would flow in such a way as to equalize the pressure.

Pressure may be measured in a number of ways, three of which are illustrated in Fig. 7–8. Usually what is directly determined is the difference between the unknown pressure and atmospheric pressure. This difference is called the *gauge pressure,* while the true pressure is called the *absolute pressure.* That is,

Absolute pressure = gauge pressure + atmospheric pressure.

Thus a tire inflated to a gauge pressure of 24 lb/in² contains air at an absolute pressure of 39 lb/in², since sea-level atmospheric pressure is 14.7 lb/in².

The metric unit of pressure is the n/m², sometimes called the *pascal,* and the British unit of pressure is the lb/ft². Unfortunately a number of other units are in more common use: the lb/in², equal to 144 lb/ft²; the *millibar* (mb), equal to 100 n/m²; the *atmosphere* (atm), representing the average pressure exerted by the atmosphere at sea level, equal to 1.013 × 10⁵ n/m² or 14.7 lb/in²; and the inch and the millimeter of mercury (mm Hg), representing the pressures exerted by columns of mercury of these heights, equal respectively to 3.39 × 10³ and 133 n/m².

7–4 Pressure and depth

The pressure inside a volume of fluid depends upon the depth below the surface, since the deeper we descend, the greater the weight of the overlying fluid. Suppose we have a tank of height h and of cross-sectional area A which is filled with a fluid of density d (Fig. 7–9). The *density* of a substance is its mass per unit volume. Densities of various substances are given in Table 7–3. The volume of the tank is

$$V = Ah,$$

so that the mass of fluid contained in it is

$$m = dV = dAh.$$

The weight of the fluid in the tank is therefore

$$w = mg = dgAh.$$

The pressure Δp the fluid exerts on the bottom

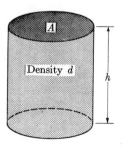

FIGURE 7–9

TABLE 7-3

Densities of various substances at atmospheric pressure and indicated temperature

Solids, 20°C	$\dfrac{10^3\,kg}{m^3}$	$\dfrac{slugs}{ft^3}$
Aluminum	2.70	5.25
Balsa wood	0.13	0.25
Brass	8.7	17
Concrete	2.3	4.5
Copper	8.89	17.3
Glass	2.6	5.0
Gold	19.3	37.5
Ice	0.92	1.8
Iron	7.8	15
Lead	11.3	22.0
Oak	0.72	1.4
Steel	7.8	15

Liquids, 20°C	$\dfrac{10^3\,kg}{m^3}$	$\dfrac{slugs}{ft^3}$
Alcohol, ethyl	0.79	1.53
Benzene	0.88	1.71
Bromine	3.19	6.18
Gasoline	0.68	1.32
Kerosene	0.80	1.55
Mercury	13.6	26.4
Oil, lubricating	0.90	1.75
Sulfuric acid	1.83	3.55
Water, pure	1.00	1.94
Water, sea	1.03	2.00

Gases, 0°C	$\dfrac{kg}{m^3}$	$\dfrac{10^{-3}\,slugs}{ft^3}$
Air	1.29	2.50
Ammonia	0.76	1.47
Argon	1.78	3.45
Carbon dioxide	1.96	3.80
Chlorine	3.16	6.13
Helium	0.18	0.35
Hydrogen	0.090	0.17
Nitrogen	1.25	2.42
Oxygen	1.43	2.78
Propane	2.02	3.92

of the tank is its weight divided by the cross-sectional area of the tank, with the result that

$$\Delta p = \frac{F}{A} = \frac{w}{A} = dgh. \tag{7-6}$$

The pressure difference between the top and the bottom of the tank is directly proportional to the height of the fluid column and to the fluid density. This result also applies to *any* depth h in a fluid, whether at the bottom or not, since the fluid beneath that depth does not contribute to the weight pressing down there.

The *total pressure* within a fluid, of course, also depends upon the pressure p_{atm} exerted on its surface by the atmosphere or, perhaps, by a piston. For example, the pressure at a depth of 8 ft in a swimming pool filled with water (density 1.94 slugs/ft³) is

$$p = p_{atm} + dgh.$$

Here

$$dgh = 1.94 \frac{slugs}{ft^3} \times 32 \frac{ft}{sec^2} \times 8 \text{ ft} \times \frac{1}{(12 \text{ in/ft})^2}$$

$$= 3.4 \frac{lb}{in^2},$$

and so

$$p = (14.7 + 3.4) \frac{lb}{in^2} = 18.1 \frac{lb}{in^2}.$$

In the event a *weight density* (such as 62.5 lb/ft³ in the case of water) is known, it is substituted for dg in Eq. (7–6). The famous and useful *Archimedes' principle* may be derived from Eq. (7–6) together with the fact that the pressure at a point in a fluid is the same in all directions. Let us consider a block of height H and of cross-sectional area A that is completely submerged in a fluid of density d so that its top is h below the surface, as in Fig. 7–10. The downward pressure p_1 on the top of the block is

$$p_1 = p_{atm} + dgh,$$

and the upward pressure p_2 on the bottom of the block is

$$p_2 = p_{atm} + dg(h + H).$$

Since the downward force the fluid exerts on the block is

$$F_1 = Ap_1,$$

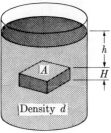

FIGURE 7–10

while the upward force is

$$F_2 = Ap_2,$$

the *net* force on the block is

$$F = F_2 - F_1 = A(p_2 - p_1) = AdgH.$$

Since the pressure on the bottom of the block is greater than that on the top, the net force is upward and is known as the *buoyant force* on the block. Because the volume of the block is

$$V = AH,$$

we have the simple result that

$$F = Vdg \qquad\qquad (7\text{-}7)$$

buoyant force = weight of displaced fluid.

The buoyant force on a submerged object is equal to the weight of fluid displaced by the object. This is Archimedes' principle.

Archimedes' principle enables us to determine only the buoyant force on a submerged object, not the resultant force on it. If the weight of the object is greater than the buoyant force on it, it will sink; if its weight is less than the buoyant force, it will rise. As an exercise let us compute the volume V a helium-filled balloon must have if it is to support a mass of 1000 kg. This mass includes the mass of the balloon itself, but *not* that of the helium inside. The total weight to be supported is therefore

$$w = mg = (1000 \text{ kg} + Vd_{He})g,$$

where d_{He} is the density of helium at atmospheric pressure. The buoyant force on the balloon is

$$F = Vd_{air}g.$$

Hence the condition for equilibrium is

$$F = w, \qquad Vd_{air}g = (1000 \text{ kg} + Vd_{He})g, \qquad V = \frac{1000 \text{ kg}}{d_{air} - d_{He}}.$$

Since

$$d_{air} = 1.29 \times 10^{-3} \frac{\text{kg}}{\text{m}^3}, \qquad d_{He} = 0.18 \times 10^{-3} \frac{\text{kg}}{\text{m}^3},$$

we find that the volume of the balloon must be

$$V = 9.1 \times 10^6 \text{ m}^3.$$

A spherical balloon of this volume has a radius of about 130 meters.

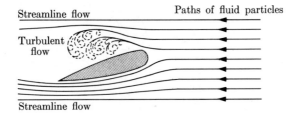

Streamline flow

Paths of fluid particles

Turbulent flow

Streamline flow

FIG. 7–11. Streamline and turbulent flows around an obstacle.

7–5 Fluid flow

The study of fluids in motion is one of the more difficult branches of mechanics because of the diversity of phenomena that may occur. However, the fundamental aspects of fluid flow can be understood on the basis of a simple model that is reasonably realistic in many cases. This model involves liquids that are incompressible and exhibit no viscosity. (Viscosity is the term we introduced earlier to describe internal friction in a fluid.) In the absense of viscosity, layers of fluid slide freely past one another and past other surfaces, so that our model applies to such liquids as water but not to such liquids as molasses. Another approximation we shall make is that the fluid undergoes *streamline flow* exclusively. In streamline flow, which is illustrated in Fig. 7–11, every particle of liquid passing a particular point follows the same path (called a *streamline*) as the particles that passed that point previously. Furthermore, the direction in which the individual fluid particles move is always the same as the direction in which the fluid as a whole moves. At the other extreme is *turbulent flow*, which is characterized by the presence of whirls and eddies, such as those in a cloud of cigarette smoke or at the foot of a waterfall. Turbulence generally occurs at high speeds and when there are obstructions or sharp bends in the path of the fluid.

The volume of liquid that flows through a pipe per unit time is easy to compute and is often of interest. If the average speed of the liquid in the pipe of Fig. 7–12 is v, each part of the stream travels the distance vt in the time interval t. Let us call the cross-sectional area of the pipe A. The volume of liquid transported the distance vt in the time t is vt multiplied by the area A, or vtA. Therefore the rate of flow R of liquid through the pipe is

$$R = \frac{vtA}{t} = vA, \qquad (7\text{–}8)$$

the product of the liquid speed and the cross-sectional area of the pipe. The rate R is the volume of liquid passing through the pipe per unit time,

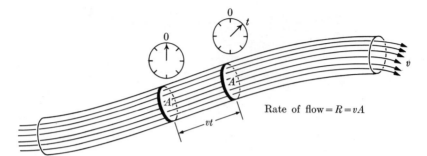

FIGURE 7–12

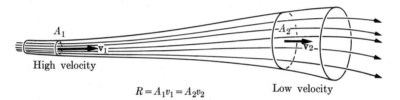

FIG. 7–13. In streamline flow, liquid speed is inversely proportional to the cross-sectional area of the pipe.

and, for convenience in many applications, is often expressed in units such as gallons/min instead of ft^3/sec or m^3/sec. If the pipe size varies, the speed of the liquid also varies so as to keep R constant. Hence a liquid flows faster through a constriction in a pipe and slower through a dilation. As in Fig. 7–13, streamlines drawn close together signify rapid motion while streamlines far apart signify slow motion.

7–6 Bernoulli's equation

When a liquid flowing through a pipe enters a region where the pipe diameter is reduced, its speed increases. A change in speed involves an acceleration; this means that a net force must be acting upon the liquid. This force can only arise from a difference in pressure between the different parts of the pipe. Evidently the pressure in the part of the pipe having a large diameter is the greater, since the liquid increases in speed on its way to the constriction. Thus we expect a relationship between the pressure in a moving liquid and its speed.

Figure 7–14 shows a curved pipe of nonuniform cross section through which a liquid flows. Let us apply the principle of conservation of energy

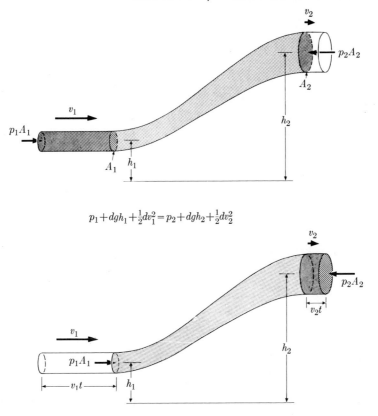

$$p_1 + dgh_1 + \tfrac{1}{2}dv_1^2 = p_2 + dgh_2 + \tfrac{1}{2}dv_2^2$$

FIG. 7–14. Bernoulli's equation.

to a parcel of the liquid of volume v_1tA_1 as it enters at the left in the time t and to the same parcel as it leaves at the right. The mass of the parcel is

$$m = dV = dv_1tA_1 = dv_2tA_2, \tag{7–9}$$

since vA has the same value at 1 and 2. The net amount of work ΔW done on the liquid parcel as it passes from 1 to 2 must be equal to the net change in its potential energy ΔPE as its height goes from h_1 to h_2, plus the net change in its kinetic energy ΔKE as its speed goes from v_1 to v_2. That is,

$$\Delta W = \Delta PE + \Delta KE. \tag{7–10}$$

The work done *on* the parcel at 1 is the force p_1A_1 on it multiplied by the distance v_1t through which the force acts. The work done *by* the parcel at 2 is the force p_2A_2 multiplied by the distance v_2t through which the force acts. The *net work done on the parcel is therefore*

$$\Delta W = p_1 A_1 v_1 t - p_2 A_2 v_2 t,$$

which we can write as

$$\Delta W = \frac{p_1 m}{d} - \frac{p_2 m}{d}$$

by virtue of Eq. (7–9). The change in the potential energy of the parcel in going from **1** to **2** is

$$\Delta PE = mgh_2 - mgh_1,$$

and the change in its kinetic energy is

$$\Delta KE = \tfrac{1}{2}mv_2^2 - \tfrac{1}{2}mv_1^2.$$

Inserting these expressions for ΔW, ΔPE, and ΔKE into Eq. (7–10) we obtain

$$\frac{p_1 m}{d} - \frac{p_2 m}{d} = mgh_2 - mgh_1 + \tfrac{1}{2}mv_2^2 - \tfrac{1}{2}mv_1^2.$$

When we divide through by the common factor m, multiply by d, and rearrange terms, the result is

$$p_1 + dgh_1 + \tfrac{1}{2}dv_1^2 = p_2 + dgh_2 + \tfrac{1}{2}dv_2^2. \qquad (7\text{–}11)$$

Equation (7–11) is known as *Bernoulli's equation* after Daniel Bernoulli (1700–1782), who first derived it. According to Bernoulli's equation the quantity $(p + dgh + \tfrac{1}{2}dv^2)$ has the same value at all points in a liquid undergoing streamline flow.

In many situations the speed, pressure, or height of a liquid is constant, and simplified forms of Bernoulli's equation hold. Thus when a liquid column is stationary, we see that the pressure difference between two depths in it is

$$p_2 - p_1 = dg(h_1 - h_2), \qquad (7\text{–}12)$$

which is just what Eq. (7–6) states. Evidently the latter formula is included in Bernoulli's equation.

Another straightforward result occurs in the event $p_1 = p_2$. As an example Fig. 7–15 illustrates a liquid emerging from an orifice at the bottom of a tank. The liquid pressure equals atmospheric pressure at the top of the tank and at the orifice. If the orifice is small compared with the cross section of the tank, the liquid level in the tank will fall slowly enough for the liquid speed at the top of the tank to be assumed zero. If the speed of the liquid as it leaves the orifices is v and the difference

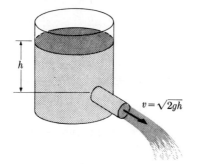

FIG. 7–15. Torricelli's theorem.

in height between the top of the liquid and the orifice is h, Bernoulli's equation reduces to

$$\tfrac{1}{2}dv^2 = dgh,$$

$$v = \sqrt{2gh}. \tag{7–13}$$

The speed with which the liquid is discharged is the same as the speed of a body falling from rest from the height h. This result is called *Torricelli's theorem*, and, like the relationship between pressure and depth, it is a special case of Bernoulli's equation. The rate at which liquid flows through the orifice may be found from Eq. (7–8) if the orifice area A is known. The volume of liquid being discharged per unit time is

$$R = vA = A\sqrt{2gh}.$$

As an illustration of this formula, we note that water will leak through a hole 1 cm² in area at the bottom of a tank in which the water level is 3 meters high at the rate of

$$R = 1\ \text{cm}^2 \times \frac{1}{10^4\ \text{m}^2/\text{cm}^2} \times \sqrt{2 \times 9.8\ \text{m}/\text{sec}^2 \times 3\ \text{m}}$$

$$= 7.7 \times 10^{-4}\ \text{m}^3/\text{sec},$$

which is a little less than a quart per second, an appreciable amount.

The most interesting special case of Bernoulli's equation occurs when there is no change in height during the motion of the liquid (Fig. 7–16). Here

$$p_1 + \tfrac{1}{2}dv_1^2 = p_2 + \tfrac{1}{2}dv_2^2, \tag{7–14}$$

which means that the pressure in the liquid is least where the speed is

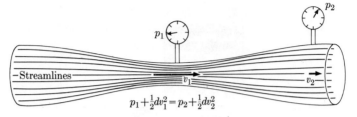

Fɪɢ. 7–16. In a horizontal pipe, the pressure is greatest when the velocity is least and vice versa.

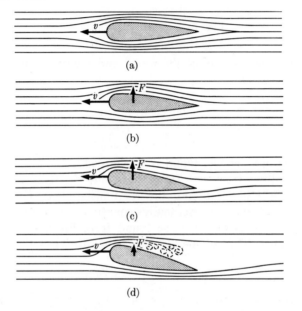

Fɪɢ. 7–17. Air flow past a wing. At (a) the flow is the same on both surfaces, so that no lift results. The lift at (c) is greater than that at (b) because of the greater pressure difference between upper and lower surfaces (the pressure is least where the streamlines are closest together). At (d) turbulence reduces the available lift.

greatest, and vice versa. There are many applications of this principle. A familiar one is the lifting force produced by the flow of air past the wing of an airplane, as in Fig. 7–17. (Air is, of course, a compressible fluid and so does not fit our model exactly, but the behavior predicted by Bernoulli's equation is not a bad approximation for gases at moderate speeds.) Air moving past the upper surface of the wing must travel faster than air moving past the lower surface; this is indicated by the closeness of the streamlines near the former. The difference in speed leads to a decreased pressure over the top of the wing, a pressure which is equivalent to a suction force that lifts the wing. The greater the difference in air speeds around the upper and lower surfaces, the greater the lift that is produced, provided the wing shape is not so extreme that turbulence results (Fig. 7–17d).

A curious effect may take place when the difference in liquid speeds at two nearby places is so great that the absolute pressure where the liquid flows most rapidly becomes negative. This can happen when the propeller of a boat rotates at a high speed. The result is a phenomenon called *cavitation* in which cavities ("holes") appear in the water. Cavitation causes the efficiency of the propeller to drop sharply.

IMPORTANT TERMS

The three categories of stress forces are *tension,* in which equal and opposite forces that act away from each other are applied to a body; *compression,* in which equal and opposite forces that act toward each other are applied to a body; and *shear,* in which equal and opposite forces that do not act along the same line of action are applied to a body. A tensile stress tends to elongate a body, a compressive stress to shorten it, and a shearing stress to change its shape without changing its volume.

Hooke's law states that the amount of deformation experienced by a body under stress is proportional to the magnitude of the stress. Thus the elongation of a wire is proportional to the tension applied to it.

The *elastic limit* is the maximum deformation a solid under stress can experience without being permanently altered. Hooke's law is only valid when the elastic limit is not exceeded.

Young's modulus Y of a particular material is equal to the tension or compression force per unit cross-sectional area applied to a sample of that material divided by the fractional change in the length of the sample. The *shear modulus* S of the material is equal to the shear force per unit cross-sectional area applied to a sample of it divided by the relative distortion of the sample. The *bulk modulus* B of the material is equal to the pressure on a sample of it divided by the fractional decrease in its volume.

The *pressure* on a surface is the perpendicular force per unit area that acts upon it. *Gauge pressure* is the difference between true pressure and atmospheric pressure.

Archimedes' principle states that the buoyant force on a submerged object is equal to the weight of fluid it displaces.

In *streamline flow* every particle of liquid passing a particular point follows the same path, whereas in *turbulent flow* irregular whirls and eddies occur. According to *Bernoulli's equation* the quantity $(p + dgh + \frac{1}{2}dv^2)$ is constant throughout a liquid undergoing streamline flow.

PROBLEMS

1. When a thin steel wire is used to support a 100-lb weight, the wire stretches by 0.02 in. What is the force constant of the wire?

2. When a coil spring is used to support a 12-kg object, the spring stretches by 4 cm. What is the force constant of the spring?

3. A coil spring has a force constant of 1000 n/m. How much will it stretch when it is used to support an object whose mass is 8 kg?

4. A coil spring has a force constant of 2 lb/ft. How much will it stretch when it is used to support an object whose mass is 0.5 slug?

5. A steel wire 1 m long and 1 mm square in cross section supports a mass of 6 kg. By how much does it stretch?

6. A copper wire 5 ft long and 0.05 in. in diameter supports a weight of 17 lb. By how much does it stretch?

7. A nylon rope $\frac{1}{2}$ in. in diameter breaks when a force of 5000 lb is applied to it. What would you estimate for the breaking strength of a nylon rope (a) $\frac{1}{4}$ in. in diameter? (b) $\frac{3}{4}$ in. in diameter?

8. A steel cable whose cross-sectional area is 0.4 in² supports an elevator weighing 1 ton. The elastic limit of the cable is 4×10^4 lb/in². What is the maximum upward acceleration that can be given the elevator if the tension in the cable is to be no more than 20% of the elastic limit?

9. An iron pipe 10 ft long is used to support a sagging floor. The inside diameter of the pipe is 4 in. and its outside diameter is 5 in. When the force on it is 3000 lb, by how much is it compressed?

10. A wall of lead bricks 1 m high is used to shield a sample of radium. Each brick was originally a cube 10 cm on an edge. What is the height of the lowest brick when the wall has been erected?

11. A 1-in. cube of raspberry flavored gelatin on a table is subjected to a shearing force of 0.1 lb. The upper surface is displaced by 0.2 in. What is the shear modulus of the gelatin?

12. A 50-kg woman balances on the heel of her left shoe, which is 1 cm in diameter. What pressure (in atm) does she exert on the ground?

13. The force on a phonograph needle whose point is 0.1 mm in radius is 1.2 n. What is the pressure it exerts on the record (in atm)?

14. A piston weighing 12 n rests on a sample of gas in a cylinder 5 cm in diameter. (a) What is the gauge pressure in the gas? (b) What is the absolute pressure in the gas?

15. An irregular nugget of gold is dropped into a filled glass of water, and 2 in³ of water overflow. Gold is worth $35 per ounce of weight. How much is the nugget worth?

16. Calculate the density of sea water at a depth of 3 mi.

17. Calculate the density of mercury at the bottom of a mercury column 1.3 m high.

18. What is the minimum area of an ice floe 3 in. thick that can support a 120-lb girl without getting her feet wet? The floe is in a fresh-water lake.

19. A barge 120 ft long and 20 ft wide weighs 20 tons. What is the depth of sea water required to float it?

20. An ice cube floats in a glass of water filled to the brim. What will happen when the ice melts?

21. What percentage of the volume of an iceberg is below the surface of the sea?

22. A sailboat has a ton of lead ballast attached to the bottom of its keel, which is 6 ft below the water surface when the boat is level. What is the torque exerted by this ballast when the boat is heeled by 20° from the vertical? (*Hint:* Why is this problem in this chapter instead of in Chapter 3?)

23. A 30-kg balloon is filled with 100 m³ of hydrogen. How much force is needed to hold it down?

24. Two spheres of the same diameter but of different mass are dropped from a tower. If air resistance is the same for both, which will reach the ground first?

25. The densities of people are slightly less than that of water. Assuming that these densities are the same, compute the buoyant force of the atmosphere on a 160-lb man.

26. Water flows through a hose whose internal diameter is 1 cm at a speed of 1 m/sec. What should the diameter of the nozzle be if the water is to emerge at 5 m/sec?

27. What gauge pressure is required in a fire hose if a stream of water is to reach a height of 60 ft?

28. A tank of height H is filled with water. A hole is made a distance y from the top. How far from the tank does the water strike the ground?

29. A horizontal stream of water leaves an orifice 1 m above the ground and strikes the ground 2 m away. (a) What is the speed of the water when it leaves the orifice? (b) What is the gauge pressure behind it?

30. A barrel filled with kerosene is 4 ft high. A crack 1 in. long and 0.07 in. wide appears at its base. How many lb of kerosene per minute flow out?

31. A man's brain is approximately $\frac{1}{3}$ m above his heart, while this distance is approximately 2 m in a giraffe. What is the pressure required to circulate blood between the heart and brain of (a) a man? (b) a giraffe? The density of blood is 1.1×10^3 kg/m³.

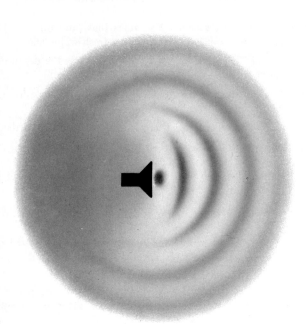

Vibrations and Waves | 8

The behavior of a vibrating body and that of a wave traveling through a medium are similar in several essential respects. In both, a particular motion repeats itself at regular intervals; in both, the properties of matter in bulk play important roles; and in both, the motions that occur represent the continuous conversion of potential energy to kinetic energy and back. The analysis of vibrations and waves will prepare us for a number of important topics we shall take up shortly as well as being a good exercise in the principles of mechanics.

8–1 Elastic potential energy

The amount of work done in stretching a body that obeys Hooke's law is easy to calculate. As we learned in Chapter 5, the work done by a force is the product of the force and the distance through which it acts. Here the force used in stretching the body is not constant, but is proportional to the elongation s at each point in the stretching process. The *average* force $\overline{F}$ applied while the body is stretched from its normal length by an amount s to some final length is

$$\overline{F} = \frac{F_{\text{initial}} + F_{\text{final}}}{2} = \frac{0 + ks}{2} = \frac{1}{2}\, ks, \qquad (8\text{–}1)$$

since the initial force is zero and the final force is ks. The work done is the product of the average force $\overline{F}$ and the total elongation s (Fig. 8–1), so that

$$W = \overline{F}s = \tfrac{1}{2}ks^2. \qquad (8\text{–}2)$$

This formula is most often used in connection with springs: to stretch (or compress) a spring whose force constant is k by an amount s from its normal length requires $\tfrac{1}{2}ks^2$ of work to be done. This work goes into *elastic potential energy.* When the spring is released, its potential energy of $\tfrac{1}{2}ks^2$ is transformed into kinetic energy or into work done on something

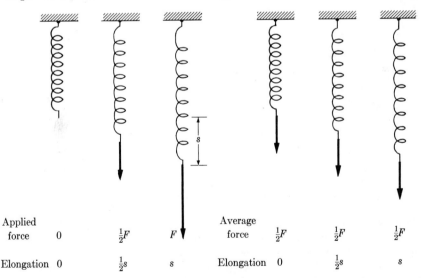

FIG. 8–1. To compute the work done in stretching a body that obey's Hooke's law, the varying force that actually acts during the expansion may be replaced by the average force.

FIG. 8–2. Examples of elastic potential energy.

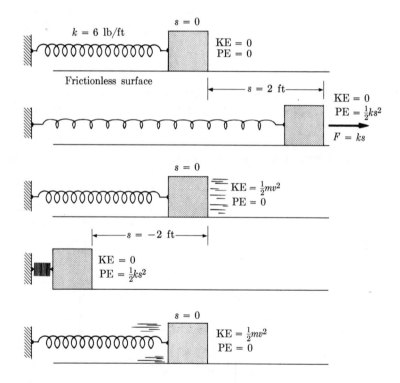

FIG. 8–3. A 0.1-slug block attached to a spring whose force constant is 6 lb/ft is pulled 2 ft from its equilibrium position. When the spring is released, its elastic potential energy of $\frac{1}{2}ks^2$ is converted into kinetic energy $\frac{1}{2}mv^2$, and as the block's momentum compresses the spring on the other side of the equilibrium position, the kinetic energy is converted back into elastic potential energy.

else (Fig. 8–2); work done against frictional forces within the spring itself always absorbs some fraction of the available potential energy. Potential energy is a useful notion whenever forces act whose magnitude depends upon position, and we shall find that gravitation and elasticity do not exhaust the areas in which it can profitably be employed.

Let us apply the ideas of this section to the horizontal spring shown in Fig. 8–3. The force constant k of the spring is 6 lb/ft. Attached to the free end of the spring is a block whose mass is 0.1 slug. If the spring is pulled out 2 ft from its equilibrium position and then released, what will the block's speed be when it returns to the equilibrium position?

When the spring is released, its elastic potential energy starts to be converted into kinetic energy of the block. We shall assume that the spring's mass is small compared with that of the block and that its internal friction may be neglected. At the equilibrium position of the spring $s = 0$, and all the initial potential energy of $\frac{1}{2}ks^2$ is now kinetic energy $\frac{1}{2}mv^2$. Hence

$$\tfrac{1}{2}mv^2 = \tfrac{1}{2}ks^2, \qquad v = \sqrt{\frac{k}{m}}\, s = \sqrt{\frac{6\text{ lb/ft}}{0.1\text{ slug}}}\, 2\text{ ft} = 15\, \frac{\text{ft}}{\text{sec}}.$$

8–2 Simple harmonic motion

When a spring is stretched and then released, it does not return to its equilibrium position and come to a stop there. Instead it continues moving until it is compressed to the point where all the kinetic energy it had is converted back into elastic potential energy (Fig. 8–3). The potential energy of the stretched spring turns into kinetic energy and then back into potential energy when the spring is compressed; neglecting friction within the spring, we find that the amount of compression $-s$ will have the same magnitude as the original extension s since

$$\tfrac{1}{2}ks^2 = \tfrac{1}{2}k(-s)^2.$$

Left to itself the spring will continue oscillating back and forth indefinitely. The behavior of a spring oscillating in this way is called *simple harmonic motion.*

Simple harmonic motion occurs whenever a force acts on a body in the opposite direction to its displacement from its normal position, with the magnitude of the force proportional to the magnitude of the displacement. The elastic force of a stretched or compressed spring always tends to restore the spring to its normal length. The third law of motion tells

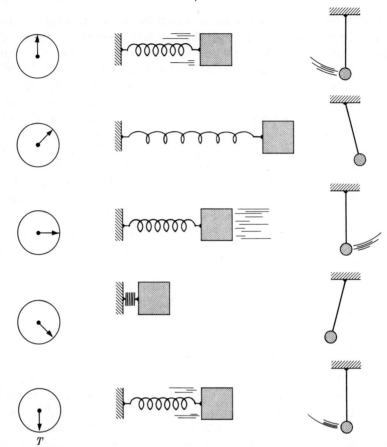

Fig. 8–4. The period T of a body undergoing simple harmonic motions is the time required for it to make one complete oscillation.

us that we can write for the *restoring force* F_r

$$F_r = -ks. \tag{8-3}$$

In other words, the restoring force *of the spring* is equal and opposite to the Hooke's law force,

$$F = ks,$$

that must be exerted *on the spring* when it is displaced by s from its equilibrium length.

The *period* of a body undergoing simple harmonic motion is the time required for it to make one complete oscillation. (A complete oscillation is often called a *cycle*). In the case of a spring, the period is the time the spring spends in going from its maximum extension, say, through its maximum compression and back to its maximum extension once more (Fig. 8–4). For all types of simple harmonic motion, the period T is given by

$$T = 2\pi \sqrt{-\frac{\text{displacement}}{\text{acceleration}}} = 2\pi \sqrt{-\frac{s}{a}}, \tag{8–4}$$

where the acceleration is that experienced by the body when it is at the specified displacement from its equilibrium position.

To calculate the acceleration of a stretched spring we start with the second law of motion,

$$F = ma,$$

and substitute the restoring force F_r given by Eq. (8–3) since it is the restoring force that causes the body to be accelerated. This procedure yields

$$F_r = ma, \qquad a = \frac{F_r}{m} = \frac{-ks}{m}.$$

With this result we find that the period of a body of mass m attached to a spring of force constant k is

$$T = 2\pi \sqrt{-\frac{s}{a}} = 2\pi \sqrt{-\frac{s}{-ks/m}} = 2\pi \sqrt{\frac{m}{k}}. \tag{8–5}$$

It is worth noting that the period T does not depend upon the maximum displacement s; no matter how much or how little the spring is initially pulled out, precisely the same amount of time is required for each cycle. If s is small, the maximum acceleration is also small and the body moves back and forth very slowly through its range, while if s is large, the acceleration is also large and the body moves correspondingly rapidly through the larger range. (The maximum displacement of a body undergoing harmonic motion on either side of its equilibrium position is called the *amplitude* of the motion.) This peculiarity of simple harmonic motion is capitalized upon in the design of clocks and watches, which use the rotational oscillations of a coil spring or the swings of a pendulum—both examples of simple harmonic motion—to maintain a constant rate independent of any fluctuations in amplitude.

In the case of the spring and mass of the previous section, we see that the period of oscillation is

$$T = 2\pi \sqrt{\frac{m}{k}} = 2\pi \sqrt{\frac{0.1 \text{ slug}}{6 \text{ lb/ft}}} = 0.81 \text{ sec.}$$

A quantity often used in describing harmonic motion is *frequency*. The frequency is the number of cycles executed per unit time. Hence frequency, whose symbol is f, is the reciprocal of period T,

$$f = \frac{1}{T}.$$ (8-6)

The frequency corresponding to the period of 0.81 sec that we found above is

$$f = \frac{1}{0.81 \text{ sec}} = 1.2 \frac{\text{cycles}}{\text{sec}}.$$

8-3 The pendulum

A pendulum executes simple harmonic motion as it swings back and forth provided that the arc through which the pendulum bob moves is a fairly small one, not more than about 5° on either side of the vertical. We shall see why this limitation arises if we use Eq. (8-4) to calculate the period of a pendulum.

Figure 8-5 shows a pendulum of length L whose bob has a mass m, together with a diagram of the forces acting on the bob. (It is assumed that the entire mass of the pendulum is concentrated in the bob.) The weight of the bob, mg, which acts vertically downward, may be resolved into two forces, $\mathbf{T}$ and $\mathbf{F}$, which act respectively parallel to and perpendicular to the supporting string L. That is, the vector sum $\mathbf{T} + \mathbf{F} = mg$. The force $\mathbf{F}$ is the restoring force that acts to return the bob to the midpoint of its motion. The space triangle hLd and the vector triangle $\mathbf{T}mg\mathbf{F}$ are similar, since each contains a right angle and two sides of one

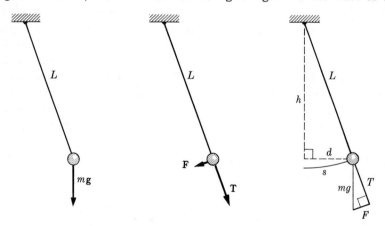

FIGURE 8-5

are parallel to the two corresponding sides of the other, and so

$$\frac{F}{d} = \frac{mg}{L}.$$

The restoring force acting on the bob is therefore

$$F = -\frac{mgd}{L},$$

where the minus sign indicates that F points in the direction of decreasing d. If the bob is not far from the midpoint of its motion, the horizontal distance d is almost exactly equal to the actual path length s, and F then is given by

$$F = -\frac{mgs}{L}$$

The acceleration of the bob that results from this force is

$$a = \frac{F}{m} = -\frac{gs}{L}.$$

Substituting in Eq. (8–4) we find that

$$T = 2\pi\sqrt{-s/a} = 2\pi\sqrt{L/g}. \tag{8–7}$$

Provided that s is small enough so that it is very close to d, then, the motion of a pendulum is simple harmonic in character with a period proportional to the square root of the pendulum's length and independent of the mass of the bob.

How long should a pendulum be for it to have a period of exactly 1 sec? We first solve Eq. (8–7) for L:

$$T = 2\pi\sqrt{\frac{L}{g}}, \qquad T^2 = \frac{4\pi^2 L}{g}, \qquad L = \frac{gT^2}{4\pi^2}.$$

Inserting the values $g = 9.8$ m/sec^2 and $T = 1$ sec, we find that

$$L = \frac{9.8 \text{ m/sec}^2 \times 1 \text{ sec}^2}{4\pi^2} = 0.25 \text{ m}.$$

8–4 Waves

Energy can be transmitted from one place to another in a variety of ways. Suppose, for instance, that we wish to supply energy to a boat in the center of a lake from a position on the shore, with the provision that the precise form in which the energy arrives does not matter. The most obvious thing to do is to throw a stone at the boat, providing it

FIGURE 8–6

with kinetic energy. Another method of approach is to build a fire, heat water drawn from the lake, and then pour it back in, providing the boat with thermal energy. A third method requires merely that we drop a stone in the water near the shore, letting the waves that move out transfer energy to the boat by causing it to move up and down (Fig. 8–6). We note in this latter case that what occurs when the stone strikes the water is not a flow of material but, instead, the spreading out of a deformation of the water surface. The energy reaching the boat arrives as such a deformation which, since the boat floats on the water's surface, appears as a combination of kinetic and potential energy. Energy propagation by means of the motion of a change in a medium, rather than the motion of the medium itself, is called *wave motion,* and it occurs in many forms in nature. We shall subsequently find *electromagnetic waves* of great interest, but in this chapter we shall restrict ourselves to the more familiar water and sound waves.

8–5 Water waves

If we were somehow to tag individual water molecules and follow them when a series of water waves pass by, we would find that their paths are similar to those shown in Fig. 8–7. Each molecule describes a roughly circular orbit about its original position, with an average velocity of zero and so never undergoing any permanent displacement. The passage of a wave across the surface of a body of water involves the motion of a pattern: energy goes from one place to another by virtue of the changing pattern, but there is no transport of mass.

Three related quantities are useful in describing wave motion (Fig. 8–8). These are the *speed v* with which the wave travels; the *wavelength* λ (Greek letter lambda), which is the distance between adjacent crests or troughs; and the *frequency f,* which is the number of waves that pass a given point per unit time, usually per second. The wave speed is equal to the distance through which a particular wave moves per second. In every second, *f* waves (by definition) go past a particular point, with

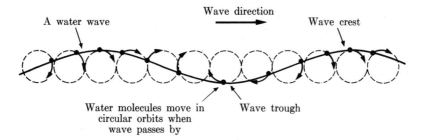

Fɪɢ. 8–7. Water molecules move in circular orbits about their original positions when a wave passes by. At the crest of a wave the molecules are moving in the direction the wave is traveling, while in the trough the molecules are moving in the opposite direction. There is no net motion of water involved in the motion of a wave.

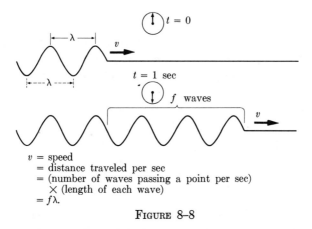

v = speed
= distance traveled per sec
= (number of waves passing a point per sec)
× (length of each wave)
= $f\lambda$.

Fɪɢᴜʀᴇ 8–8

each wave representing a distance of λ. Therefore a wave travels a total distance of $f\lambda$ per second, which is the wave speed v. That is,

$$v = f\lambda, \tag{8–8}$$

a basic formula which may be applied to all wave phenomena.

Sometimes it is more useful to consider the *period* T of a wave, which is the time required for one complete wave to pass a given point (Fig. 8–9). Since f waves pass by per second, the period of each wave is

$$T = \frac{1}{f}. \tag{8–9}$$

If there are five waves per second passing by, for example, each wave

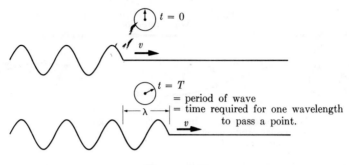

FIGURE 8–9

has a period of $\frac{1}{5}$ sec. In terms of period T, Eq. (8–8) becomes

$$v = \frac{\lambda}{T}.$$ (8–10)

Suppose that we find the distance between adjacent wave crests on the ocean one day to be 160 ft, with a crest passing underneath us every 4.5 sec. This means that, since $\lambda = 160$ ft and $T = 4.5$ sec, the wave velocity is

$$v = \frac{\lambda}{T} = \frac{160 \text{ ft}}{4.5 \text{ sec}} = 36 \frac{\text{ft}}{\text{sec}}.$$

The frequency of the waves is

$$f = \frac{1}{T} = \frac{1}{4.5 \text{ sec}} = 0.22 \frac{\text{waves}}{\text{sec}}.$$

The motion of the pattern that constitutes a wave in a medium can be much more rapid than the motion of the individual particles of the medium. The waves we have just discussed might have an *amplitude A* of 4 ft, that is, the crests are 4 ft above the normal sea level and the troughs 4 ft below (Fig. 8–10). Thus the water particles are moving in circles whose radii are 4 ft, so that as each wave passes by, the molecules

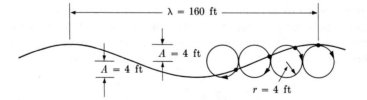

FIG. 8–10. The amplitude of a water wave is the height of its crest above normal sea level, which is equal to the depth its trough lies below sea level.

travel a distance d equal to the circumference $2\pi r$ of the circle. Hence

$$d = 2\pi r = 2\pi A = 25 \text{ ft.}$$

It takes 4.5 sec for a wave to go past a given point, which means that the particles must cover the 25-ft circumference of their circular orbits in 4.5 sec. The speed v' of the particles is therefore

$$v' = \frac{d}{T} = \frac{25 \text{ ft}}{4.5 \text{ sec}} = 5.6 \frac{\text{ft}}{\text{sec}}.$$

The speed of the *wave*, however, is **36** ft/sec, over six times greater. Thus energy can be transported by wave motion at a faster rate than might be possible through direct mass motion.

8–6 Sound waves

Water waves are actually a combination of two more fundamental types of waves, *longitudinal* and *transverse*. Longitudinal waves occur when the individual particles of a medium vibrate back and forth *in the direction* in which the waves travel. Figure 8–11 shows longitudinal waves in a coil spring; each portion of the spring is alternately compressed and extended as the waves pass by. Longitudinal waves, then, are essentially density fluctuations. Transverse waves occur when the individual particles of a medium vibrate from side to side *perpendicular to the direction* in which the waves travel. Figure 8–12 shows transverse waves in a stretched wire; each section of the wire moves from one side to the other and back again as the patterns constituting the waves travel down the wire.

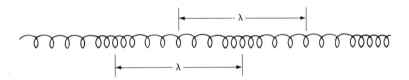

FIG. 8–11. Longitudinal waves in a coil spring.

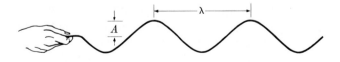

FIG. 8–12. Transverse waves in a stretched wire.

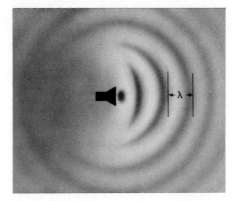

FIG. 8–13. Sound consists of longitudinal waves, representing condensations and rarefactions in the air in its path.

When a periodic disturbance takes place in air, longitudinal *sound* waves spread out from it. The air in the path of a sound wave becomes alternately denser and rarer, as in Fig. 8–13. The resulting changes in pressure cause our eardrums to vibrate with the same frequency, producing the physiological sensation of sound.

The speed V of sound in air at 0°C is 332 m/sec, which is 1090 ft/sec. (V increases about 0.2% for each °C increase in air temperature.) Hence the musical note A, whose frequency is 440 cycles/sec, represents a wavelength λ in air of

$$\lambda = \frac{V}{f} = \frac{1090 \text{ ft/sec}}{440 \text{ cycles/sec}} = 2.5 \text{ ft.}$$

Sound waves are transmitted by media other than air. Table 8–1 gives the speed of sound in various solids, liquids, and gases. Note that, generally, the stiffer the material, the faster the speed, which is reasonable when we reflect that stiffness implies particles tightly coupled together and therefore more immediately responsive to one another's motions.

The wavelength of the musical note A in sea water is

$$\lambda = \frac{V}{f} = \frac{3424 \text{ ft/sec}}{440 \text{ cycles/sec}} = 7.8 \text{ ft,}$$

over three times longer than the wavelength of the same note in air. When a sound wave produced in one medium enters another in which its speed is different, the frequency of the wave remains the same while its wavelength changes.

TABLE 8-1

The speed of sound in various media at the specified temperatures

Medium	Speed		Temperature, °C
	m/sec	ft/sec	
Air	332	1090	0
Carbon dioxide	259	676	0
Chlorine	206	658	0
Water, distilled	1404	3413	0
Water, sea	1440	3424	0
Acetone	1146	3624	30
Paraffin	1304	4266	15
Copper	3560	11,680	20
Iron	5130	16,830	20

8–7 Harmonic motion and waves

We might suspect at this point that there is a connection between the properties of waves and the behavior of particles undergoing simple harmonic motion. This suspicion is reinforced when we pluck one end of a long coil spring and watch what happens. If we gradually pull out an end of the spring, the entire spring lengthens uniformly; here, however, we pull it out quickly and then release it, so that the distortion has not had time to move down the entire spring when we let go. The result is a wave moving down the spring, exactly like the longitudinal waves of Fig. 8–11. We conclude that longitudinal waves in a spring actually consist of a series of coupled harmonic oscillations, a conclusion that can be extended to longitudinal waves in all media.

Transverse waves can be analyzed in a similar way. As Fig. 8–14 shows, a transverse wave in a stretched string consists of harmonic oscillations of the particles making up the string. These oscillations are per-

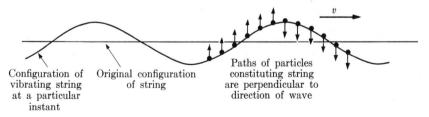

Configuration of vibrating string at a particular instant

Original configuration of string

Paths of particles constituting string are perpendicular to direction of wave

FIG. 8–14. A transverse wave in a stretched string.

pendicular to the direction in which the wave moves, while those in longitudinal waves are parallel to the wave direction.

The correspondence between harmonic and wave motion permits us to obtain an interesting result. As we found in the early part of this chapter, the work done in extending (or compressing) a spring by a distance s is $\frac{1}{2}ks^2$, where k is the force constant of the spring. If this spring is then released, it oscillates back and forth, with its energy going from potential to kinetic and back to potential, but its total energy at all times equals the $\frac{1}{2}ks^2$ it was originally. In wave terminology, s is the amplitude A, and since a wave is a series of harmonic oscillations, its total energy must be proportional to A^2, the square of its amplitude. This dependence of wave energy on A^2 is true for waves of all kinds, longitudinal and transverse, and even applies to electromagnetic waves, which are nonmechanical in character, provided that the amplitude A is properly interpreted. Ocean waves illustrate this relationship: a wave 10 ft high possesses four times the energy of a wave 5 ft high, something no sailor would doubt.

IMPORTANT TERMS

A body under stress possesses *elastic potential energy,* which is equal to the work done in deforming it.

Simple harmonic motion is an oscillatory motion that occurs whenever a force acts on a body in the opposite direction to its displacement from its normal position, with the magnitude of the force proportional to the magnitude of the displacement. Thus the force on a body in simple harmonic motion always tends to return it to its normal position.

The *period T* of a body undergoing simple harmonic motion is the time required for it to make one complete oscillation. The *frequency f* of such a body is the number of complete oscillations it makes per unit time. Thus $f = 1/T$.

The *amplitude* of a body undergoing simple harmonic motion is its maximum displacement on either side of its equilibrium position.

Wave motion is characterized by the propagation of a change in a medium, rather than by the net motion of the medium itself. The passage of a wave across the surface of a body of water, for instance, involves the motion of a pattern of alternate crests and troughs, with the individual water molecules themselves ideally executing uniform circular motion.

The *frequency f* of a series of waves is the number of waves that pass a particular point per unit time, while their *wavelength* λ is the distance between adjacent crests or troughs. The *period T* is the time required for one complete wave to pass a particular point.

Longitudinal waves occur when the individual particles of a medium vibrate back and forth in the direction in which the waves travel. *Sound* consists of longitudinal waves.

Transverse waves occur when the individual particles of a medium vibrate from side to side perpendicular to the direction in which the waves travel. The vibrations of a stretched string are transverse waves.

PROBLEMS

1. A force of 0.5 lb is required to push a Jack-in-the-box into its box, an operation in which the spring is compressed 4 in. If the Jack-in-the-box weighs 0.2 lb, what will its maximum speed be when it pops out?

2. A toy rifle employs a spring whose force constant is 30 lb/ft. In use, the spring is compressed 2 in., and when released, it propels a lead ball weighing 0.01 lb. What is the speed with which the ball leaves the rifle?

3. A 5-kg object is dropped on a vertical spring from a height of 2 m. The force constant of the spring is 1500 n/m. What is the maximum compression the spring will suffer?

4. The periods of the earth's rotation on its axis and revolution about the sun are 24 hr and 365 days respectively. What is the frequency of each of these motions?

5. When a 2-lb weight is suspended from a spring, the spring stretches by 3 in. If the weight oscillates up and down, what is its period? What is its frequency?

6. A body whose mass is 1 kg hangs from a spring. When the body is pulled down 5 cm from its equilibrium position and released, it oscillates once per second. (a) What is the force constant of the spring? (b) What is the body's speed when it passes through its equilibrium position? (c) What is the maximum acceleration of the body?

7. A body whose mass is 0.2 slug is suspended from a spring and oscillates with a frequency of 5 cycles/sec and an amplitude of 0.5 ft. (a) What is the total energy of the motion? (b) What is the maximum acceleration of the body?

8. A body whose mass is 0.4 kg is suspended from a spring and oscillates with a period of 2 sec. By how much will the spring contract when the body is removed?

9. A body whose mass is 0.005 kg is in simple harmonic motion with a period of 0.04 sec and an amplitude of 0.01 m. (a) What is its maximum acceleration? (b) What is the maximum force on the body? (c) What is its acceleration when it is 0.005 m from its equilibrium position? (d) What is the force on it at that point?

10. A chandelier is suspended from a high ceiling with a cable 20 ft long. What is its period of oscillation?

11. A pendulum whose length is 1.53 m oscillates 24 times per minute in a particular location. What is the acceleration of gravity there?

12. What is the wavelength in meters of sound waves of frequency 8000 cycles/sec in (a) air, (b) sea water, (c) iron? (Use the values for the speed of sound in these media given in Table 8–1.)

13. Water waves are observed approaching a lighthouse at a speed of 18 ft/sec. There is a distance of 20 ft between adjacent crests. (a) What is the frequency of the waves? (b) What is their period?

14. A violin string is set in vibration at a frequency of 440 cycles/sec. How many vibrations does it make while its sound travels 500 ft?

15. A particular groove in a phonograph record moves past the needle at a speed of 0.4 m/sec. The sound produced has a frequency of 3000 cycles/sec. What is the wavelength of the wavy indentations of the groove?

16. A tuning fork vibrating 440 times per sec is placed in distilled water. (a) What are the frequency and wavelength in meters of the waves produced within the water? (b) What are the frequency and wavelength in meters of the waves in the surrounding air the water waves produce when they reach the surface?

Heat | 9

Because we are so familiar with the idea that heat is a form of energy, it may not be easy for us to sympathize with the struggles our predecessors had in understanding its nature. In particular, the hypothesis that heat is an invisible fluid called *caloric* is hard for us to take seriously, although it was not a bad notion when first proposed. While we will not take the time to discuss the rise and subsequent downfall of the caloric theory of heat, we should note that, like so many other unsuccessful physical theories, it began as a simplifying idea, became complicated as it was modified to account for observations that did not agree with it, and ultimately perished in favor of another idea that provided in its turn a simpler, more consistent picture of what was going on. In this chapter we shall consider various aspects of heat and thermal phenomena from a macroscopic point of view, leaving the ultimate explanation for heat in terms of the microscopic structure and behavior of matter for the next chapter.

9–1 Temperature

Temperature, like force, is a key concept in physics which, while we have a clear idea of its meaning in terms of our sense impressions, is difficult to define with precision. We shall attain such precision later, but for the moment we shall dodge the issue and accept temperature merely as something responsible for sensations of hot and cold.

There are a number of properties of matter that vary with temperature, and these can be used to construct *thermometers,* devices for measuring temperature. For example, when an object is heated sufficiently, it glows, at first a dull red, then bright red, and finally, at a high enough temperature, it becomes "white hot." By measuring the color of the light it gives off, we can accurately determine the temperature of an object. This method can only be used at rather high temperatures, however.

Of wider application is the fact that matter usually expands when its temperature is increased and contracts when its temperature is decreased. Railroad tracks must be laid with gaps between successive rails to allow for expansion in the summer; heated air above a radiator rises as it expands and becomes lighter than the surrounding air; a column of mercury in a glass tube changes length with a change in temperature. All three of these types of observation have resulted in practical thermometers. Two strips of different metals that are joined together bend to one side with a change in temperature owing to different rates of expansion in the two metals, a fact employed in constructing household oven thermometers (Fig. 9–1a). In a constant-volume gas thermometer (Fig. 9–1b), a very sensitive laboratory instrument, the height of the mercury column at the left is adjusted until the mercury column at the right just touches the gas bulb. The difference in heights of the two mercury columns is a measure of the pressure needed to maintain the gas in a fixed volume, and hence a measure of the temperature. And, of course, the most common thermometer (Fig. 9–1c) uses the expansion of a liquid, usually mercury or colored alcohol, to measure temperature.

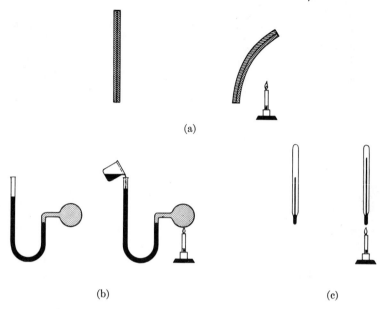

(a)

(b)   (c)

FIG. 9–1. (a) A bimetallic strip thermometer. (b) A constant-volume gas thermometer. (c) A liquid-in-glass thermometer.

Before we can use any of these or other thermal properties of matter to construct a practical thermometer, we must begin by specifying a temperature scale and the method by which we shall calibrate the thermometer in terms of this scale. Water is a readily available liquid which freezes into a solid, ice, and vaporizes into a gas, steam, at definite temperatures at sea level atmospheric pressure. We can establish a temperature scale by defining the freezing point of water (or, more exactly, the point at which a mixture of ice and water is in equilibrium, with exactly as much ice melting as water freezes) as 0° and the boiling point of water (or, more exactly, the point at which a mixture of steam and water is in equilibrium) as 100°. This scale is called the *celsius**
scale, and temperatures measured in it are written, for example, "40°C."

To calibrate a thermometer, say an ordinary mercury thermometer, we first plunge it into a mixture of ice and water. When the mercury column has come to rest we mark the position of its top 0°C on the glass (Fig. 9–2). Then we plunge it into a mixture of steam and water, and when the mercury column has again come to rest, we mark the new position of the top of the mercury column 100°C. Finally we divide the interval between the 0°C and 100°C markings into 100 equal parts, each

---

* Also called the *centigrade* scale.

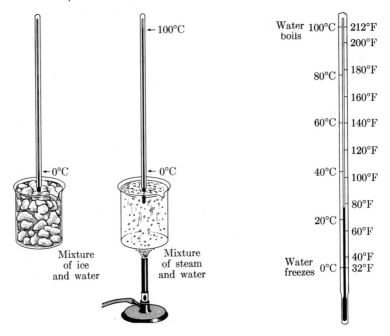

FIG. 9–2. Calibrating a thermometer on the centigrade scale. A mixture of ice and water is, by definition, at 0°C and a mixture of steam and water is, again by definition, at 100°C.

FIG. 9–3. A comparison of the fahrenheit and celsius temperature scales.

representing a change in temperature of 1°C, and extend the scale with divisions of the same length beyond 0°C and 100°C as far as is convenient. In doing this we have, of course, assumed that changes in the length of the mercury column are always directly proportional to the changes in temperature that brought them about.

Another temperature scale in common use is the *fahrenheit* scale. Fahrenheit degrees are five-ninths as large as celsius degrees, since on the former scale the freezing point of water is 32°F and its boiling point is 212°F (Fig. 9–3). With the formulas

$$°F = \tfrac{9}{5}°C + 32° \quad \text{and} \quad °C = \tfrac{5}{9}(°F - 32°), \qquad (9–1)$$

we can convert temperatures from one scale to the other. For instance, a fahrenheit temperature of 70°F, which is normal room temperature, on the celsius scale is

$$\tfrac{5}{9}(70° - 32°) = 21°C.$$

9–2 Heat

Heat and temperature are related, though different, physical quantities, and the distinction between them is both subtle and important. A simple example will illustrate this distinction. Suppose that we have a block of ice we want to melt, and that we can use either a small amount of water at 200°F or a large amount of water at 50°F for this purpose (Fig. 9–4). The 200°F water is quite hot, nearly boiling in fact, while the 50°F water is definitely cool to the touch. Yet we find experimentally that the large mass of cool water melts more of the ice than the small mass of hot water! To interpret this result we say that something we shall call *heat* is involved in melting ice, and that the above amount of hot water has a higher temperature but a lower heat content than the much larger amount of cool water. The hot water contains more heat than an identical volume of cool water, but there is in this case so much more of the latter that its total heat content is the greater.

There are a number of other striking examples that can be given which bear upon heat and temperature. Before we consider any of them, though, it is appropriate to make a preliminary definition of heat and

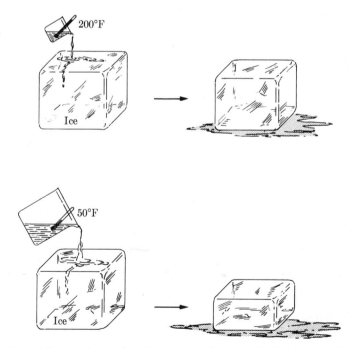

FIG. 9–4. A large volume of cool water contains more heat than a small volume of hot water.

to use it in defining a unit of heat. Such a definition might state that

Heat is a quantity that causes an increase in the temperature of a body of matter to which it is added and a decrease in the temperature of a body of matter from which it is removed, provided that the matter does not change state during the process.

The latter part of the definition is required because, as we shall see, changes of state (for instance, from ice to water or water to steam) involve the transfer of heat to or from a body without any change in temperature.

In the metric system the unit of heat, called the *kilocalorie* (abbreviated kcal), is that amount of heat required to raise the temperature of 1 kg of water through 1°C. Similarly, 1 kcal of heat must be removed from 1 kg of water to reduce its temperature by 1°C. Because this amount of heat actually varies slightly with temperature, the kilocalorie is formally defined as the amount of heat involved in changing the temperature of 1 kg of water from 14.5°C to 15.5°C; the difference is insignificant for most purposes, however.

Although we shall not employ them here, two other heat units are in common use, the *calorie* and the *British thermal unit,* abbreviated cal and Btu respectively. The calorie is equal to 10^{-3} kcal (0.001 kcal). The "calorie" that dieticians speak of is the same as the kilocalorie. The Btu is the amount of heat required to raise the temperature of 1 lb of water by 1°F; it is equal to 0.252 kcal.

9–3 Specific heat

Other substances respond to the addition or removal of heat with temperature changes greater than that of an equal mass of water. One kcal of heat increases the temperature of 1 kg of helium (its volume held constant) by 1.3°C, of 1 kg of ice by 2°C, and of 1 kg of gold by 33°C; 1 kg of water by definition increases in temperature by 1°C when 1 kcal of heat is added to it. The *specific heat* (symbol c) of a substance refers to what we might think of as its thermal inertia; c is the amount of heat required to change the temperature of 1 kg of a particular substance by 1°C. Table 9–1 is a list of specific heats for various substances. A high specific heat implies a relatively small change in temperature for a given change in heat, just as a large inertial mass implies a relatively small acceleration when a given force is applied.

With the help of specific heat we can obtain a formula for the quantity of heat Q involved when a mass m of a substance undergoes a

TABLE 9-1

Specific heats of various substances*

Substance	Specific heat, kcal/kg·°C
Alcohol (ethyl)	0.58
Aluminum	0.22
Copper	0.093
Glass	0.20
Gold	0.030
Granite	0.19
Ice	0.50
Iron	0.11
Lead	0.030
Marble	0.21
Mercury	0.033
Silver	0.056
Steam	0.48
Sulfuric acid	0.27
Turpentine	0.42
Water	1.00
Wood	0.42
Zinc	0.092

* The actual values vary somewhat with temperature; the ones
given in the table represent averages.

change in temperature of ΔT. The formula is simply

$$Q = mc\, \Delta T. \tag{9-2}$$

To cool 14 kg of aluminum from 80°C to 15°C, for instance, involves the
quantity of heat

$$Q = mc\, \Delta T = 14 \text{ kg} \times 0.22\, \frac{\text{kcal}}{\text{kg·°C}} \times (-65°\text{C}) = -200 \text{ kcal};$$

the minus sign means that this quantity of heat is to be removed from
the aluminum to achieve the temperature change of −65°C.

Let us try a somewhat more elaborate calculation. We pour 0.2 kg of
coffee at 90°C into an 0.05-kg cup at 20°C. Assuming that no heat is
transferred to or from the outside, what is the final temperature of the
coffee? We shall assume that the specific heat of coffee is that of water

and that the specific heat of the cup is that of glass. To solve this problem, we begin by noting that

$$\text{Heat gained by cup} = \text{heat lost by coffee.}$$

If the final temperature of both coffee and cup is T,

$$\text{Heat gained by cup} = m_{\text{cup}}c_{\text{cup}}\,(T - 20°C)$$

$$= 0.05 \text{ kg} \times 0.20\,\frac{\text{kcal}}{\text{kg·°C}} \times (T - 20°C)$$

$$= (0.01T - 0.2)\text{ kcal}$$

and

$$\text{Heat lost by coffee} = m_{\text{coffee}}c_{\text{coffee}}\,(90°C - T)$$

$$= 0.2 \text{ kg} \times 1.0\,\frac{\text{kcal}}{\text{kg·°C}} \times (90°C - T)$$

$$= (18 - 0.2T)\text{ kcal.}$$

Hence, equating the heat gained by the cup with the heat lost by the coffee,

$$0.01T - 0.2 = 18 - 0.2T,$$

$$0.21T = 18.2,$$

$$T = 87°C.$$

The temperature of the coffee will drop 3°C as it warms the cup.

9–4 Change of state

Not always does the addition or removal of heat from a sample of matter lead to a change in its temperature. Figure 9–5 is a graph showing how the temperature of 1 kg of ice varies as we add more and more heat to it. The ice was initially at −50°C, and, since its specific heat is 0.5 kcal/kg·°C, each kilocalorie of heat that is supplied increases its temperature by 2°C. At 0°C, however, the steady temperature rise ceases, and the temperature remains at 0°C while we add 80 kcal more. This is the amount of heat needed to melt 1 kg of ice at 0°C to water at 0°C. Then, when all the ice has turned to water, the temperature goes up once more as further heat is supplied. Since the specific heat of water is 1 kcal/kg·°C, there is now a rise of 1°C per kilocalorie of heat. This

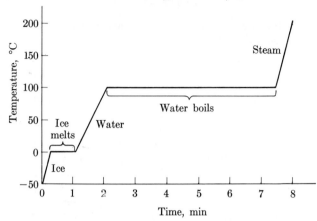

Rate of heating = 100 kcal/min

FIG. 9–5. A graph of temperature versus time for 1 kg of water, initially ice at −50°C, to which heat is being added at the constant rate of 100 kcal/min.

rate of change is less than that of ice, since the specific heat of water is greater than that of ice, and so the slope of the graph is less. When 100°C is reached, the temperature again stays constant despite further heating until a total of 540 kcal is added. This is the amount of heat needed to vaporize 1 kg of water at 100°C to steam at 100°C. After all the water has become steam, its temperature increases again. The specific heat of steam is 0.48 kcal/kg·°C, so the temperature increase is 2.1°C per kilocalorie of heat, a steeper slope than the ones corresponding to ice or water.

The *heat of fusion* of a substance is the amount of heat that must be supplied to change 1 kg of it at its melting point from the solid to the liquid state; the same amount of heat must be removed from 1 kg of the substance in the liquid state at its melting point to change it to a solid. The usual symbol for heat of fusion is L_f.

The *heat of vaporization* of a substance is the amount of heat that must be supplied to change 1 kg of it at its boiling point from the liquid to the gaseous (or vapor) state; the same amount of heat must be removed from 1 kg of the substance in the gaseous state at its boiling point to change it into a liquid. The usual symbol for heat of vaporization is L_v. The heats of fusion and vaporization for a number of substances are listed in Table 9–2 together with their melting and boiling points.

Under certain circumstances most substances can change directly from the solid to the vapor state, or vice versa, processes both called *sublimation*. For example, "dry ice" (solid carbon dioxide) evaporates

TABLE 9-2

Heats of fusion and vaporization and melting and boiling points of various substances

Substance	Melting point, °C	L_f, kcal/kg	Boiling point, °C	L_v, kcal/kg
Alcohol (ethyl)	−114	25	78	204
Bismuth	271	12.5	920	190
Bromine	−7	16	60	43
Lead	330	5.9	1170	175
Lithium	186	160	1336	511
Mercury	−39	2.8	358	71
Nitrogen	−210	6.1	−196	48
Oxygen	−219	3.3	−183	51
Sulfuric acid	8.6	29	326	122
Water	0	80	100	540
Zinc	420	24	918	475

directly to gaseous carbon dioxide at temperatures above −78.5°C, and does not pass through the liquid state. With the exception of carbon dioxide and a few other substances, however, sublimation does not occur except at pressures well below that of the atmosphere.

Let us see how a typical problem involving a change of state may be solved. What is the minimum amount of ice at −10°C that must be added to 0.5 kg of water at 20°C in order to bring the water down to 0°C? We begin, as before, with the statement of heat conservation,

$$\text{Heat gained by ice} = \text{heat lost by water.}$$

The heat Q_1 gained by the unknown mass of ice in going from −10°C to its melting point at 0°C is

$$Q_1 = m_{ice}c_{ice}\,\Delta T_{ice} = m_{ice} \times 0.50\,\frac{\text{kcal}}{\text{kg·°C}} \times 10°C,$$

and the heat Q_2 gained by the ice in melting at 0°C is

$$Q_2 = m_{ice}L_{f\ ice} = m_{ice} \times 80\,\frac{\text{kcal}}{\text{kg}}.$$

Hence

$$\text{Heat gained by ice} = Q_1 + Q_2 = m_{ice}(5 + 80)\,\frac{\text{kcal}}{\text{kg}}.$$

The heat lost by the water in cooling to 0°C is

$$\text{Heat lost by water} = m_{\text{water}} c_{\text{water}} \, \Delta T_{\text{water}}$$

$$= 0.5 \text{ kg} \times 1 \, \frac{\text{kcal}}{\text{kg} \cdot °\text{C}} \times 20°\text{C}$$

$$= 10 \text{ kcal}.$$

Equating the heat gained with the heat lost,

$$85 m_{\text{ice}} \, \frac{\text{kcal}}{\text{kg}} = 10 \text{ kcal}, \qquad m_{\text{ice}} = 0.118 \text{ kg}.$$

9–5 The triple point

Let us examine the effect of pressure on changes of state. Figure 9–6 shows how the boiling point of water varies with pressure, behavior typical of other liquids as well. The upper limit of this *vaporization curve*, which occurs at a temperature of 374°C and a pressure of 218 atm, is known as the *critical point*. A substance cannot exist in the liquid state at a temperature above that of its critical point, regardless of how great the pressure may be. Helium has the lowest critical temperature, −268°C, and is therefore a gas at all temperatures above that.

The melting points of solids also depend upon pressure, although to a smaller extent than the boiling points. The variation of the melting point of ice with pressure is shown in the *fusion curve* of Fig. 9–7. Ice, together with gallium and bismuth, is unique in that its melting point *decreases*

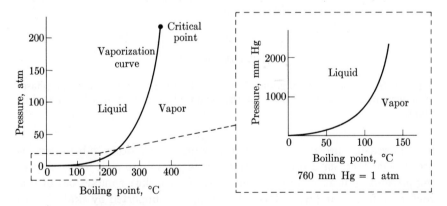

FIG. 9–6. The vaporization curve of water, showing how the boiling point of water varies with pressure.

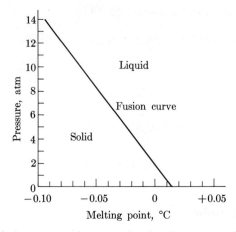

FIG. 9–7. The fusion curve of water, showing how the melting point of ice varies with pressure.

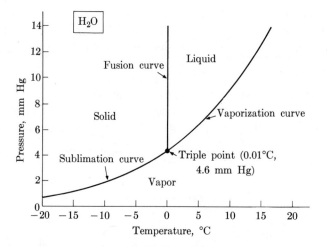

FIG. 9–8. Triple-point diagram of water. The solid, liquid, and vapor phases of water can exist simultaneously at the temperature and pressure of the triple point.

with increasing pressure; the melting points of all other substances increase with increasing pressure. Hence it is possible to melt ice by applying pressure to it as well as by heating it. An ice skater makes use of this fact in an interesting way. His entire weight is supported by skate blades of very small area, and the resulting pressure on the ice may exceed 1000 atm. The ice under the blades melts because of the great pressure, creating a thin film of water that acts as an efficient lubricant. On unusually

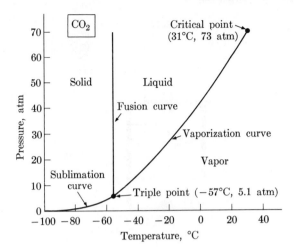

Fɪɢ. 9–9. Triple-point diagram of carbon dioxide.

cold days even such pressures may not be sufficient to melt the ice, and skating then becomes impossible.

The fusion and vaporization curves of water intersect at a temperature of 0.01°C and a pressure of 4.6 mm Hg, as shown on the combined plot of Fig. 9–8. Along the fusion curve both ice and water can simultaneously exist, and along the vaporization curve both water and water vapor can simultaneously exist; hence under conditions corresponding to those of the intersection of the two curves, the solid, liquid, and vapor states of water can all exist together. This intersection is accordingly called the *triple point* of water.

At pressures below that of its triple point, no substance can exist as a liquid. The dividing line on a pressure-temperature graph between the solid and vapor states is called the *sublimation curve*, since it represents the conditions required for a solid to vaporize directly or a vapor to solidify directly. At atmospheric pressure the addition of heat causes ordinary ice to melt, since the triple point of water lies well below 1 atm, but the addition of heat causes solid carbon dioxide to sublime, since its triple point lies above 1 atm (Fig. 9–9).

9–6 Mechanical equivalent of heat

As we noted in Chapter 5, the energy dissipated when work is done against frictional forces appears as heat. This observation by itself, however, is not enough to persuade us that energy is being conserved. What

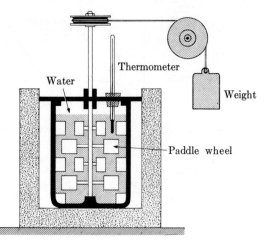

FIG. 9–10. Joule's method for measuring the mechanical equivalent of heat. As the weight descends it turns paddle wheels inside an insulated container of water, so that its potential energy is converted first into the kinetic energy of the paddle wheels and then into heat energy of the water. When the energy lost by the descending weight is divided by the increase in the heat content of the water, the mechanical equivalent of heat is found to be 4185 j/kcal.

we must also know is how much heat is produced when a given amount of energy is dissipated. Is it always the same quantity of heat, or does it vary in some way with the circumstances involved in its production? This fundamental question did not receive an adequate answer until little more than a century ago.

The first definite evidence that the ratio between the energy lost in some mechanical process and the heat that appears as a result of the process is a constant came from the experiments of James Prescott Joule (1818–1889), an English brewer. Joule used an apparatus similar to that sketched in Fig. 9–10: a descending weight causes paddle wheels inside an insulated container to rotate, thereby stirring the liquid within the container. The stirring must be done against the frictional resistance of the liquid (the viscosity of a liquid is a measure of this resistance), and the heat produced can be determined by measuring the increase in the liquid's temperature. The results are always the same, no matter what the nature of the liquid, its volume, or the total amount of work done on it by the falling weight: 1 kcal of heat appears for every 4185 joules of work done. The quantity

$$J = 4185 \ \frac{\text{j}}{\text{kcal}}$$

is accordingly known as the *mechanical equivalent of heat.*

As a simple application of the mechanical equivalent of heat we can compute the difference in temperature of water at the top and at the bottom of a waterfall. In the course of dropping from a height h, a mass m of water loses the potential energy

$$PE = mgh.$$

This potential energy appears as kinetic energy in the falling water, and we shall assume that it is all dissipated as the heat

$$Q = mc \, \Delta T$$

at the bottom of the fall. To convert Q, which is in kilocalories, to joules, we multiply it by J, since the units of J are j/kcal. Hence

$$PE = JQ, \qquad mgh = Jmc \, \Delta T,$$

and the rise in temperature of the water is

$$\Delta T = \frac{gh}{Jc} = 0.0023h \, \frac{°C}{m}.$$

We note that the mass m of water that we have been considering turns out to be irrelevant, as it should. We can apply the general formula we have obtained to any waterfall whose height h we know. For example, the Victoria Falls on the Zambesi River are 108 meters high, so that the water at the foot of the falls is

$$\Delta T = 0.0023 \, \frac{°C}{m} \times 108 \text{ m} = 0.25°C$$

warmer than at the top, hardly an appreciable amount.

IMPORTANT TERMS

That property of a body of matter that causes us to experience a sensation of hot or cold when we touch it is called its *temperature*.

A *thermometer* is a device for measuring temperature. The two temperature scales in common use are the *celsius* (centigrade) scale, in which the freezing point of water is assigned the value 0°C and its boiling point the value 100°C, and the *fahrenheit* scale, in which these points are assigned the values 32°F and 212°F respectively.

Heat is a quantity that causes an increase in the temperature of a body of matter to which it is added and a decrease in the temperature of a body of matter from which it is removed, provided that the matter does not change state during the process. The unit of heat in the metric system is the *kilocalorie* (kcal), which is that amount of heat required to change the temperature of 1 kg of water by 1°C.

The *specific heat* of a substance is the amount of heat required to change the temperature of 1 kg of the substance by 1°C.

The *heat of fusion* of a substance is the amount of heat that must be supplied to change 1 kg of it at its melting point from the solid to the liquid state; the same amount of heat must be removed from 1 kg of the substance in the liquid state at its melting point to change it to a solid.

The *heat of vaporization* of a substance is the amount of heat that must be supplied to change 1 kg of it at its boiling point from the liquid to the gaseous (or vapor) state; the same amount of heat must be removed from 1 kg of the substance at its boiling point to change it into a liquid.

Sublimation is the direct conversion of a substance from the solid to the vapor state, or vice versa, without it first becoming a liquid.

The *critical point* is the upper limit of the vaporization curve of a substance; a substance cannot exist in the liquid state at a temperature above that of its critical point.

The *triple point* is the intersection of the vaporization, fusion, and sublimation curves of a substance. All three states of a substance may exist in equilibrium at the temperature and pressure of its triple point.

The *mechanical equivalent of heat* is the constant ratio between the energy dissipated in some mechanical process and the heat that appears as a result of the process. It is equal to 4185 j/kcal.

PROBLEMS

1. The melting point of lead is 330°C and its boiling point is 1170°C. Express these temperatures on the fahrenheit scale.

2. The normal temperature of the human body is 98.6°F. What is this temperature on the celsius scale?

3. At what temperature would celsius and fahrenheit thermometers give the same reading?

4. Mercury freezes at −40°C. What is this temperature on the fahrenheit scale?

5. Dry ice (solid carbon dioxide) vaporizes at −112°F. What is this temperature on the celsius scale?

6. How much heat must be added to 1 kg of copper to raise its temperature from 20°C to 100°C?

7. Why will the engine of a car whose cooling system is filled with an alcohol antifreeze be more likely to overheat in the summer than one whose cooling system is filled with water?

8. A 60-kg woman is on a diet that provides her with 2500 kcal daily. If a corresponding amount of heat were added to 60 kg of water at 37°C, what would its final temperature be?

9. A 0.6-kg copper container holds 1.5 kg of water at 20°C. A 0.1-kg iron ball at 120°C is dropped into the water. What is the final temperature of the water?

10. A 0.1-kg piece of silver is taken from a bath of hot oil and placed in a 0.08-kg glass jar containing 0.2 kg of water at 15°C. The temperature of the water increases by 8°C. What was the temperature of the oil?

11. How much more heat must be added to 1 kg of ice at 0°C to convert it to steam at 100°C than is required to raise the temperature of 1 kg of water from 0°C to 100°C?

12. How much steam at 120°C is required to melt 0.5 kg of ice at 0°C?

13. How much ice at −10°C is required to cool a mixture of 0.1 kg ethyl alchohol and 0.1 kg water from 20°C to 5°C?

14. A 5-kg iron bar is taken from a forge at a temperature of 1000°C and plunged into a pail containing 10 kg of water at 60°C. How much steam is produced?

15. By mistake, 0.2 kg of water at 0°C is poured into a vessel containing liquid nitrogen at −196°C. How much nitrogen vaporizes?

16. A 1-kg block of ice at 0°C falls into a lake whose water is also at 0°C, and 0.01 kg of ice melts. What was the minimum altitude from which the ice fell?

17. A lead bullet at 100°C strikes a steel plate and melts. What was its minimum speed?

18. A 100-kg wooden beam is pushed across a stone floor by a force just sufficient to overcome friction. The coefficient of friction is 0.4. Assuming that all the work done against friction goes into heating the beam, what is its rise in temperature for each foot that it is pushed?

19. A 50-kg block of ice at 0°C is pushed across a wooden floor also at 0°C for a distance of 20 m. A total of 25 gm of ice melts as a result of the friction of the block on the floor. What is the coefficient of friction in this case?

20. The rate at which work is being done is often specified in *watts*, one watt equaling one joule per second. (a) How many kcal/sec are equivalent to 1 watt? (b) How many kcal are evolved by a 60-watt light bulb per hour? (c) Radiant energy from the sun arrives at the earth's surface at the approximate rate of 20 kcal/m² per minute. Express this rate in watts/m².

21. Find the average pressure exerted on the ground by a 180-lb man each of whose feet has an area of 20 in².

22. A vertical cylinder 5 cm in radius has a 12-kg piston. How much pressure does the piston exert on the contents of the cylinder?

23. The statement that atmospheric pressure equals 760 mm Hg means that a column of mercury 760 mm high exerts a pressure of 1 atm on its base. A given volume of mercury weighs 13.6 times as much as the same volume of water. How high would a column of water have to be in order that the pressure on its base equal 1 atm?

24. The gauge on an air compressor reads 150 lb/in². To what absolute pressure does this correspond (a) when the gauge is at sea level; (b) when the gauge is at an altitude of 10,000 ft, where atmospheric pressure is 10 lb/in²?

25. A small amount of water is placed in a can and brought to a boil. The can is then tightly covered and removed from the stove. Explain why the can collapses after it has cooled for a few minutes.

26. The pressure and temperature of the atmosphere at 115,000 ft are 4 mm Hg and $-23°C$ respectively. What is the state of water under those conditions? of carbon dioxide?

27. Why does food cook faster in a pressure cooker?

28. Water expands slightly when it freezes. What connection might you suspect between this fact and the fact that the melting point of ice decreases with increasing pressure?

29. Carbon dioxide is usually shipped in tanks under a pressure of approximately 1000 lb/in². At 20°C, is the carbon dioxide a solid, liquid, or gas?

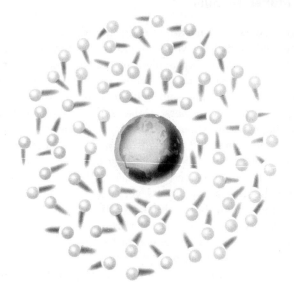

Structure of Matter | 10

Thus far we have been outlining the laws and principles that underlie the more conspicuous phenomena associated with macroscopic aggregates of matter. However, we have not touched upon the microscopic properties of the aggregates themselves—what they are composed of, what their ultimate structure is, and how their internal character is related to their interactions with the external world. To answer these questions we must, of course, first make further observations and conduct additional experiments. On the basis of the results we then find, we may anticipate having to introduce new laws and principles into the scheme we are developing, under the name *physics*, to describe the nature and behavior of matter. In this chapter we shall descend into the realm of molecules, the first step in our exploration of the microscopic universe.

10–1 Matter in bulk

Matter occurs in four *states:* solid, liquid, gas, and plasma. Solids are characterized by definite volumes and shapes that vary only slightly with environmental changes (provided, of course, that they remain as solids during the changes). Liquids, while possessing definite volumes in the same sense as solids, have no definite shapes, instead merely taking on the shapes of their containers. Gases exhibit neither definite volumes nor definite shapes: they expand to fill large containers, may be compressed to fit into small ones, and escape from open ones (Fig. 10–1). Liquids and gases are jointly called fluids because of their ability to flow more or less readily from one place to another. The line of demarcation between solids and liquids is not always very clear; a material like pitch seems quite hard and rigid, but a chunk of it will eventually lose its shape and spread out on whatever surface it is resting.

Plasmas are much like gases except that their constituent particles have become electrically charged, and their behavior in consequence depends strongly upon electromagnetic forces in addition to the forces influencing the other states of matter. Because plasmas do not exist under ordinary terrestrial conditions, they are very strange to us, but, since nearly all the matter in the universe as a whole is in the plasma state, they have received a great deal of study in recent years. We shall confine ourselves in this chapter to the more familiar solids, liquids, and gases.

Liquids and gases are almost always *homogeneous*, which means that every portion of a particular sample is exactly like every other portion. Solids may be either homogeneous or *heterogeneous;* if the latter, some portions of a particular sample may be different from others. A bar of gold, for example, is a homogeneous solid, while a piece of wood is a heterogeneous one. A heterogeneous solid is not always easy to recognize as such, and instruments such as the microscope (or even more sophisticated devices) may be required for definite identification.

Solid

Liquid   Gas

FIGURE 10–1

Homogeneous substances may be further classified into *elements, compounds,* and *solutions.* *Elements* are the simplest substances we encounter in bulk; they cannot be decomposed or transformed into one another by ordinary chemical or physical means. There are 105 known elements, listed in Appendix IV together with their symbols and certain of their properties, of which 92 have been found in nature and 13 artificially prepared. At room temperature and sea-level atmospheric pressure, 10 elements are in the gaseous state, namely argon, chlorine, fluorine, helium, hydrogen, krypton, nitrogen, oxygen, radon, and xenon, and two in the liquid state, namely bromine and mercury. The rest are in the solid state, the majority being metals.

Two or more elements may combine chemically to form a *compound,* a new substance whose properties are different from those of the elements that compose it. Each constituent of a *solution,* in contrast, retains its characteristic properties (except, of course, for the mechanical properties of solids and gases dissolved in liquids), and may be separated from the other constituents by relatively simple procedures. Boiling and freezing are examples of such procedures, since the temperatures at which these changes of state occur have specific values for each element or compound. Air, for example, is a solution of several gases, chiefly nitrogen and oxygen. Oxygen boils at $-183°C$ while nitrogen boils at $-196°C$, $13°$ lower; hence if we heat a sample of liquid air to a temperature over $-196°C$ but under $-183°C$, the nitrogen will vaporize and we will be left ideally with oxygen alone. Under certain circumstances, nitrogen and oxygen unite to form the compound nitric oxide; the boiling point of nitric oxide is $-152°C$, and heating a sample of liquid nitric oxide to this temperature will result in the vaporization of the entire sample. The constituents of a solution may be elements or compounds or both.

Another distinction between compounds and solutions is that the elements in a compound are present in certain definite proportions, always

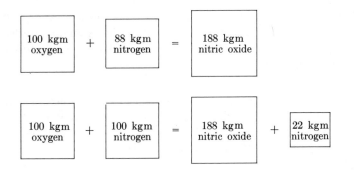

FIG. 10–2. The law of definite proportions.

the same for a particular compound, while the constituents of a mixture may be present in a wide range of proportions. At sea level the ratio by weight of the nitrogen and oxygen in the atmosphere varies slightly about an average of 3.2:1, and is several percent greater at high elevations; the ratio by weight of the nitrogen and oxygen in nitric oxide is invariably 0.88:1. If there is an excess of either nitrogen or oxygen when nitric oxide is being prepared, the excess amount will not combine but will be left over and can be separated out at an appropriate temperature (Fig. 10–2).

10–2 Thermal expansion

When the temperature of a sample of matter changes, the dimensions of the sample also change. This is a familiar effect: sidewalks buckle on a hot summer day, the mercury column rises in a heated thermometer, warm air rises because, owing to its expansion, it is lighter than surrounding cool air. The change ΔL in the length of a solid rod is, to a good approximation, proportional to both the original length L_0 of the rod and the temperature change ΔT. That is,

$$\Delta L = aL_0 \, \Delta T, \tag{10-1}$$

where a, the *coefficient of linear expansion*, is a constant whose value depends upon the nature of the rod. Table 10–1 is a list of coefficients of linear expansion for various materials.

As an example of Eq. (10–1), let us consider a steel girder 50 ft long whose temperature changes from 5°C to 25°C, an increase of 20°C. The coefficient of linear expansion of steel is 1.2×10^{-5}°C, and so

$$\Delta L = aL_0 \, \Delta T = 1.2 \times \frac{10^{-5}}{°C} \times 50 \text{ ft} \times 20°C = 0.012 \text{ ft},$$

which is a little over $\frac{1}{8}$ inch. This does not seem a very significant change until we compute the force exerted by the expanding girder, which is equal to the force required to stretch it by 0.012 ft. To do this we refer to Eq. (7–1), namely

$$F = ks,$$

relating the force F and the increase in length s it produces when applied to an object (Fig. 10–3). If the girder has a cross-sectional area of 40 in², the corresponding value of the force constant is probably about 3×10^7 lb/ft. Substituting 0.012 ft for s, we see that

$$F = ks = 3 \times 10^7 \, \frac{\text{lb}}{\text{ft}} \times 0.012 \text{ ft} = 3.6 \times 10^5 \text{ lb};$$

TABLE 10-1		TABLE 10-2	
Coefficients of linear expansion		Coefficients of volume expansion	
Substance	*Coefficient,* $\times 10^{-5}/°C$	*Substance*	*Coefficient,* $\times 10^{-4}/°C$
Aluminum	2.4	Ethyl alcohol	11
Brass	1.8	Glass (average)	0.2
Concrete	0.7–1.2	Glycerin	5.1
Copper	1.7	Ice	0.5
Iron	1.2	Mercury	1.8
Lead	3.0	Pyrex glass	0.09
Quartz	0.05	Water	2.1
Silver	2.0		
Steel	1.2		

a force of 360,000 lb, which is 180 tons, is associated with the expansion of the girder!

A formula similar to Eq. (10–1) holds for the changes in volume ΔV of a solid or liquid whose temperature changes by an amount ΔT. Here we have

$$\Delta V = bV_0\,\Delta T, \tag{10-2}$$

where V_0 is the original volume and b is the *coefficient of volume expansion*. Table 10–2 is a list of coefficients of volume expansion for various substances. In general the coefficients of linear and volume expansion are related by

$$b = 3a, \tag{10-3}$$

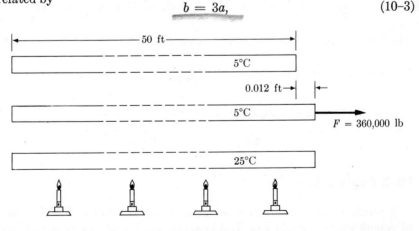

FIG. 10–3. A 50-ft steel girder expands 0.012 in. when its temperature is increased by 20°C. At constant temperature, a force of 360,000 lb would be required to produce the same increase in length.

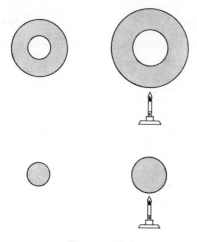

FIGURE 10–4

so that we can readily determine the values of b for the materials of Table 10–1.

As an exercise we shall calculate the volume of water that overflows when a Pyrex beaker filled to the brim with 250 cm³ of water at 20°C is heated to 60°C. First, we note that a cavity in a body expands or contracts by precisely as much as a solid object having the same volume as the cavity and having the composition of the body (Fig. 10–4). This means that we can write for the increase in capacity of the beaker

$$V_b = b_P V_b \, \Delta T = 0.09 \times \frac{10^{-4}}{°C} \times 250 \text{ cm}^3 \times 40°C = 0.09 \text{ cm}^3.$$

The increase in the volume of the water is

$$\Delta V_w = b_w V_w \, \Delta T = 2.1 \times \frac{10^{-4}}{°C} \times 250 \text{ cm}^3 \times 40°C = 2.1 \text{ cm}^3,$$

and so the volume of water that overflows is

$$\Delta V_w - \Delta V_b = 2.0 \text{ cm}^3.$$

10–3 Boyle's law

A peculiar difficulty arises when we attempt to measure the coefficient of volume expansion of a gas. Unlike solids and liquids, gases do not have specific volumes at a particular temperature, but expand to fill their containers. The only way to change the volume of a gas is to change the

$p = 20$ lb/in^2 $p = 40$ lb/in^2 $p = 120$ lb/in^2

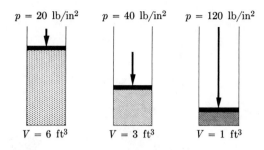

$V = 6$ ft^3 $V = 3$ ft^3 $V = 1$ ft^3

FIG. 10–5. Boyle's law.

capacity of its container. However, even though its volume may remain the same, another property of a confined gas varies with its temperature, namely the pressure it exerts on the container walls. The air pressure in an automobile tire drops in cold weather and increases in warm, an illustration of this effect.

When the temperature of a sample of gas is held constant, the absolute pressure it exerts on its container is inversely proportional to the volume of the container. Expanding the container lowers the pressure; shrinking the container raises the pressure. This property of gases is called *Boyle's law* (Fig. 10–5) after its discoverer, Robert Boyle (1627–1691). We can write Boyle's law as

$$pV = \text{constant}, \qquad (10\text{–}4)$$

or, alternatively, as

$$p_1 V_1 = p_2 V_2. \qquad (10\text{–}5)$$

In the latter case p_1 is the absolute gas pressure when its volume is V_1, and p_2 is the gas pressure when its volume is V_2; the temperature must be the same in states 1 and 2 for Eqs. (10–4) and (10–5) to hold.

We can use Boyle's law to find how much air at the sea-level atmospheric pressure of 15 lb/in^2 can be stored in the 12-ft^3 tank of an air compressor which can withstand an absolute pressure of 100 lb/in^2. Letting state 1 represent the air in the compressor tank and state 2 the air at atmospheric pressure,

$$V_2 = \frac{p_1}{p_2} V_1 = \frac{100 \text{ lb/in}^2 \times 12 \text{ ft}^3}{15 \text{ lb/in}^2} = 80 \text{ ft}^3.$$

A total of 80 ft^3 of air at atmospheric pressure can be stored in the 12-ft^3 compressor tank at the high pressure.

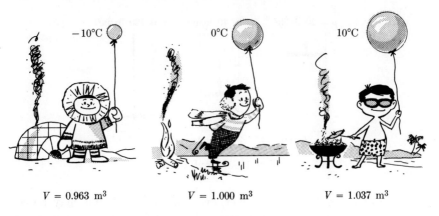

$$V = 0.963 \text{ m}^3 \qquad V = 1.000 \text{ m}^3 \qquad V = 1.037 \text{ m}^3$$

FIG. 10–6. Charles's law.

10–4 Charles's law

Now let us see what happens to a gas when its temperature is changed. As we said earlier, if the volume of a gas is held constant, the pressure it exerts on its container depends upon its temperature. According to Boyle's law, then, if we hold the gas pressure constant, its volume should vary with temperature. When this prediction is experimentally tested, which was first done over 150 years ago by Charles and Gay-Lussac, it is found that the change in volume ΔV of a gas sample is related to a change ΔT in its temperature by the same formula, Eq. (10–2),

$$\Delta V = bV_0 \, \Delta T$$

that holds for solids and liquids. The significant thing about gases at constant pressure is that they *all* have very nearly the same coefficient of volume expansion b; by contrast, as Tables 10–1 and 10–2 indicate, the thermal coefficients for solids and liquids may have markedly different values for different substances.

At $0°C$ the coefficient of volume expansion b_0 of all gases is

$$b_0 = \frac{0.0037}{°C} = \frac{1/273}{°C}.$$

If we vary the temperature of a gas sample while holding its pressure constant, its volume changes by $1/273$ of its volume at $0°C$ for each $1°C$ temperature change. A child's large balloon filled with air whose volume at $0°C$ is 1.000 m^3 has a volume of 1.037 m^3 at $10°C$ and 0.963 m^3 at $-10°C$ (Fig. 10–6).

A natural question to ask here is, what happens when the balloon is cooled to $-273°C$? At that temperature the air in the balloon should have lost $273/273$ of its volume at $0°C$, and therefore vanish entirely! Actually, all gases condense into the liquid state at temperatures above $-273°C$, so the question has no physical meaning. But $-273°C$ is still a significant temperature. If we set up a new temperature scale, the *absolute temperature scale*, and designate $-273°C$ as the zero point on this scale, the volume of a gas is always proportional to its absolute temperature. To help in understanding this statement we refer to Fig. 10–7, which is a plot of the volume of a particular sample of a gas at constant pressure versus its temperature in both the celsius and absolute scales. The proportionality between volume and absolute temperature is evident. At constant pressure, then, *Charles's law* states that

$$V/T = \text{constant}, \tag{10–6}$$

or, alternatively,

$$\frac{V_1}{T_1} = \frac{V_2}{T_2}, \tag{10–7}$$

where 1 and 2 refer to different states of the gas.

Temperatures in the absolute scale are designated °K, after Lord Kelvin, a noted British physicist of the last century. To convert temperatures from one scale to the other we note that

$$°K = °C + 273° \quad \text{and} \quad °C = °K - 273°. \tag{10–8}$$

If there were a gas that did not liquify before reaching $0°K$, at $0°K$ its volume would shrink to zero. Since it is impossible to conceive of a negative volume, it is natural to think of this temperature as *absolute zero*. Actually, $0°K$ is indeed the lower limit to temperatures capable of being attained, but on the basis of a stronger argument than one based on imaginary gases.

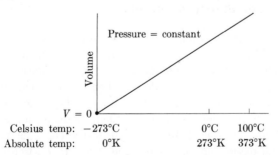

| Celsius temp: | $-273°C$ | | $0°C$ | $100°C$ |
| Absolute temp: | $0°K$ | | $273°K$ | $373°K$ |

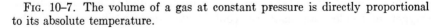

FIG. 10–7. The volume of a gas at constant pressure is directly proportional to its absolute temperature.

10–5 The ideal gas law

We can combine Boyle's law, Eq. (10–5),

$$p_1V_1 = p_2V_2 \qquad (T = \text{constant}),$$

and Charles's law, Eq. (10–7),

$$\frac{V_1}{T_1} = \frac{V_2}{T_2} \qquad (p = \text{constant})$$

(in which it is understood that T is to be expressed in °K), into a single law,

$$\frac{p_1V_1}{T_1} = \frac{p_2V_2}{T_2}. \tag{10–9}$$

Obviously Eq. (10–9) becomes (10–5) when T is the same in states 1 and 2, and it becomes (10–7) when p is the same in both states. Another way of expressing Eq. (10–9) is

$$\frac{pV}{T} = \text{constant.} \tag{10–10}$$

Equation (10–10) is obeyed approximately by all gases; the significant thing is not that the agreement with experiment is never perfect, but that *all* gases, of whatever kind, exhibit virtually identical behavior. An *ideal gas* is defined as one that obeys Eq. (10–10) exactly. While no ideal gases actually exist, they do provide a target for theories of the gaseous state to aim at. It is reasonable to suppose that Eq. (10–10) is a consequence of the essential nature of gases; hence the next step is to account for Eq. (10–10) and only afterward to seek reasons for its failure to be completely correct. Equation (10–10) is called the *ideal gas law*.

10–6 Molecular theory of matter

The idea that matter is not infinitely divisible, that all substances are composed of characteristic individual particles, is an ancient one. The ultimate particles of any substance are called *molecules*. Although molecules may be further broken down, when this happens they no longer are representative of the original substance. The molecules of a compound consist of the *atoms* of its constituent elements joined together in a definite ratio. Thus each molecule of water contains two hydrogen atoms and one oxygen atom. While the ultimate particles of elements are atoms, many elemental gases consist of molecules rather than atoms. Oxygen molecules, for instance, contain two oxygen atoms each. The

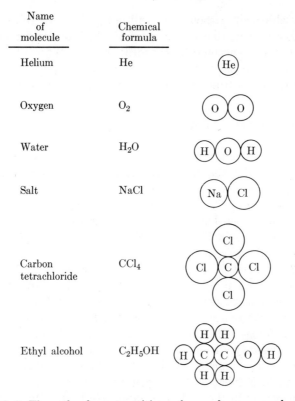

Name of molecule	Chemical formula	
Helium	He	
Oxygen	O_2	
Water	H_2O	
Salt	NaCl	
Carbon tetrachloride	CCl_4	
Ethyl alcohol	C_2H_5OH	

FIG. 10–8. The molecular compositions of several common substances.

molecules of other gases, such as helium and argon, are single atoms. Figure 10–8 shows schematically the composition of some common molecules.

The masses of atoms and molecules are usually expressed in units such that the oxygen atom has a mass of precisely 16. It is customary, though illogical, to refer to masses in these units as *atomic weights* and *molecular weights*. Appendix IV contains a list of the atomic weights of the elements; if we know the composition of a compound, we can determine the corresponding molecular weight. The oxygen atom has an actual mass of 2.656×10^{-26} kg; hence

$$1 \text{ atomic weight unit} = \frac{2.656 \times 10^{-26} \text{ kg}}{16}$$

$$= 1.660 \times 10^{-27} \text{ kg}.$$

A considerable amount of experimentation and ingenious reasoning had to be carried out before the reality of atoms and molecules became definitely established. Although we will not go into the full story of the molecular theory of matter, a large part of which involves chemistry, we shall show that it can account for the ideal gas law. We shall also discuss briefly how the physical properties of solids and liquids, and the deviations of actual gases from the ideal gas, fit into the molecular theory.

10–7 Kinetic theory of gases

According to the assumptions of the *kinetic theory of gases,* a development of the molecular theory of matter, a gas consists of a great many tiny individual molecules that do not interact with one another except when collisions occur. The molecules are supposed to be far apart compared with their dimensions and to be in constant motion, incessantly hurtling to and fro as in Fig. 10–9, being kept from escaping into space only by the solid walls of a container (or, in the case of the earth's atmosphere, by gravity). A natural consequence of the random motion and large molecular separation is the tendency of a gas to completely fill its container and to be readily compressed or expanded. In a solid, on the other hand, the molecules are close together, and mutual attractive and repulsive forces hold them in place to provide the solid with its characteristic rigidity. In a liquid the intermolecular forces are sufficient to keep the volume of the liquid constant; however, they are not strong enough to prevent adjacent molecules from sliding past one another, which results in the ability of liquids to flow.

Let us see how the kinetic theory of gases accounts for the ideal gas law. At first glance it may not seem that there *is* any straightforward connection between them. But it is also clear that if the kinetic theory is to have any meaning at all, it must yield the same behavior that we obtain from experiment. To bridge the gap between the microscopic picture of a gas as an aggregate of molecules in random motion and the macroscopic picture of a gas as a continuous fluid with certain physical properties, we shall use the laws of mechanics. The derivation is worth following in detail, since it illustrates how a theory whose assumptions are not based upon direct observation may be compared with data obtained by such observation.

We begin with a model situation: a box L long on each side filled with N identical molecules each having the same mass m. While these molecules are actually traveling about in all directions, the effects of their collisions with the walls of the box are the same as if one-third of them were moving back and forth between each pair of opposite walls (Fig. 10–10). We therefore imagine, for the sake of convenience, that one-third of our N molecules bounce between the top and bottom walls

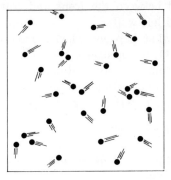

FIGURE 10–9

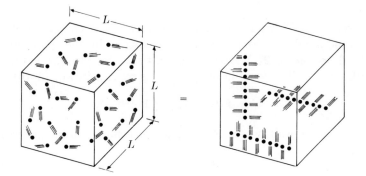

FIGURE 10–10

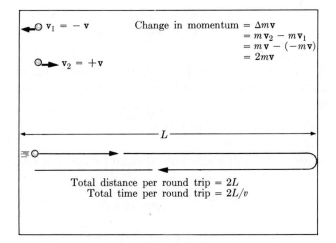

FIGURE 10–11

of the box, one-third between the front and rear walls, and one-third between the right and left walls.

Next we consider what happens when a molecule strikes one of the walls. As in Fig. 10–11, it approaches the wall with the velocity $-\mathbf{v}$, so that its initial momentum is $-m\mathbf{v}$, and bounces off with the velocity $+\mathbf{v}$ (same speed but opposite direction), so that its final momentum is $+m\mathbf{v}$. Hence the molecule has experienced a *change in momentum* of

$$\Delta m\mathbf{v} = m\mathbf{v}_2 - m\mathbf{v}_1 = m\mathbf{v} - (-m\mathbf{v}) = 2m\mathbf{v}.$$

The distance between the walls of the box is L, and it takes a molecule whose speed is v the time

$$\Delta t = t_2 - t_1 = \frac{2L}{v}$$

to make the round trip from one wall to the opposite one and back again. (The time required is the total distance divided by the speed, and here the total distance is $2L$.) The change in the molecule's momentum when it strikes the wall is equal in magnitude, as we saw in Eq. (5–10), to the impulse

$$\mathbf{F}(t_2 - t_1) = \mathbf{F}\,\Delta t$$

it exerts on the wall, where $\mathbf{F}$ is the average force during the period of time Δt. Hence

$$\mathbf{F}\,\Delta t = \Delta m\mathbf{v},$$

and the magnitude of the average force on the wall caused by the successive impacts of the molecule we are considering is

$$F = \frac{\Delta mv}{\Delta t} = \frac{2mv}{2L/v} = \frac{mv^2}{L}. \qquad (10\text{–}11)$$

Now we inquire as to the pressure on the wall due to the impacts of all the $N/3$ molecules that strike it. If the average value of v^2 for the different molecules is designated by $\overline{v^2}$, the sum of the individual forces they exert is

$$F_{\text{total}} = \frac{N}{3}\frac{m\overline{v^2}}{L}. \qquad (10\text{–}12)$$

The pressure on the wall is this force divided by the area L^2 of the wall:

$$\text{Pressure} = \frac{\text{total force on wall}}{\text{area of wall}},$$

$$p = \frac{F_{\text{total}}}{L^2} = \frac{1}{3}\frac{N m\overline{v^2}}{L^3}.$$

But L^3 is the volume V of the box! Hence

$$p = \frac{1}{3} \frac{N\overline{mv^2}}{V},$$

or, multiplying both sides by V,

$$pV = \frac{1}{3} N\overline{mv^2}. \tag{10–13}$$

10–8 Molecular kinetic energy

The significance of Eq. (10–13) becomes clearer when we rewrite it in the form

$$pV = \tfrac{2}{3}N(\overline{\tfrac{1}{2}mv^2}) = \tfrac{2}{3}N \,\overline{KE}, \tag{10–14}$$

where $\overline{KE} = (\overline{\tfrac{1}{2}mv^2})$ is the average kinetic energy per gas molecule. This is the closest that the kinetic theory of gases can come to the ideal gas law, which we may write

$$pV = cT, \tag{10–15}$$

where c is a number whose value depends upon the particular gas sample under scrutiny and T is its absolute temperature. The two formulas agree, however, if we assume that

The average kinetic energy of the molecules of a gas is proportional to the absolute temperature of the gas.

This assumption means that we set Eqs. (10–14) and (10–15) equal to each other. Then we have

$$\tfrac{2}{3} N \,\overline{KE} = cT, \qquad \overline{KE} = \tfrac{3}{2} (c/N) \, T.$$

The quantity c/N turns out, on the basis of evidence we have not considered here, to be a universal constant of nature. It is known as *Boltzmann's constant*, after the 19th-century Austrian physicist Ludwig Boltzmann. The symbol for Boltzmann's constant is k, and its value is

$$k = 1.38 \times 10^{-23} \, \text{j/}^\circ\text{K}.$$

In terms of Boltzmann's constant,

$$\overline{KE} = \tfrac{3}{2}kT; \tag{10–16}$$

$\tfrac{3}{2}kT$ *is the average kinetic energy of the molecules of all gases at the absolute temperature* T.

We can use Eq. (10–16) to compute the average speeds of various gas molecules at particular temperatures by solving it for v:

$$\overline{\tfrac{1}{2}mv^2} = \tfrac{3}{2}\,kT,$$

$$v = \sqrt{\frac{3kT}{m}}. \tag{10–17}$$

Let us apply this formula to the molecules of oxygen, one of the major constituents of air. Oxygen molecules are composed of two oxygen atoms each. The molecular weight of oxygen is therefore 2×16 or 32. As we learned earlier, one atomic weight unit is equivalent to a mass of 1.660×10^{-27} kg, and so each oxygen molecule has a mass of

$$m_{\text{oxygen}} = 32 \times 1.660 \times 10^{-27} \text{ kg}$$

$$= 5.31 \times 10^{-26} \text{ kg}.$$

At an absolute temperature of 273°K (corresponding to 0°C), the average speed of oxygen molecules is therefore

$$v = \sqrt{\frac{3 \times 1.38 \times 10^{-23}\,\text{j/°K} \times 273\text{°K}}{5.31 \times 10^{-26}\,\text{kg}}}$$

$$= 4.61 \times 10^2\,\frac{\text{m}}{\text{sec}},$$

which is a little over 1000 mi/hr! Evidently molecular speeds are very large compared with those of the macroscopic bodies familiar to us.

It is important to keep in mind that the v of Eq. (10–17) is an *average;* the actual molecular speeds vary considerably on either side of this average value. Figure 10–12 is a graph showing the actual distribution of molecular speeds in oxygen at 73°K and at 273°K and in hydrogen at 273°K.

The interpretation of absolute zero in terms of the elementary kinetic theory of gases is a simple one: it is that temperature at which all molecular movement ceases.

The kinetic theory of gases leads us directly to the ideal gas law, but, as we said earlier, the ideal gas law is only a good approximation to reality. If we examine the initial assumptions we made in order to obtain the kinetic theory, it is easy to see why we should expect discrepancies between theory and experiment. For example, we assumed tacitly that gas molecules have volumes so small as to be negligible, that they exert no forces upon one another except in actual collisions, that these col-

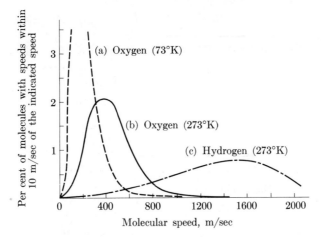

Fig. 10–12. The distribution of molecular speeds in oxygen at 73°K (−200°C), in oxygen at 273°K (0°C), and in hydrogen at 273°K. The average molecular speed increases with temperature and decreases with molecular weight; hence molecular speeds in oxygen at 73°K (curve a) are on the average less than at 273°K (curve b), and at the same temperature molecular speeds in hydrogen (curve c) are on the average greater than in oxygen (curve b), since oxygen molecules are the heavier. At the same temperature the average molecular *energy* is the same for all gases.

lisions conserve kinetic energy, and that the molecules have no kinetic energy of rotation or internal vibration. When the kinetic theory is worked out starting from more realistic assumptions, the results are in excellent agreement with observational data.

10–9 Kinetic theory of matter

The kinetic theory of gases does not meet with quite as much success when its notions are applied to the solid and liquid states. However, the concept that the heat energy of a body resides in the kinetic energy of its molecules helps in understanding a variety of phenomena in these states. Thermal expansion in a solid, for example, now has a straightforward explanation. From the point of view of elasticity, a solid is composed of molecules which behave as though joined together by tiny springs, as in Fig. 10–13, thereby accounting for Hooke's law (Chapter 7). The molecules, according to kinetic theory, constantly oscillate about their equilibrium positions. An increase in temperature means an increase in the energies of these molecular oscillators, larger amplitudes of motion, and more space occupied by each molecule. Hence the solid expands when its temperature is raised.

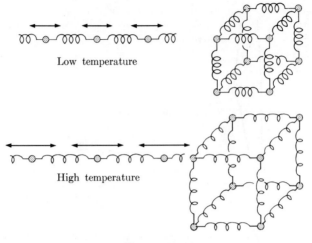

Low temperature

High temperature

FIGURE 10–13

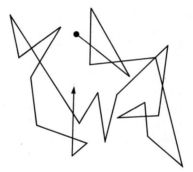

FIGURE 10–14

The random motion of water molecules led to an important event in the history of science. In **1827** the British botanist Robert Brown noticed that pollen grains in water are in continual, agitated movement. Similar *Brownian motion* is apparent whenever very small particles are suspended in a fluid medium, for example smoke particles in air (Fig. 10–14). According to kinetic theory, Brownian motion originates in the bombardment of the particles by molecules of the fluid. This bombardment is completely random, with successive molecular impacts coming from different directions and contributing different impulses to the particles. Albert Einstein, in 1905, found that he could account for Brownian motion quantitatively by assuming that, as a result of continual collisions with fluid molecules, the particles themselves have an average kinetic energy of $\frac{3}{2}kT$. Surprising as it may seem, this was the first direct verification of the reality of molecules, and it convinced many distinguished scientists who had previously been reluctant to believe that such things actually exist.

IMPORTANT TERMS

Solids are characterized by definite volumes and shapes that vary only slightly with environmental changes.

Liquids, while possessing definite volumes in the same sense as solids, have no definite shapes, but take on the shapes of their containers.

Gases exhibit neither definite volumes nor definite shapes, but instead expand to fill large containers, may be compressed to fit into small ones, and escape from open ones.

Plasmas are gases composed of electrically charged particles, and their behavior depends strongly upon electromagnetic forces. Most of the matter in the universe is in the plasma state.

Elements are the simplest substances encountered in bulk. They cannot be decomposed nor transformed into one another by ordinary chemical or physical means.

Two or more elements may combine chemically to form a *compound*, a new substance whose properties are different from those of the elements that compose it.

The constituents of a *solution* may be elements or solutions or both, and each retains its characteristic properties (except for the mechanical properties of solids and gases dissolved in liquids).

The *coefficient of linear expansion* is the ratio between the change in length of a solid rod of a particular material and its original length per 1° change in temperature. The *coefficient of thermal expansion* is the ratio between the change in volume of a sample of a particular solid or liquid and its original volume per 1° change in temperature.

Boyle's law states that, at constant temperature, the absolute pressure of a sample of a gas is inversely proportional to its volume, so that pV = constant at that temperature regardless of changes in either p or V individually.

Charles's law states that, at constant pressure, the volume of a sample of a gas is directly proportional to its absolute temperature, so that V/T = constant at the pressure regardless of changes in either V or T individually.

The *absolute temperature scale* has its zero point at $-273°$C. Temperatures in the absolute scale are designated °K; the conversion formulas are $°K = °C + 273°$ and $°C = °K - 273°$. *Absolute zero* is $0°K$, which is $-273°C$.

The equation

$$\frac{pV}{T} = \text{constant},$$

a combination of Boyle's and Charles's laws, is called the *ideal gas law* and is obeyed approximately, though not exactly, by all gases.

The ultimate particles of any substance are called *molecules*. The molecules of a compound consist of *atoms* of its constituent elements joined together in a definite ratio.

According to the *kinetic theory of gases*, a gas consists of a great many tiny individual molecules that do not interact with one another except when collisions occur. The molecules are far apart compared with their dimensions and are in constant random motion. The ideal gas law may be derived from the kinetic theory of gases.

Brownian motion is the continual agitated movement of tiny particles when suspended in a fluid medium; it is a consequence of the bombardment of the particles by molecules of the fluid.

PROBLEMS

1. Classify the following substances as homogeneous or heterogeneous: salt, leather, stone, diamond, iron, blood, solid carbon dioxide, gaseous carbon dioxide, helium, rust.

2. Classify the following homogeneous liquids as elements, compounds, or solutions: mercury, alcohol, gin, pure water, sea water, bromine, tea, glycerin, liquid oxygen, liquid air.

3. A steel tape measure is calibrated at 20°C. A reading of 27.1 m is obtained when it is used to determine the width of a building at 5°C. What is the true width of the building?

4. How large a gap should be left between steel rails that are 10 m long when laid at 20°C if they are to just barely touch at 30°C?

5. A rod 2 m long expands by 1 mm when heated from 8°C to 72°C. What is the coefficient of linear expansion of the material from which the rod is made?

6. The outside diameter of a wheel is 1.000 m. An iron tire for this wheel has an inside diameter of 0.992 m at 20°C. To what temperature must the tire be heated in order for it to fit over the wheel?

7. In the construction of a light bulb, wires are led through the glass at the base by means of air-tight seals. If the wires were made of copper, what would happen when the light is turned on and the bulb heats up? What must be true for a wire to be successfully used for this purpose?

8. A Pyrex beaker is filled to the brim with 250 cm³ of glycerin at 15°C. How much glycerin overflows at 25°C?

9. A Pyrex flask holds 500 cm³ of mercury at 0°C. How much mercury will run out when it is heated to 80°C?

10. Vodka that is "100 proof" is a mixture of half ethyl alcohol and half water. How much profit per quart will a merchant make if he buys it at $5.00 per quart at −5°C and sells it at $5.00 per quart at 25°C?

11. A sample of gas occupies 2 m³ at 300°K and an absolute pressure of 2×10^5 n/m². (a) What is its pressure at the same temperature when it has been compressed to a volume of 1 m³? (b) What is its volume at the same temperature when its pressure has been decreased to 1.5×10^5 n/m²? (c) What is its volume at a temperature of 400°K and a pressure of 2×10^5 n/m²?

12. Starting from the ideal gas law, obtain an equation relating the pressure and temperature of a gas at constant volume.

13. A sample of gas occupies 100 cm³ at 0°C and 1 atm pressure. What is its volume (a) at 50°C and 1 atm pressure; (b) at 0°C and 2.2 atm pressure; (c) at 50°C and 2.2 atm pressure?

14. An automobile tire contains air at a gauge pressure of 24 lb/in² at 0°C. If the volume of the tire is unchanged, what will the pressure be when the temperature has increased to 27°C?

15. The absolute temperature scale described in the text is based upon the celsius scale. An absolute temperature scale, called the rankine scale, may also be constructed based upon the fahrenheit scale. Temperatures in this scale are designated °R. (a) What is the fahrenheit temperature of absolute zero? (b) Devise equations for converting temperatures in °F to °R and vice versa, similar to Eq. (10–8). (c) What is the rankine temperature of the freezing and boiling points of water?

16. A sample of gas occupies a volume of 8 ft³ at a temperature of 400°R and 1 atm pressure. What is its volume at 500°R and 1 atm pressure?

17. According to the kinetic theory of gases, molecular motion ceases only at absolute zero. How can this be reconciled with the definite shape and volume of a solid at temperatures well above absolute zero?

18. (a) What is the average kinetic energy of the molecules of a gas at 0°C? (b) at 100°C?

19. Actual molecules attract one another slightly. Does this tend to increase or decrease gas pressures from values computed from the ideal gas law? Why?

20. In terms of the kinetic theory, explain why the pressure of a gas in a closed container increases when the gas is heated.

21. Which of the following gases has (a) the highest average molecular speed at a given temperature? (b) the lowest average molecular speed? CO_2, UF_6, H_2, He, Xe, NH_3. (Compute molecular weights from the atomic weights given in Appendix IV.)

22. The average speed of a hydrogen molecule at room temperature and atmospheric pressure is 1 mi/sec. (a) What happens to the average speed of such a molecule if the temperature remains constant but the pressure is doubled? (b) What is the average speed of a nitrogen molecule (N_2) at room temperature and atmospheric pressure? Compute this from the above figure for hydrogen.

23. (a) Find the average speed of carbon dioxide (CO_2) molecules at 0°C. (b) At what temperature would this speed be doubled?

24. The minimum speed which a body must have if it is to leave the earth permanently is 11,200 m/sec (36,800 ft/sec). This speed is called the *escape velocity*. At the top of the earth's atmosphere the temperature is approximately 0°C. With the help of Fig. 10–12, explain why hydrogen escapes from the atmosphere more readily than oxygen.

25. One of the assumptions of the kinetic theory is that the average distance between molecules is much greater than the dimensions of the molecules themselves. Oxygen and nitrogen molecules are roughly 2×10^{-10} m in diameter, and there are 2.7×10^{25} molecules in a cubic meter of air at room temperature and atmospheric pressure. (a) On the average, how far apart are the molecules in air? (b) How many molecular diameters is their average separation?

26. The average speed of air molecules is roughly 4×10^2 m/sec, and the average distance an air molecule goes between collisions with other molecules is about 10^{-7} m. What is the average number of collisions an air molecule makes per second?

27. A vessel contains 1 kg of chlorine at 0°C and a pressure of 1 atm. When another kg of chlorine is added to the vessel at the same temperature, the pressure rises to 2 atm. Explain the change in pressure on the basis of the kinetic theory of gases.

28. Silver is a vapor at 1500°K. What is the average speed of silver atoms in a vapor at this temperature?

Thermodynamics | 11

The science of thermodynamics has as its basic concern the transformation of heat energy into mechanical energy. Thermodynamics thus plays a central role in technology, since almost all of the "raw" energy available for our use is liberated in the form of heat. Even so recent an innovation as the nuclear reactor evolves energy as heat, and its utilization poses problems almost identical with those that arise in the utilization of heat from a coal fire. A device that converts heat into mechanical energy is called a *heat engine,* and the principles that govern its operation are the same whether its heat input ultimately originates in the fission of uranium nuclei or in the oxidation of hydrocarbon molecules. In this chapter we shall study these principles, their significance, and how they are applied.

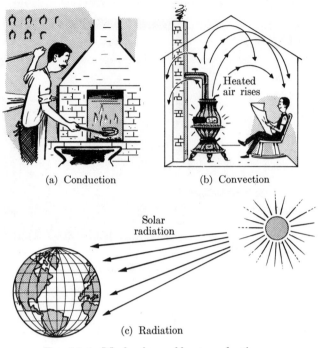

(a) Conduction   (b) Convection

(c) Radiation

FIG. 11–1. Mechanisms of heat conduction.

11–1 Heat transfer

Before we take up thermodynamics itself, it is appropriate for us to look into the three different mechanisms by which heat may be transferred from one place to another. An example of each is shown in Fig. 11–1. When we place one end of an iron rod in a fire, the other end becomes warm as a result of the *conduction* of heat through the iron. Conduction is a very slow process in air; a stove warms a room chiefly through the actual movement of heated air, a process called *convection*. Neither conduction nor convection can take place appreciably in the virtual void of interplanetary space. Instead, the heat the earth receives from the sun arrives in the form of *radiation*. These mechanisms all illustrate a fundamental fact: the natural direction of heat flow is from hot bodies to cold ones.

In most materials, conduction is a simple consequence of the kinetic theory of matter. Molecules at the hot end of a rod vibrate faster and faster as the temperature there increases. When these molecules collide with their less energetic neighbors, some of their kinetic energy is transferred to the latter (Fig. 11–2). Through successive molecular collisions

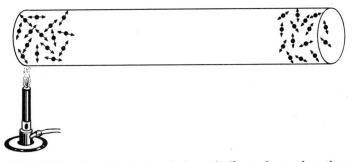

Fig. 11–2. Molecules at the hot end of a rod vibrate faster than those at the cold end.

energy travels down the rod, and, since we perceive random molecular motion as heat, we equally well describe the situation by saying that heat travels down the rod. The average positions of the molecules themselves do not change in conduction.

For heat to be conducted through a body, its ends must be at different temperatures. If the entire body is at the same temperature, all its molecules have the same average energy, and any molecule has as much chance of losing energy in a collision with a nearby molecule as it has of gaining energy. Hence there is no net flow of energy from a region of rapidly moving molecules to an adjacent one of slowly moving molecules. In fact, we drew upon the necessity of a temperature difference for heat transport in defining temperature: one body has a higher temperature than another if, when they are placed in contact, heat flows from the former to the latter.

There are wide differences in the ability of various substances to conduct heat. Gases are poor conductors, because their molecules are relatively far apart and collisions between them correspondingly infrequent. The molecules of liquids and nonmetallic solids are closer together, leading to somewhat higher thermal conductivities. Metals exhibit by far the greatest ability to conduct heat; this is why saucepans are made of metal but have handles of wood or plastic. The reason for the exceptional thermal conductivity of metals is the same as that for their exceptional electrical conductivity: the presence of a significant number of electrons that are able to move about freely instead of being bound permanently to particular atoms. Acquiring kinetic energy at the hot end of a metal object, the free electrons can travel past many atoms before giving up their energy in collisions, and thereby can speed up the rate of energy transport toward the cold end of the object. Heat conduction by free electrons in a metal compares with heat conduction by molecular interactions as travel by express train compares with travel by local train. We shall further explore the interesting and important subject of free electrons in metals in Chapter **22**.

Convection is a much simpler physical process than conduction, since it merely consists of the actual motion of a volume of hot fluid from one place to another. Convection may be either *natural* or *forced*. In natural convection, the buoyancy of a heated fluid leads to its motion; when a portion of a fluid (either gas or liquid) is heated, it expands to become of lower density than the surrounding, cooled fluid, and hence rises upward. A steam or hot-water heating system employs radiators in each room which heat the rooms with the help of the convection currents they set up. In forced convection, a blower or pump directs the heated fluid to its destination.

11–2 Radiation

In the process of radiation, energy is transported by means of *electromagnetic waves*. These waves travel at the speed of light (3 × 10^8 m/sec = 186,000 mi/sec), and require no material medium for their passage. Radio and radar waves, light waves, and x- and gamma-rays are all electromagnetic waves; they differ only in their wavelength. Figure 11–3 shows the classification of electromagnetic waves according to wavelength. A hot body radiates electromagnetic waves of all wavelengths, though the intensities of the different wavelengths may be very different. We are all familiar with, say, the glowing embers of a wood fire, where enough radiation is emitted as visible light for our eyes to respond, but other wavelengths as well are given off.

Three features of the radiation from a hot body are worth noting: (1) dark surfaces are the best emitters of radiant energy; (2) the rate of radiation from a body increases rapidly as its temperature increases (in fact, this rate is pro-

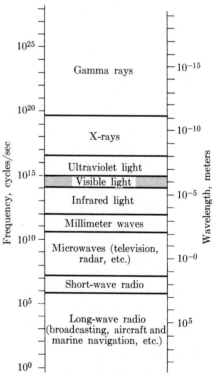

Fig. 11–3. The electromagnetic wave spectrum.

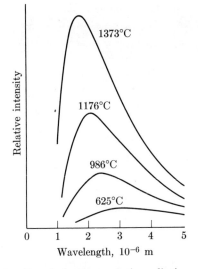

Fɪɢ. 11–4. The intensity of electromagnetic radiation as a function of wavelength emitted by a heated object at various temperatures. The total emitted radiation (which is proportional to the area under each curve) increases with increasing temperature, while the predominant wavelength becomes shorter with increasing temperature.

portional to T^4, where T is the absolute temperature of the body); and (3) the predominant wavelength in the radiation becomes shorter as the temperature of the body increases. A body that glows red is cooler than one that glows bluish-white. The latter effects are illustrated in Fig. 11–4.

A body need not be so hot that it gives off visible light for it to be radiating electromagnetic energy—*all* bodies radiate such energy continuously, whatever their temperature. However, a body also absorbs electromagnetic energy from its surroundings, and if both are at the same temperature, the rates of emission and absorption are the same. When a body is at a higher temperature than its surroundings, it radiates more energy than it absorbs, and this excess we detect.

11–3 The first law of thermodynamics

In discussing heat engines it is convenient to think of them as abstract systems whose interactions with their surroundings can be completely controlled. A particular system might actually be a gasoline engine, a human being, or the earth's atmosphere, to give a few examples, but in every case there are three characteristic processes: (1) the absorption of heat from an external source; (2) the performance of mechanical work;

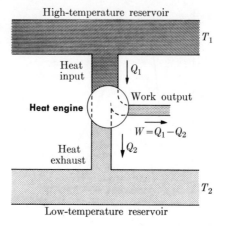

High-temperature reservoir

Heat input

Q_1

Work output

Heat engine

$W = Q_1 - Q_2$

Heat exhaust

Q_2

T_1

T_2

Low-temperature reservoir

FIGURE 11-5

and (3) the release of heat to the external environment (Fig. 11–5). Different heat engines carry out these processes in different ways, but the general pattern of operation is always the same.

There are two general principles that have been found to govern the operation of all heat engines. The *first law of thermodynamics* expresses the conservation of energy:

Energy cannot be created or destroyed, but may be converted from one form to another.

When applied to a heat engine, this law states that

$$\text{Work output} = \text{net heat input.}$$

The net heat input equals the heat supplied to the engine minus the heat it exhausts to the outside. In the case of an engine operating in a cycle that is continuously repeated, energy may be alternately stored and released from storage, but during a complete cycle the work performed by such an engine cannot exceed the amount of heat energy supplied to it in that cycle.

11–4 The second law of thermodynamics

The first law of thermodynamics prohibits an engine from operating without a source of energy, but it does not tell us anything about the character of possible sources of energy. For instance, there is an immense amount of heat energy in the atmosphere and the oceans, yet we know that more work must be done to extract this energy than can be performed with its help. Or, to give an extreme case, it is energetically possible for a pailful of water to rise spontaneously into the air, cooling and freezing into ice as its heat energy changes into potential energy. After all, a block of ice dropped from a sufficient height melts when it strikes the ground, its initial potential energy first being converted to kinetic energy and then into heat. Needless to say, water does not rise upward of its own accord, and we must find an appropriate way of expressing this conclusion.

The *second law of thermodynamics* is the physical principle, independent of the first law and not derivable from it, that supplements the first law in limiting our choice of heat sources for our engines. It can be stated in a number of equivalent ways, a common one being as follows:

It is impossible to construct an engine, operating in a cycle (that is, continuously), which does nothing other than take heat from a source and perform an equivalent amount of work.

According to the second law of thermodynamics, then, no engine can be completely efficient—some of its heat input *must* be ejected. As we shall see, the greatest efficiency any heat engine is capable of depends upon the temperatures of its heat source and of the reservoir to which it exhausts heat. The greater the difference between these temperatures, the more efficient the engine. The second law is a consequence of the empirical fact we have already noted:

The natural direction of heat flow is from a reservoir of heat at a high temperature to a reservoir of heat at a low temperature, regardless of the total heat content of each reservoir.

The latter statement, in fact, may be regarded as an alternative expression of the second law.

If we are to utilize the heat content of the atmosphere or the oceans, we must first provide a reservoir at a lower temperature than theirs in order to extract heat from them. There is no reservoir in nature suitable for this purpose, for if there were, heat would flow into it until its temperature reached that of its surroundings. To establish a low-temperature reservoir, we must employ a refrigerator, which is a heat engine running in reverse by using up energy to extract heat, and in so doing we will perform more work than we can successfully obtain from the heat of the atmosphere or oceans.

We might summarize the laws of thermodynamics by saying that the first law prohibits the work output of a heat engine from exceeding its heat input, while the second law prohibits the engine from even doing that well.

11–5 The Carnot engine

All cyclic heat engines exhibit the same characteristic behavior: they absorb heat at a high temperature, convert some of it into mechanical energy, and exhaust the rest at a low temperature. We may therefore simplify the task of analyzing their principles of operation by referring to an idealized heat engine free of the complications involved in actual engines. A suitably simple and straightforward engine was suggested for this purpose by Sadi Carnot in 1824. The *Carnot engine* is not subject to such practical difficulties as friction and the loss of stored heat by conduction or radiation, but naturally must obey all physical laws. Our chief concern is the efficiency of the Carnot engine: what proportion of the heat supplied to it can it transform into mechanical energy? Of course, all the effects we are neglecting reduce the performance of actual engines below that of the Carnot engine, but it is of great interest to find out the ultimate limits on engine efficiency.

A Carnot engine turns heat into work without itself undergoing a permanent change. It is distinct from, say, a dynamite blast, which is a one-shot process rather than a cyclic one that can continue indefinitely. A Carnot engine consists of a cylinder that is filled with an ideal gas and has a movable piston at one end. The four stages in its operating cycle are shown in Fig. 11–6, together with a graph of each stage on a pressure-volume diagram. In the first stage, an amount of heat Q_1 is added to the gas, which expands from its initial volume V_a to the final volume V_b while remaining at its original temperature T_1. During this expansion

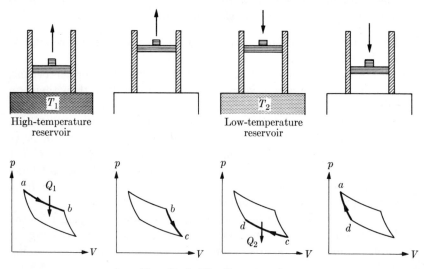

Fig. 11–6. The Carnot cycle.

the pressure drops from p_a to p_b. Then the heat source is removed, and the gas is allowed to expand further to V_c, which involves a further pressure drop to p_c. This second expansion takes place at the expense of the energy stored in the gas as heat, and so its temperature falls from T_1 to T_2. During these expansions the piston exerts a force on whatever it is attached to, and thereby performs work.

Having done work in pushing the piston outward, the engine must now be returned to its initial state in order for it to be able to do further work. The third stage in the operation of a Carnot engine involves a compression of the gas at the constant temperature T_2, during which an amount of heat Q_2 is given off. The volume of the gas is reduced to V_c while its pressure increases to p_c. In the final stage, the gas is returned to its original temperature, pressure, and volume by a compression during which heat is neither added to it nor removed from it. In compressing the gas a force had to be exerted on the piston, which means that work had to be done by some external agency on the engine.

In its cycle the Carnot engine performs some net amount of work W, which is the difference between the work it does during the two expansions and the work done on it during the two compressions. It has taken in the heat Q_1 and ejected the heat Q_2; we observe that the heat Q_2 *must* be ejected if the engine is to return to its initial state to await another cycle. The efficiency of any engine is

$$\text{Eff} = \frac{\text{work output}}{\text{energy input}}, \tag{11-1}$$

so that for a Carnot engine

$$\text{Eff} = \frac{W}{Q_1}. \tag{11-2}$$

According to the first law of thermodynamics,

$$W = Q_1 - Q_2, \tag{11-3}$$

since at the end of the cycle the gas has the same properties it began with. Hence

$$\text{Eff} = \frac{Q_1 - Q_2}{Q_1} = 1 - \frac{Q_2}{Q_1}. \tag{11-4}$$

The smaller the ratio of the ejected heat Q_2 to the absorbed heat Q_1, the more efficient the engine.

The heat Q transferred to or from a Carnot engine is directly proportional to the absolute temperature T of the reservoir with which it is in contact. That is,

$$\frac{T}{Q} = \text{constant.} \tag{11-5}$$

The derivation of Eq. (11–5) is not given here, because, although it does not involve any physical principles we do not already know, a knowledge of advanced mathematics is necessary. As a consequence of Eq. (11–5), the ratio Q_2/Q_1 between the amounts of heat ejected and absorbed per cycle by a Carnot engine is equal to the ratio T_2/T_1 between the temperatures of the respective reservoirs:

$$\frac{Q_2}{Q_1} = \frac{T_2}{T_1}. \tag{11–6}$$

The efficiency of a Carnot engine may therefore be written

$$\text{Eff} = 1 - \frac{T_2}{T_1}. \tag{11–7}$$

The smaller the ratio between T_2 and T_1, the more efficient the engine.

The second law of thermodynamics may be used to prove that a Carnot engine operating between two heat reservoirs is the most efficient one. Let us assume for the sake of argument that a more efficient engine does exist. We can operate this engine in reverse as a refrigerator to take the ejected heat from a Carnot engine and return it to the high-temperature reservoir, using the engine itself as the source of power (Fig. 11–7). Because of its greater efficiency, the imaginary engine needs less energy to return heat to the high-temperature reservoir than the Carnot engine provides. We can employ the extra energy that is left over in any way we

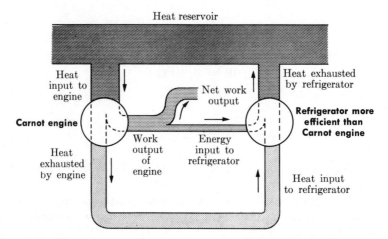

FIG. 11–7. The existence of an engine more efficient than a Carnot engine violates the second law of thermodynamics. Here a hypothetical superefficient engine is shown operating in reverse as a refrigerator coupled to a Carnot engine.

choose. The net result of the entire process, then, has been the removal of heat from a reservoir and the performance of an equivalent amount of work, which violates the second law of thermodynamics. We conclude that the Carnot engine is the most efficient, so that Eq. (11–7) represents the upper limit of engine efficiency.

Essentially the same argument can also be used to show that *all* reversible engines have the same efficiency when operated between the same two reservoirs. Hence a Carnot engine that employs an ideal gas as its working substance is just one representative of a whole class of hypothetical reversible engines. For example, analogs of the simple Carnot cycle can be devised based upon the electrochemical changes that occur in a storage battery or upon the magnetic changes that occur in a paramagnetic substance. A real engine is never exactly reversible because of such irreversible transformations as those involved in friction and in heat losses through the engine walls.

11–6 Practical engines

Practical heat engines have low efficiencies. One reason for their inability to make more use of their energy inputs is the unavailability of heat reservoirs at sufficiently high and low absolute temperatures for the ratio T_2/T_1 to be small. Steam enters a typical steam engine at about 450°K and is discharged at about 373°K. A Carnot engine operating between these temperatures has an efficiency of only 17%, which is therefore the upper limit to the efficiency of *any* engine operating between 450°K and 373°K. And, of course, the inevitable presence of friction and heat losses to the atmosphere reduces the efficiency even further.

An internal combustion engine is able to achieve a relatively high operating efficiency by generating the input heat within the engine itself. In a gasoline engine a mixture of air and gasoline vapor is ignited in each cylinder by a spark plug, and the evolved heat is converted into mechanical energy by the pressure of the hot gases on a piston. The greater the ratio between initial and final volumes of the expanding gases, the greater the engine efficiency. In a gasoline engine this ratio is limited to 7 or 8 to 1, since the gasoline-air mixture in the cylinder will otherwise spontaneously ignite during its compression before the end of the stroke is reached. The more efficient Diesel engine circumvents this difficulty by compressing only air and injecting fuel oil into the hot, compressed air at the instant the piston has reached the end of its travel. No spark plug is required. The compression ratio in a Diesel engine might be 16 to 1. The effective values of T_1 and T_2 in a modern Diesel engine are perhaps 1800°K and 800°K, respectively, for a Carnot efficiency of 56%.

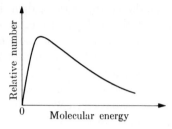

Fɪɢ. 11–8. The distribution of molecular energies in a gas at a particular temperature.

11–7 Statistical mechanics

It is interesting to consider the meaning of the second law of thermodynamics on the molecular level. As we learned in the previous chapter, heat is molecular kinetic energy, and the temperature of a body is a measure of the average kinetic energy of each of its constituent molecules. The molecules of a gas are in constant random motion and undergo frequent collisions with one another. While we cannot hope to follow an individual gas molecule in its vicissitudes, it is possible to predict on the basis of statistical arguments what fraction of the time it will have any specified amount of kinetic energy. Hence we can calculate the distribution of molecular energies in a gas sample at a particular temperature. This distribution, which has been confirmed by experiment, has the form shown in Fig. 11–8, and holds for all equilibrium conditions in which each molecule has the same average energy over a period of time. A molecule that moves more swiftly than usual at one instant will move less swiftly at a later instant after a number of collisions have taken place.

An equilibrium condition is the most probable one, according to *statistical mechanics,* a branch of physics which uses abstract mathematical methods to deduce the behavior of assemblies of so many particles that deviations from statistically probable behavior are not significant. If we toss a coin a dozen times, it is unlikely that heads and tails will come up equally often, but if we toss it a million times, the percentage deviation from an equal number of heads and tails will be minute. Let us consider a heat reservoir at a high temperature and a heat reservoir at a low temperature. The molecules of each are in equilibrium with the molecular energy distributions shown in Fig. 11–9(a) and (b). If we consider the two reservoirs as a single system, the molecular energy distribution in the system is like that of Fig. 11–9(c). This distribution is, in a statistical sense, very improbable; if they are mixed together, the molecules of the two reservoirs would soon blend their energies in collisions to attain the

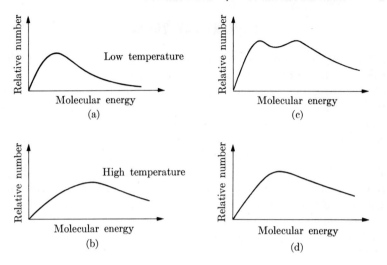

FIGURE 11–9

equilibrium distribution of Fig. 11–9(d), which corresponds to a temperature intermediate between the initial ones of the two reservoirs. We note
the important fact that the total energy contents of the distributions of
Fig. 11–9(c) and (d) are identical; the only distinction between them
is the manner in which the energy is allotted to the molecules on the
average. However, it is just this distinction with which the second law
of thermodynamics is concerned, because this law states that a system of
two heat reservoirs at different temperatures can be made to yield a net
work output while a single heat reservoir, no matter how much energy it
contains, cannot be made to perform any net work. A system of molecules
whose energies are distributed in the most probable way is "dead" thermo-
dynamically, while a system having a different distribution of molecular
energies is capable of doing mechanical work. The universe may be
thought of as a single system of molecules, and its evolution is powered
by the flow of energy from high-temperature reservoirs (the stars) to low-
temperature reservoirs (everything else). Ultimately the entire universe
will be at the same temperature and all of its constituent particles will
have the same average energy, a condition sometimes called the "heat
death" of the universe.

IMPORTANT TERMS

A *heat engine* is any device that converts heat into mechanical energy.

Heat can be transferred from one place to another by means of *conduction, convection,* or *radiation.* In conduction heat is transported by successive molecular collisions, and in convection by the motion of a volume of hot fluid from one place to another. The vehicle for heat transfer by radiation consists of *electromagnetic waves* which require no material medium for their passage.

The *first law of thermodynamics* states that the work output of a heat engine is equal to its net heat input plus any decrease in its stored energy.

The *second law of thermodynamics* states that it is impossible to construct an engine, operating in a repeatable cycle, which does nothing other than take energy from a source and perform an equivalent amount of work.

A *Carnot engine* is an idealized engine which is not subject to such practical difficulties as friction or heat losses by conduction or radiation but which obeys all physical laws. The efficiency of a Carnot engine which absorbs heat at the absolute temperature T_1 and exhausts heat at the absolute temperature T_2 is equal to $1 - T_2/T_1$; no engine operating between the same two temperatures can be more efficient than a Carnot engine operating between them.

Statistical mechanics attempts to deduce the behavior of assemblies of particles by considering statistically the most probable behavior of their constituent particles.

PROBLEMS

1. In the winter why does the steel blade of a shovel seem colder than its wooden handle?

2. A Thermos bottle consists of two glass vessels, one inside the other, with the space between them evacuated. The vessels are both coated with thin films of silver. Why is this device so effective in keeping its contents at a constant temperature?

3. Why is the specific heat of a gas when its pressure is held constant always greater than its specific heat when its volume is held constant?

4. What is the maximum possible efficiency of an engine that obtains heat at 200°C and exhausts heat at 80°C?

5. An engine operating between 300°C and 60°C is 15% efficient. What would its efficiency be if it were a Carnot engine?

6. A Carnot engine takes in 10^3 kcal of heat from a reservoir at 327°C and exhausts heat to a reservoir at 127°C. How much work does it do?

7. One of the most efficient engines ever developed operates between about 2000°K and 700°K. Its actual efficiency is 40%. What percentage of its maximum possible efficiency is this?

8. Three designs for a heat engine to operate between 450°K and 300°K are proposed. Design A is claimed to require a heat input of 0.2 kcal for each 1000 j of work output, design B a heat input of 0.6 kcal, and design C a heat input of 0.8 kcal. Which design would you choose and why?

9. The sun's corona is a very dilute gas at a temperature of about 10^6 °K that is believed to extend into interplanetary space at least as far as the earth's orbit. Why can we not use the corona as the high-temperature reservoir of a heat engine in an earth satellite?

10. An attempt is made to cool a kitchen during the summertime by leaving the refrigerator door open and closing the kitchen door and windows. What will happen and why?

11. In an attempt to cool a room in the summer, a man turns on an electric fan and leaves the room. Will the room be cooler when he returns?

12. A Carnot refrigerator (which is a Carnot engine operating in reverse) extracts heat from a freezer at −5°C and exhausts it at 25°C. How much work per kcal of heat extracted is required?

13. A Carnot refrigerator is used to make 1 kg of ice at −10°C from 1 kg of water at 20°C, which is also the temperature of the kitchen. How many joules of work must be done?

Electricity | 12

The success of the laws of motion and of gravitation and the kinetic-molecular theory of matter might tempt us into thinking that we now have, at least in outline, a complete picture of the workings of the physical universe. To dispel this notion all we need do is perform a simple experiment: on a dry day, we run a hard rubber comb through our hair, and find that the comb is now able to attract small bits of paper and lint. The attraction is surely not gravitational, because the gravitational force is far too small and should not, in any case, depend upon whether the comb is run through our hair. What we have discovered in this experiment is the entirely new realm of *electrical* phenomena, so called after *elektron*, the Greek word for amber, a substance used in the earliest studies of electricity.

12-1 Electric charge

Let us begin our study of electricity by examining the results of a few basic experiments. The first experiment consists of stroking a rubber rod with some fur and touching two small pith balls suspended from strings in turn with the rod. The pith balls then swing apart, indicating that a repulsive force now acts between them (Fig. 12–1). Next, we stroke a glass rod with a silk cloth and touch two different suspended pith balls in turn with it. Once more the pith balls swing apart, indicating a repulsive force (Fig. 12–2). Now we bring near each other one of the pith balls touched with the rubber rod and one touched with the glass rod; they swing *toward each other* this time, indicating an attractive force (Fig. 12–3).

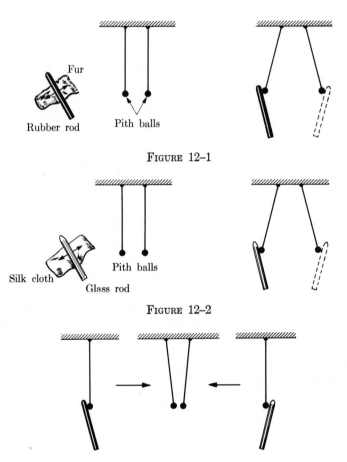

Fur

Rubber rod   Pith balls

FIGURE 12–1

Silk cloth   Pith balls

Glass rod

FIGURE 12–2

FIGURE 12–3

FIGURE 12-4

By convention, we call whatever it is that the rubber rod possesses by virtue of having been stroked with the fur *negative electrical charge* and whatever it is that the glass rod possesses by virtue of having been stroked with silk *positive electrical charge*. We can summarize the results of the above experiments by saying that *like charges repel; unlike charges attract*.

Where do the charges come from when one substance is stroked with another? When we charge one pith ball with a rubber rod and another with the fur the rod was stroked with, we find (Fig. 12-4) that the two balls *attract;* since the rubber rod is negatively charged, this experiment indicates that the fur is positively charged. A similar experiment with a glass rod and a silk cloth indicates that the cloth acquires a negative charge during the stroking. Evidently the process of stroking serves to *separate* charges. We might infer that rubber has an affinity for negative charges and fur an affinity for positive charges, so that, when rubbed together, each tends to acquire a different kind of charge.

A great many experiments with a variety of substances have shown that there are only the two kinds of electrical charge, positive and negative, that we have spoken of. All electrical phenomena involve either or both kinds of charge. An "uncharged" body actually possesses equal amounts of positive and negative charge, so that appropriate treatment—mere rubbing is sufficient for some substances—can leave a net excess of either kind on the body and thereby cause it to exhibit electrical effects.

12-2 Coulomb's law

As we saw in Chapter 4, Newton was able to determine the law of gravitation,

$$F = G\,\frac{m_1 m_2}{r^2},$$

by combining Kepler's analysis of planetary motion with the fact that

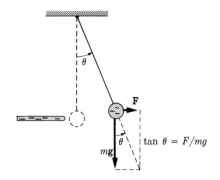

$$\tan \theta = F/mg$$

Fig. 12–5. A negatively charged rod exerts a repulsive force **F** on a negatively charged pith ball suspended from a string. The ball swings to the side until the vector sum of the electric force **F** on the ball and the ball's weight mg is directed along the string. The ball is then in equilibrium. The condition for equilibrium is therefore $\tan \theta = F/mg$.

the acceleration of gravity does not depend upon the mass of the falling body. The law that electrical forces obey can be readily determined in the laboratory because these forces are so much greater in magnitude than gravitational ones.

To establish the variation of the force with distance, we could use an arrangement similar to that shown in Fig. 12–5. A pith ball, let us say negatively charged, is suspended on a string, and a negatively charged body is brought near it. The pith ball, of course, swings off to the side so that its string makes an angle of θ with the vertical. This means that, as the diagram shows, the horizontal electrical force **F** on the ball and the vertical downward force of gravity mg (where m is the mass of the ball) are related by the equation

$$\tan \theta = \frac{F}{mg}, \qquad F = mg \tan \theta.$$

By placing the body at various distances from the pith ball, we can evaluate the variation of F with the separation r. The result is that F is found to be *inversely proportional to* r^2. Repeating the experiment with a positively charged body yields exactly the same result, except that, of course, the force is now attractive and the ball will swing toward the body.

The same technique can be used to find out how the electrical force between two charged bodies depends upon the amount of charge on each. We begin by charging a metal sphere and then placing it in contact with another, identical metal sphere. Electric charge is able to flow freely in metals (in fact, this is one of the defining properties of metals), and so

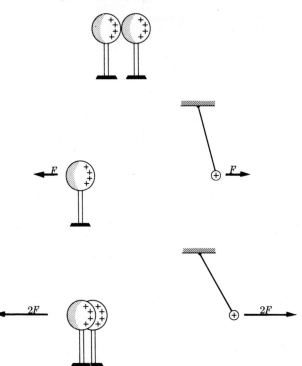

Fɪɢ. 12–6. The electrified force between charged bodies is directly proportional to the magnitude of the charges.

the charge originally on the first sphere distributes itself over both spheres until they have identical charges (Fig. 12–6). We bring one of the spheres near a suspended charged pith ball and note its deflection, and then bring the other sphere alongside the first to determine the change in force when the charge is doubled. What we find is that the force is doubled, too. If we repeat the experiment using three identically charged spheres, we find that the force between all three spheres together and the pith ball is three times greater than the force between one of the spheres and the pith ball. The electrical force F is therefore *directly proportional to the magnitude of the charges.*

The law of force between charges that results from the above experiments was first obtained by the 18th-century French scientist Charles Coulomb, and is called Coulomb's law in his honor. If we use the symbol q for electric charge, Coulomb's law for the magnitude of the force between the charges q_1 and q_2 states that

$$F = \frac{1}{4\pi\epsilon_0} \frac{q_1 q_2}{r^2} ; \qquad (12\text{–}1)$$

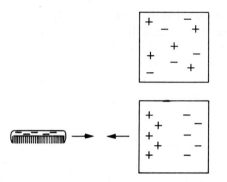

FIG. 12-7. How a charged body attracts an uncharged one.

the force between two charges is proportional to the product of the charges and inversely proportional to the square of the distance between them. In the metric system the unit of charge is the *coulomb* (abbreviated coul). When q_1 and q_2 are expressed in coulombs, the constant $1/4\pi\epsilon_0$ in Eq. (12-1) has the approximate value

$$\frac{1}{4\pi\epsilon_0} = 9 \times 10^9 \; \frac{\text{n·m}^2}{\text{coul}^2}.$$

Electric force is, of course, a vector quantity, with Eq. (12-1) giving only its magnitude. The direction of the vector $\mathbf{F}$ is always along the line joining q_1 and q_2, and the force is attractive if q_1 and q_2 have opposite signs and repulsive if they have the same sign.

The coulomb is an extremely large unit. The force between two charges of 1 coul each a distance of 1 meter apart is

$$F = \frac{9 \times 10^9 \, \text{n·m}^2/\text{coul}^2 \times 1 \, \text{coul} \times 1 \, \text{coul}}{1 \, \text{m}^2}$$
$$= 9 \times 10^9 \, \text{n},$$

which is about 2 billion lb! Because electrostatic forces are so powerful, even the most highly charged bodies that can be produced seldom contain more than a minute fraction of a coulomb.

Let us return for a moment to the first experiment we mentioned, in which a charged rubber comb attracted bits of paper. Since the paper bits were originally uncharged, how could the comb exert a force on them? The explanation is shown in Fig. 12-7. When the negatively charged comb is brought near the paper, some of the negative charges in the paper which are not tightly bound in place move as far away as they can from the rod, while some of the positive charges which are not tightly bound move toward the rod. Because electrical forces vary inversely with distance, the attraction between the rod and the closer positive charges is greater than the repulsion between the rod and the farther negative charges, and so the paper moves toward the rod. Only a small amount of charge separation actually occurs, and so, with little force available, only very light objects can be attracted in this way.

12–3 The electric field

Electric forces, like the gravitational forces we discussed in Chapter 4, act between bodies that are not in contact with each other. The proper way to regard such forces involves the concept of a force *field*. When a mass is present somewhere, the properties of space in its vicinity can be considered to be so altered that another mass brought to this region will experience a force there. The "alteration in space" caused by a mass is called its *gravitational field*, and any other mass is thought of as interacting with the field and not directly with the mass responsible for it. Similarly, an electric charge produces an *electric field* around it that interacts with any other charges present. One reason it is preferable not to think of two charges as exerting forces upon each other directly is that if one of them is changed in magnitude or position, the consequent change in the forces each experiences does *not* occur immediately, but takes a definite time to be established (Fig. 12–8). This delay cannot be understood on the basis of Coulomb's law, but is easily accounted for

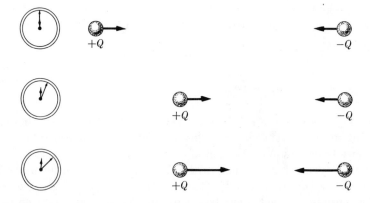

Fig. 12–8. Changes in an electric field travel with a finite speed. Hence, when one charge is moved closer to another one, the increase in the force between them does not appear instantaneously. The effect is greatly exaggerated in the figure, since the speed with which the field travels is the speed of light, 3×10^8 m/sec.

in terms of the field concept by assuming that changes in the field travel with a finite speed. We will learn later that this speed is the speed of light, 3×10^8 m/sec.

We would like to specify electric field in such a way that it will be possible to determine the force acting on any arbitrary charge at any point in the field. Accordingly, we define the *electric field intensity* **E** at a point as the ratio between the force **F** on an arbitrary positive

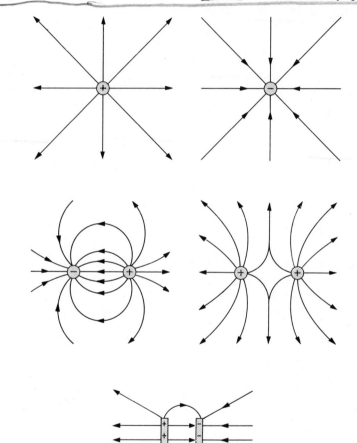

FIG. 12–9. Lines of force of the electric fields near various combinations of charges.

charge q_0 at that point and the magnitude of q_0:

$$\mathbf{E} = \frac{\mathbf{F}}{q_0}. \qquad (12\text{--}2)$$

The units of $\mathbf{E}$ are newtons per coulomb. Electric field intensity is a vector quantity that possesses both magnitude and direction.

Once we know what the intensity $\mathbf{E}$ is somewhere, from its definition we see that the force the field exerts on a charge q there is

$$\mathbf{F} = q\mathbf{E}. \qquad (12\text{--}3)$$

We can use Coulomb's law to determine the magnitude of the electric field surrounding a single charge q. First we find the force F that q exerts upon a test charge q_0 at a distance r away, which is

$$F = \frac{1}{4\pi\epsilon_0}\frac{qq_0}{r^2}.$$

Next we substitute this expression for F into Eq. (12–2), yielding

$$E = \frac{1}{4\pi\epsilon_0}\frac{qq_0}{q_0r^2} = \frac{1}{4\pi\epsilon_0}\frac{q}{r^2}. \qquad (12\text{--}4)$$

Equation (12–4) tells us what the electric field intensity is at the distance r from a charge of magnitude q, namely that it falls off inversely as r^2.

The representation of a force field using *lines of force* that we spoke of in Chapter 4 is very convenient in visualizing electric fields. In drawing the lines of force corresponding to a particular electric field we follow two rules: (1) the direction of a line of force at any point is that in which a positive charge would move if placed at that point, and (2) the spacing of lines of force is such that they are close together where the field is strong, far apart where the field is weak. Rule (1) means that lines of force are to be thought of as leaving positive charges and entering negative ones. Figure 12–9 shows the pattern of lines of force near various combinations of charges.

12–4 The electron

Is electric charge continuous, so that it can be divided into smaller portions indefinitely, or do the charges we find in nature and in the laboratory actually consist of multiples of some minimum, indivisible charge? Beginning in 1906 the American physicist Robert A. Millikan performed a series of careful experiments to find out. The question had been studied earlier by a number of investigators, but their results were

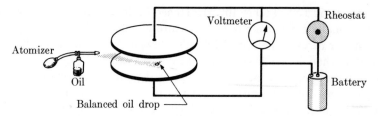

FIG. 12–10. The Millikan oil-drop experiment.

not conclusive. Millikan's apparatus is shown schematically in Fig. 12–10. An atomizer is used to spray tiny oil drops between two metal plates. In passing through the atomizer nozzle, the drops become electrically charged. A uniform electric field is established by charging the upper and lower plates oppositely, in this case with the help of a battery. If this field has the proper magnitude and direction, it can exert an upward electrical force on a charged oil drop that exactly counterbalances the downward gravitational force on the drop. The electric force on an oil drop whose charge is q is

$$F = qE$$

in an electric field of intensity E, while the gravitational force on the drop is just its weight

$$W = mg.$$

When $F = W$, the drop is acted upon by no net force and remains stationary. If this is true,

$$qE = mg,$$

and the charge on the drop is

$$q = \frac{mg}{E}. \qquad (12\text{–}5)$$

Millikan could measure directly or indirectly all the quantities on the right-hand side of Eq. (12–5), which permitted him to determine the precise magnitudes of a great many charges. The advantage of this method is that extremely small values of q can be determined, while the suspended pith ball method we spoke of at the beginning of this chapter can only be used for relatively large values of q.

Millikan found many different charges on the various oil drops he studied, but they shared a singular property: all the charges were multiples of a single value, 1.60×10^{-19} coul. The measurements he made of q ranged from 1.60×10^{-19} coul up to 27.2×10^{-19} coul, in every case equal to 1.60×10^{-19} coul multiplied by a whole number (27.2×10^{-19}, for instance, is $17 \times 1.60 \times 10^{-19}$). Hence he felt justified in concluding that the smallest electric charge in nature is 1.60×10^{-19} coul.

A number of years before Millikan's work, J. J. Thomson (1856–1940) in England had found that when a strong electric field is established in a tube containing a gas at low pressure, a stream of fast, submicroscopic particles goes from the negative to the positive plate. These particles, called *electrons*, are negatively charged. Thomson was able to establish by an ingenious experiment that electrons all have the same ratio of charge to mass:

$$\frac{q}{m} = 1.76 \times 10^{11} \frac{\text{coul}}{\text{kg}}.$$

However, he was *not* able to establish that all electrons are alike; all he could say was that each of them has the same relative amounts of charge and mass.

The discovery that the basic unit of charge in nature has a magnitude of 1.60×10^{-19} coul meant that Thomson's electrons *are* all identical, with the mass

$$m = \frac{1.60 \times 10^{-19} \text{ coul}}{1.76 \times 10^{11} \text{ coul/kg}} = 9.1 \times 10^{-31} \text{kg}.$$

The electron is the lightest particle known and is one of the constituents of all atoms.

It is easy to show that the electrical forces subatomic particles exert upon one another are so much stronger than their mutual gravitational ones that the latter can be neglected completely. Let us consider the hydrogen atom, which has the simplest structure of all atoms. It consists of an electron and a *proton* (a particle of mass 1.7×10^{-27} kg and a positive charge of 1.6×10^{-19} coul) separated by an average of 5.3×10^{-11} meters. The electrical force between the electron and proton is

$$F_e = \frac{1}{4\pi\epsilon_0} \frac{q_e q_p}{r^2}$$

$$= \frac{9 \times 10^9 \text{ n·m}^2/\text{coul}^2 \times (1.6 \times 10^{-19} \text{ coul})^2}{(5.3 \times 10^{-11} \text{ m})^2}$$

$$= 8.1 \times 10^{-8} \text{ n},$$

while the gravitational force between them is

$$F_g = G \frac{m_e m_p}{r^2}$$

$$= \frac{6.7 \times 10^{-11} \text{ n·m}^2/\text{kg}^2 \times 9.1 \times 10^{-31} \text{ kg} \times 1.7 \times 10^{-27} \text{ kg}}{(5.3 \times 10^{-11} \text{ m})^2}$$

$$= 3.7 \times 10^{-47} \text{ n}.$$

The electrical force is over 10^{39} times greater than the gravitational force!

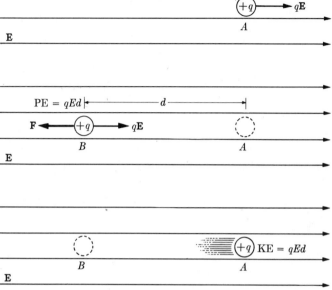

FIGURE 12–11

12–5 Potential difference

In our study of mechanics we found the related concepts of work and potential energy to be useful in analyzing a wide variety of situations. These concepts are equally useful in the study of electrical phenomena, but we shall introduce them in a slightly different form for the sake of convenience.

Let us examine a charge q in a uniform electric field $\mathbf{E}$, as in Fig. 12–11. To take the charge from point A to point B, we must apply a force of

$$\mathbf{F} = q\mathbf{E} \tag{12–6}$$

to it because we have to push against a force of the same magnitude exerted by the field. When the charge is at B, we have performed the amount of work

$$W = Fd = qEd \tag{12–7}$$

on it, where d is the distance between A and B. When it is at B, the charge therefore has the potential energy

$$PE = W = qEd \tag{12–8}$$

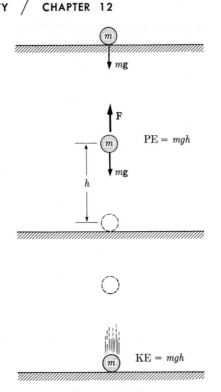

$$PE = mgh$$

$$KE = mgh$$

FIGURE 12–12

with respect to point A; if we release it, this potential energy will become kinetic energy of motion as the electric field **E** accelerates the charge, so that when it is back at A it will have a kinetic energy of qEd.

An exact analog of the charge q in the electric field **E** is the case of a stone of mass m in the earth's gravitational field. To lift the stone to a height h above the ground (Fig. 12–12) requires that we perform the amount of work mgh. When the stone is at h it has the potential energy mgh *with respect to the ground.* If we release the stone, it will fall to the ground with a final kinetic energy of mgh.

To describe the situation of a charge in an electric field, the quantity *potential difference* is introduced. The potential difference V_{AB} between two points A and B is defined as the ratio between the work that must be done to take a charge q from A to B and the value of q:

$$V_{AB} = \frac{W_{AB}}{q}. \qquad (12\text{–}9)$$

The units of potential difference are joules per coulomb. Because this quantity is so frequently employed, its units have been given the name *volts.* Thus

$$1 \text{ volt} = 1 \text{ j/coul.}$$

In a uniform electric field, W_{AB} is given by Eq. (12–7), with the result that the potential difference between A and B is

$$V_{AB} = \frac{qEd}{q} = Ed. \qquad (12\text{–}10)$$

In a uniform electric field, the potential difference between two points parallel to the direction of the field is the product of the field intensity and their separation. A positive potential difference means that the energy of the charge is *greater* at B than at A; a negative potential difference means that its energy is *less* at B than at A. If V_{AB} is positive, then, a charge at B tends to return to A, while if V_{AB} is negative it tends to move in the opposite direction from A.

One advantage of specifying the potential difference between two points in an electric field rather than the field intensity between them is that we normally create an electric field by imposing a difference of potential between two points in space. A *battery* is a device that uses chemical means to produce a potential difference between two terminals. A "six-volt" battery is one that has a potential difference of six volts between its terminals. When a charge q goes from one terminal of a battery whose potential difference is V to the other, the work

$$W = qV \qquad (12\text{–}11)$$

is done on it, according to Eq. (12–9), regardless of the path taken by the charge and regardless of whether the actual electric field that caused the motion of the charge is strong or weak; given V we can find W at once, no matter what the details of the process are. Hence the notion of potential difference permits us to simplify our analyses of electrical phenomena just as the notion of potential energy permitted us to simplify our analyses of mechanical phenomena.

We shall shortly consider a particularly important application of potential difference, but a simple example to show how it may be employed is appropriate here. Suppose that we have a tube, like that shown in Fig. 12–13, in which we have a source of electrons at one end and a metal plate at the other. A 100-volt battery is connected between the electron source and the metal plate, so that there is a potential difference of 100 volts between them. The negative terminal of the battery is connected to the electron source. What is the velocity of the electrons when they

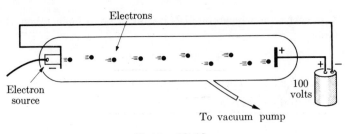

FIGURE 12–13

arrive at the metal plate? (The tube is evacuated to prevent collisions between the electrons and air molecules.)

To determine the kinetic energy, we note that the work done by the electric field within the tube on an electron is

$$W = qV$$
$$= 1.6 \times 10^{-19} \, \text{coul} \times 100 \, \text{volts}$$
$$= 1.6 \times 10^{-17} \, \text{j};$$

since the kinetic energy of the electron is equal to the work done on it, its kinetic energy after passing through the entire field is 1.6×10^{-17} joules. Hence

$$\text{KE} = \tfrac{1}{2}mv^2 = W,$$

$$v = \sqrt{\frac{2W}{m}} = \sqrt{\frac{2 \times 1.6 \times 10^{-17} \, \text{j}}{9.1 \times 10^{-31} \, \text{kg}}} = 5.9 \times 10^6 \, \frac{\text{m}}{\text{sec}} \cdot$$

It is not necessary for us to know anything specific about the electric field that actually accelerates the electron or about its actual path in the field.

The *electron volt* (abbreviated ev), a widely used unit in atomic physics, is the energy acquired by an electron that has been accelerated by a potential difference of 1 volt. Hence

$$1 \, \text{ev} = 1.6 \times 10^{-19} \, \text{j}.$$

The above electron had an energy of 100 ev when it reached the positive plate.

IMPORTANT TERMS

An example of an object possessing a *negative electric charge* is a rubber rod that has been stroked with fur. An example of an object possessing a *positive electric charge* is a glass rod that has been stroked with silk. The unit of electric charge is the *coulomb*.

Coulomb's law states that the force one charge exerts upon another is directly proportional to the magnitudes of the charges and inversely proportional to the square of the distance between them. The force between like charges is repulsive, and that between unlike charges is attractive.

The *electric field intensity* at a point is the force that would act on a positive charge of 1 coul at that point.

The electrical *potential difference* between two points is the work that must be done to take a charge of 1 coul from one point to the other. The unit of potential difference is the *volt*, which is equal to 1 j/coul. The *electron volt* is the energy acquired by an electron that has been accelerated by a potential difference of 1 volt. It is equal to 1.6×10^{-19} joules.

PROBLEMS

1. Electricity was once regarded as a weightless fluid, an excess of which was "positive" and a deficiency of which was "negative." What phenomena can this hypothesis still explain? What phenomena can it not explain?

2. An insulating rod has a positive charge at one end and a negative charge at the other. Both charges have the same magnitude. How will the rod move when it is placed in a uniform electric field (a) whose direction is parallel to the rod; (b) whose direction is perpendicular to the rod?

3. Two charges of unknown magnitude and sign are observed to repel one another with a force of 0.1 n when they are 5 cm apart. What will the force be when they are (a) 10 cm apart? (b) 50 cm apart? (c) 1 cm apart?

4. Why do people sometimes get small electric shocks when they leave automobiles?

5. How do we know that the inverse square force holding the earth in its orbit around the sun is not an electrical force?

6. (a) When two bodies attract each other electrically, must both of them be charged? (b) When two bodies repel each other electrically, must both of them be charged?

7. A particle carrying a charge of $+6 \times 10^{-5}$ coul is located halfway between two other charges, one of $+1 \times 10^{-4}$ coul and the other of -1×10^{-4} coul, that are 40 cm apart. All three charges lie on the same straight line. What is the magnitude and direction of the force on the $+6 \times 10^{-5}$ coul charge?

8. A small sphere carrying a charge of $+2 \times 10^{-4}$ coul is 0.1 m from another small sphere carrying a charge of -5×10^{-4} coul. What is the magnitude and direction of the force exerted by the -5×10^{-4} coul charge on the $+2 \times 10^{-4}$ coul charge?

9. How far apart should two electrons be if the force each exerts on the other is to equal the weight of an electron?

10. A test charge of -5×10^{-5} coul is placed between two other charges so that it is 5 cm from a charge of -3×10^{-5} coul and 10 cm from a charge of -6×10^{-5} coul. The three charges lie along a straight line. What is the magnitude and direction of the force on the test charge?

11. Two metal spheres, one with a charge of $+2 \times 10^{-5}$ coul and the other with a charge of -1×10^{-5} coul, are 10 cm apart. (a) What is the force between them? (b) The two spheres are brought into contact, and then separated again by 10 cm. What is the force between them now?

12. According to one model of the hydrogen atom, it consists of a proton circled by an electron whose orbit has a radius of 5.3×10^{-11} m. How fast must the electron be moving if the orbit is to be a stable one?

13. Two charges, one of 1.5×10^{-6} coul and the other of 3×10^{-6} coul, are 0.2 m apart. Where is the electric field in their vicinity equal to zero?

14. What is the electric field intensity 0.4 m from a charge of $+7 \times 10^{-5}$ coul?

15. Two charges of $+4 \times 10^{-6}$ coul and $+8 \times 10^{-6}$ coul are 2 m apart. What is the electric field intensity halfway between them?

16. A body whose mass is 10^{-6} kg carries a charge of $+10^{-6}$ coul. What is the magnitude of an electric field that can hold the body suspended in equilibrium?

17. A particle carrying a charge of 10^{-5} coul starts moving from rest in a uniform electric field whose intensity is 50 volts/m. (a) What is the force on the particle? (b) How much kinetic energy will the particle have after it has moved 1 m?

18. It is necessary to do 13.6 ev of work to remove the electron from a hydrogen atom. (a) How much energy in joules would be required to remove the electrons from the 6×10^{26} hydrogen atoms in a kilogram of hydrogen? (b) How many kilocalories is this?

19. What is the energy in electron volts of an electron whose speed is 10^6 m/sec?

20. What is the energy in electron volts of a potassium atom whose speed is 10^6 m/sec?

21. What is the speed of an electron whose energy is 50 ev?

22. How much energy is expended when a charge of 30 coul is sent through an electric motor if the potential difference is 220 volts?

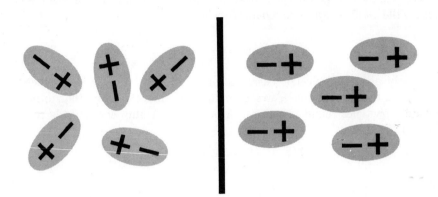

Electrical Properties of Matter

13

Different substances vary considerably in their electrical properties. In this chapter we shall examine several significant aspects of the behavior of matter when under the influence of electric fields.

13–1 Electric current

A flow of electric charge from one place to another is called an *electric current*. The magnitude of an electric current, i, is the amount of charge passing a given point per second; that is,

$$i = q/t \qquad (13-1)$$

where q is the charge that goes past in a time t. The unit of electric current is the *ampere* (abbreviated amp), where

$$1 \text{ amp} = \frac{1 \text{ coul}}{\text{sec}}.$$

Nearly all actual electric currents consist of flows of electrons, which, owing to their small mass, are able to move more readily than other charged particles found in nature. A current of 1 amp corresponds to a flow of

$$\frac{1 \text{ coul/sec}}{1.6 \times 10^{-19} \text{ coul/electron}} = 6.3 \times 10^{18} \frac{\text{electrons}}{\text{sec}}.$$

This figure does not mean, if we have a copper wire in which there is a current of 1 amp, that 6.3×10^{18} electrons flow from one end of the wire to the other each second. What it does signify is that this many electrons enter one end of the wire and the same number leave the other end each second, but they do not have to be the same electrons. A legitimate analogy is with the flow of water in a pipe.

Nearly all substances fall into two categories, *conductors*, through which electric currents can flow easily, and *insulators*, through which currents have great difficulty in flowing. Metals, many liquids, and plasmas (gases whose molecules are charged) are conductors, while nonmetallic solids, certain liquids, and gases whose molecules are electrically neutral are insulators. Several substances, called *semiconductors*, are intermediate in their ability to conduct current.

Just as the rate of flow of water between two points depends upon the difference of height between them, the rate of flow of electric current between two points depends upon the difference of potential between them (Fig. 13–1). In a metal, the current i that flows in a conductor is proportional to the potential difference V between its ends:

$$i = \frac{1}{R} V. \tag{13–2}$$

The constant of proportionality is written $1/R$, so that a large value of R corresponds to a small current for a given potential difference, and a small value of R to a large current. We call R the *resistance* of the conductor.

Equation (13–2) is called *Ohm's law*, since it was first verified experimentally by the German physicist Georg Ohm (1787–1854). The unit of electrical resistance is the *ohm*, where

$$1 \text{ ohm} = 1 \frac{\text{volt}}{\text{amp}}.$$

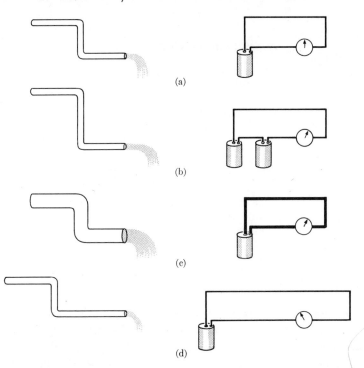

FIG. 13–1. Analogy between the flow of water in a pipe and the flow of electric current in a wire. The rate of flow of water in a pipe may be increased over its value at (a) by having the water fall from a greater height (b) or by using a larger pipe (c). The rate of flow is decreased when a longer pipe is used (d). The rate of flow of electric current in a wire may be increased over its value at (a) by increasing the potential difference between the ends of the wire (b) or by using larger wire (c). The current is decreased when a longer wire is used (d).

When a potential difference of 1 volt is applied across the ends of a conductor whose resistance is 1 ohm, 1 amp of current flows. For instance, a potential difference of 120 volts applied across a resistance of 240 ohms in a light bulb results in a current of

$$i = \frac{V}{R} = \frac{120 \text{ volts}}{240 \text{ ohms}} = 0.5 \text{ amp.}$$

The resistance of a conductor depends upon (1) the material of which it is composed (metals vary widely in their ability to conduct currents, with silver and copper among the best and iron and lead among the worst), (2) its length L (the greater L is, the greater the value of R), and (3) the cross-sectional area A of the conductor (the greater A is, the smaller the value of R). This again corresponds to water in a pipe: the

TABLE 13–1

Approximate resistivities at 20°C

Substance	Resistivity, ohm·m
Aluminum	2.63×10^{-8}
Carbon	3500×10^{-8}
Copper	1.62×10^{-8}
Germanium	2×10^{8}
Gold	2.22×10^{-8}
Iron	$9\text{–}13 \times 10^{-8}$
Lead	20.4×10^{-8}
Mercury	94×10^{-8}
Phosphorus	1×10^{12}
Platinum	11.1×10^{-8}
Silver	1.60×10^{-8}
Sulfur	1×10^{17}

longer the pipe, the more chance friction against the pipe wall has to slow down the flow of water, and the larger the cross-sectional area of the pipe, the greater the volume of water that can flow through per second when everything else is the same. This analogy is illustrated in Fig. 13–1. The simple formula

$$R = \rho \, \frac{L}{A} \tag{13–3}$$

has been found to hold, where ρ (Greek letter *rho*) is called the *resistivity* of the conducting material. Table 13–1 gives a list of the resistivities of several elements at room temperature, 20°C. There is evidently a substantial variation in resistivity; sulfur, for example, has a resistivity almost 10^{25} greater than that of silver. We shall look into the origin of this variation in Chapter 22.

13–2 Energy and power

Owing to the resistance that all conductors offer to the flow of current through them, work must be done continuously to maintain a current. Electrical resistance is analogous to friction, and so the work that is done in causing a flow of current is dissipated as heat. The heating effect of electric currents is made use of in such devices as electric stoves, toasters, and light bulbs.

To calculate the energy dissipated by a wire of resistance R carrying a current i during the time t, we begin with the work done in taking a charge q through a potential difference V. From Eq. (12–11) we see that this amount of work is

$$W = qV.$$

Now

$$i = \frac{q}{t}, \quad q = it \quad \text{and} \quad i = \frac{V}{R}, \quad V = iR;$$

hence

$$W = qV = i^2 Rt. \tag{13–4}$$

The energy dissipated is the product of the square of the current, the resistance, and the time. In the case of the light bulb we mentioned earlier, the energy dissipated per second is

$$W = i^2 Rt$$
$$= (0.5 \text{ amp})^2 \times 240 \text{ ohms} \times 1 \text{ sec}$$
$$= 60 \text{ j}.$$

The rate at which work is being done (or at which energy is being dissipated) turns out to be a useful quantity in many applications. This rate is called *power*, whose symbol is P. That is,

$$P = W/t, \tag{13–5}$$

where W is the amount of work done in a period of time t. The unit of power in the mks (metric) system is the *watt:*

$$1 \text{ watt} = 1 \text{ j/sec}.$$

The above light bulb evidently dissipates energy at the rate of 60 watts.

While power is a perfectly general concept, it is perhaps most familiar in electrical applications. From Eqs. (13–4) and (13–5) we see that

$$P = \frac{W}{t} = \frac{i^2 Rt}{t} = i^2 R. \tag{13–6}$$

This expression for electrical power may be rewritten with the help of Ohm's law to give

$$P = iV = \frac{V^2}{R}.$$

Depending upon which quantities are known in a specific case, any of these formulas may be used.

A further example may be helpful. The *horsepower* (abbreviated hp) is a traditional unit of power in engineering, and is equal to 746 watts. An electric motor rated at $\frac{1}{2}$ hp and operated from a 120-volt source of electricity draws a current of

$$i = \frac{P}{V} = \frac{\frac{1}{2} \text{ hp} \times 746 \text{ watts/hp}}{120 \text{ volts}} = 3.1 \text{ amp.}$$

13–3 Capacitance

A *capacitor* is a device that stores electrical energy in the form of electric field. The simplest capacitor consists of a pair of parallel metal plates separated by air or other insulating material. When a potential difference V is placed across the plates, each acquires a charge Q of opposite sign to the other (Fig. 13–2), and an electric field E appears between them. The charge Q is always directly proportional to the impressed potential difference V, so that the ratio Q/V is a constant for any capacitor. The value of this ratio for a given capacitor is known as its *capacitance*. The symbol for capacitance is C, and so

$$C = \frac{Q}{V}. \tag{13–7}$$

The unit of capacitance is the *farad*, abbreviated f, where

$$1 \text{ f} = 1 \frac{\text{coul}}{\text{volt}}.$$

The capacitance of a pair of separated conductors depends solely upon their geometrical configuration and upon the material between them.

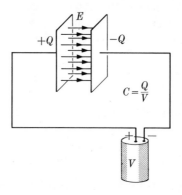

FIGURE 13–2

A parallel-plate capacitor, for instance, has a capacitance of

$$C = K\epsilon_0 \frac{A}{d},$$ (13–8)

where A is the area of either plate, d is their separation, K is a factor that varies with the nature of the material between the plates, and ϵ_0 is the electrostatic constant found in Coulomb's law. The factor K is exactly 1 for vacuum and is very close to 1 for air. The plates of such a capacitor might be 10 cm square and 1 mm apart, so that $A = 10^{-2}\,\text{m}^2$ and $d = 10^{-3}\,\text{m}$. In air its capacitance is

$$C = 1 \times 8.85 \times 10^{-12}\,\frac{\text{coul}^2}{\text{n·m}^2} \times \frac{10^{-2}\,\text{m}^2}{10^{-3}\,\text{m}}$$

$$= 8.85 \times 10^{-11}\,\text{f.}$$

The farad is so large a unit that, for practical purposes, it is usually replaced by the *microfarad* (μf) or *micromicrofarad* ($\mu\mu$f) whose values are

$$1\,\mu\text{f} = 10^{-6}\,\text{f}, \qquad 1\,\mu\mu\text{f} = 10^{-12}\,\text{f.}$$

The above capacitor therefore has a capacitance of $88.5\,\mu\mu$f. If a potential difference of 100 volts is placed across the plates of this capacitor, they will acquire charges of

$$Q = CV$$

$$= 8.85 \times 10^{-11}\,\text{f} \times 100\,\text{volts}$$

$$= 8.85 \times 10^{-9}\,\text{coul.}$$

13–4 Electrostatic potential energy

We can calculate the electrostatic potential energy of a charged capacitor by analogy with the method we used in the case of a stretched spring. Figure 13–3 shows various stages in the transfer of electrons from one plate of a capacitor to the others. Initially the potential difference between the plates is small, and little work is necessary to move the electrons. As the charge builds up, the potential difference becomes greater, and more work is needed per electron. If the final potential difference is V, the average potential difference $\overline{V}$ during the charge transfer is

$$\overline{V} = \frac{V_{\text{final}} + V_{\text{initial}}}{2} = \frac{V + 0}{2} = \frac{1}{2}\,V.$$

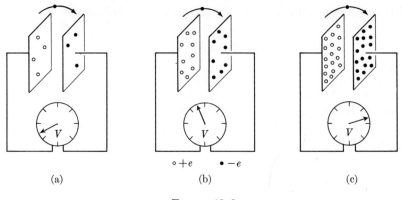

$\circ + e \qquad \bullet - e$

(a)   (b)   (c)

FIGURE 13–3

Since the total charge transferred is Q, the work W that was done is the product of Q and the average potential difference $\overline{V}$, or

$$W = Q\overline{V} = \tfrac{1}{2}QV. \tag{13–9}$$

Hence $\tfrac{1}{2}QV$ is the electrostatic potential energy of a capacitor whose plates have the charge Q and a potential difference of V. Equation (13–9) may be written in the alternative forms

$$W = \tfrac{1}{2}CV^2 \tag{13–10}$$

$$= \frac{1}{2}\frac{Q^2}{C} \tag{13–11}$$

by making use of the definition $C = Q/V$. Equations (13–9) through (13–11) are true for *all* capacitors regardless of this construction.

When an 88.5-$\mu\mu$f capacitor is charged to a potential difference of 100 volts, the energy stored in it is

$$W = \tfrac{1}{2}CV^2$$

$$= \tfrac{1}{2} \times 8.85 \times 10^{-11}\,\text{f} \times (100\ \text{volts})^2$$

$$= 4.43 \times 10^{-7}\,\text{j}.$$

The capacitance of a parallel-plate capacitor is

$$C = \epsilon_0\,\frac{A}{d}$$

when it is in vacuum. The energy of such a capacitor when the potential

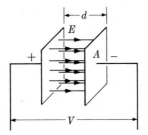

FIGURE 13–4

difference across it is V is therefore

$$W = \frac{1}{2} CV^2 = \frac{1}{2} \epsilon_0 \frac{A}{d} V^2. \tag{13-12}$$

The electric-field intensity E between the plates is

$$E = \frac{V}{d},$$

according to Eq. (12–10). By substituting Ed for V in Eq. (13–12), we have

$$W = \tfrac{1}{2}\epsilon_0 Ad\, E^2.$$

We note that Ad, the product of the area of the plates and their separation distance, is the volume occupied by the electric field E (Fig. 13–4). If we define the *energy density* w of an electric field as the electrostatic potential energy per unit volume associated with it, we have here

$$w = \frac{\text{total energy of electric field}}{\text{volume occupied by electric field}}$$

$$= \frac{W}{Ad} = \frac{1}{2} \epsilon_0 E^2. \tag{13-13}$$

This important formula states that the energy density of an electric field is directly proportional to the square of its intensity E. While we have derived Eq. (13–13) for the specific case of a parallel-plate capacitor, it is a completely general result.

The capacitor we have been using as an example has an electric field of

$$E = \frac{V}{d} = \frac{100 \text{ volts}}{10^{-3} \text{ m}} = 10^5 \frac{\text{volts}}{\text{m}}$$

between its plates. The energy density of this field is

$$w = \frac{1}{2} \, \epsilon_0 E^2 = \frac{1}{2} \times 8.85 \times 10^{-12} \, \frac{coul^2}{n \cdot m^2} \times \left(10^5 \, \frac{volts}{m} \right)^2$$

$$= 0.0443 \, j/m^3.$$

Until now we have been content to discuss fields of force as though they are merely convenient ways to regard the transmission through space of such forces as those of gravitation and electricity. As we just learned for the case of an electric field, these fields contain energy, and we must revise our previous casual approach to take seriously the existence of force fields as definite physical entities. The energy contents of electric and magnetic fields are essential to the analysis of electromagnetic waves given in Chapter 16.

13–5 Dielectric constant

Let us now examine what happens when a slab of an insulating material is placed between the plates of a capacitor. While the slab cannot conduct electric current, it can respond to an electric field in two other possible ways. The molecules of all substances either normally have an asymmetrical distribution of electric charge within them or assume such a distribution under the influence of an electric field. A molecule of the former kind is known as a *polar molecule,* and behaves as though one end is positively charged and the other negatively charged. In an assembly of polar molecules when there is no external electric field, the molecules are randomly oriented as in Fig. 13–5(a). When an electric field is present, it acts to align the molecules opposite to the field, as in Fig. 13–5(b). While nonpolar molecules ordinarily have symmetric charge distributions, an electric field is able to distort their inner structures so that an effective separation of charge takes place (Fig. 13–6). Again the molecules have their charged ends aligned opposite to the external field. In either case, then, the net electric field between the plates of the capacitor is *less* than it would be with nothing between them.

In Eq. (13–8), the factor K, called the *dielectric constant,* is a measure of how effective a particular substance is in reducing an electric field set up across a sample of it. Table 13–2 is a list of dielectric constants for various substances. Water and alcohol molecules are highly polar, and the values of K for water, ice, and ethyl alcohol are accordingly high.

A charged capacitor has an energy of

$$W = \frac{1}{2} \, \frac{Q^2}{C}.$$

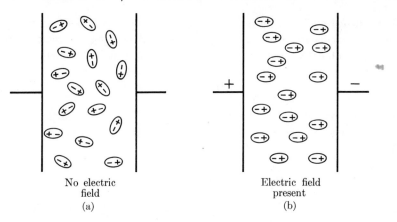

FIG. 13–5. An electric field tends to align polar molecules opposite to the field.

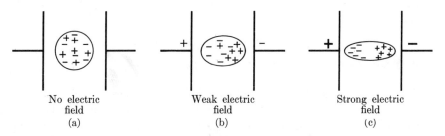

FIG. 13–6. An electric field tends to distort the charge distributions in nonpolar molecules.

TABLE 13–2

Dielectric constants at 20°C except where noted

Substance	K
Air	1.0006
Air, liquid (−191°C)	1.4
Alcohol, ethyl	26
Amber	2.9
Benzene	2.3
Glass	5–8
Granite	7–9
Ice (−2°C)	94
Mica	2.5–7
Sulfur	3.9
Water	80

If we insert a slab of material of dielectric constant K between the plates of a charged capacitor which originally had nothing between them, the charge Q on each plate does not change. However, the capacitance of the capacitor is now

$$C = KC_0, \tag{13-14}$$

where C_0 was its initial value. The energy W of the charged capacitor with the slab in place is therefore related to its former energy W_0 by the formula

$$W = \frac{W_0}{K}. \tag{13-15}$$

Because K is greater than one, energy has been lost, with the missing energy having gone into the alignment and/or distortion of the molecules of the material between the plates. To remove the slab of material requires the performance of enough work to make up the difference between W and W_0.

13–6 Charging a capacitor

When a capacitor is connected to a battery, it does not immediately become fully charged. At first the only limit to the current that flows to the capacitor is the resistance R in the circuit, so that the initial current is $i = V_b/R$, where V_b is the potential difference across the terminals of the battery. As the capacitor becomes charged, however, a potential difference appears across it whose polarity is such as to tend to oppose the further flow of current. When the charge on the capacitor has built up to some value q, this opposing potential difference is $V = q/C$. Hence the net potential difference is $V_b - q/C$, and the current is

$$i = \frac{V_b - q/C}{R}. \tag{13-16}$$

As q increases, then, its *rate* of increase drops. Figure 13–7 is a graph showing how q varies with time when a capacitor is being charged; the capacitor is connected to the battery at $t = 0$. A mathematical analysis of Eq. (13–16) shows that after a time interval of RC (the product of the resistance R in the circuit and the capacitance C of the capacitor), the charge on the capacitor reaches 63% of its ultimate value of $Q = CV_b$. The time RC is therefore a convenient measure of how rapidly the capacitor becomes charged; it is accordingly called the *time constant* of the circuit. In principle the capacitor acquires its ultimate charge Q only after an infinite time has elapsed, but, as we can see from Fig. 13–7,

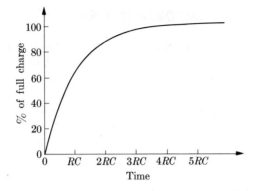

FIG. 13–7. The growth of charge in a capacitor.

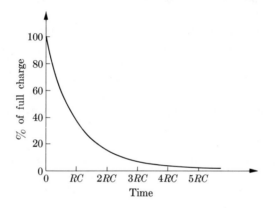

FIG. 13–8. The decay of charge in a short-circuited capacitor.

this value is very nearly reached after only 3 or 4 RC. It is easy to verify that RC has the dimensions of time. From their definitions,

$$R = \frac{V}{i} = \frac{V}{q/t} = \frac{Vt}{q}, \qquad C = \frac{q}{V},$$

and so

$$RC = \frac{Vt}{q} \times \frac{q}{V} = t.$$

Figure 13–8 shows the gradual discharge of the capacitor when its terminals are short-circuited by a wire.

IMPORTANT TERMS

A flow of electric charge from one place to another is called an *electric current*. The unit of electric current is the *ampere*, which is equal to a flow of 1 coul/sec.

Ohm's law states that the current that flows in a conductor is proportional to the potential difference between its ends. The *resistance* of a conductor is the ratio between the potential difference across its ends and the current that flows. The unit of resistance is the *ohm*, which is equal to 1 volt/amp. The resistance of a conductor is proportional to its length and to the *resistivity* ρ of the material of which it is made, and inversely proportional to its cross-sectional area.

The rate at which work is being done is called *power*. The unit of power is the *watt*, which is equal to 1 joule/sec.

A *capacitor* is a device that stores electrical energy in the form of an electric field. The ratio between the charge on either plate of a capacitor and the potential difference between the plates is called its *capacitance*. The unit of capacitance is the *farad*, which is equal to 1 coul/volt.

The *energy density* of an electric field is the electrostatic potential energy per unit volume associated with it. The energy density of an electric field in free space of intensity $\mathbf{E}$ is $\frac{1}{2}\epsilon_0 E^2$.

The *dielectric constant K* of a particular material is a measure of how effective it is in reducing an electric field set up across a sample of it.

The *time constant* of a circuit containing resistance and capacitance is a measure of how rapidly the capacitor can be charged or discharged.

PROBLEMS

1. A certain 12-volt storage battery is rated at 80 amp·hr, which means that when it is fully charged it can deliver a current of 1 amp for 80 hr, 2 amp for 40 hr, 80 amp for 1 hr, etc. (a) How many coulombs of charge can this battery deliver? (b) How much energy is stored in it?

2. In a Van de Graaff generator an insulating belt is used to carry charges to a large metal sphere. In a typical generator of this kind, the potential difference between the sphere and the source of the charges is 5×10^6 volts. (a) If the belt carries charge to the sphere at a rate of 10^{-3} amp, how much power is required? (b) How much energy in ev will an electron have if it is accelerated by such a potential difference? (c) Express the answer to (b) in joules.

3. A current of 3 amp flows through a wire whose ends are at a potential difference of 12 volts. How much charge flows through the wire per minute?

4. Approximately 10^{20} electrons/cm participate in conducting electric current in a certain wire. (That is, 10^{20} electrons in each centimeter of the wire are

in motion when a current is being carried by the wire.) What is the average speed of the electrons when there is a current of 1 amp in the wire?

5. Do bends in a wire affect its electrical resistance? Explain your answer.

6. How many electrons flow through the filament of a 120-volt, 60-watt electric light bulb per second?

7. It is sometimes said that an electrical appliance "uses up" electricity. What does such an appliance actually use up in its operation?

8. The starter of a certain car requires a current of 100 amp from a 12-volt battery to turn over the motor. How much horsepower does this represent?

9. An electric water heater has a resistance of 12 ohms and is operated from a 120-volt power line. If no heat escapes from it, how much time is required for it to raise the temperature of 40 kg of water from 15°C to 80°C?

10. Prove that when two resistors R_1 and R_2 are connected together in a circuit so that current flows consecutively from one through the other, their combined resistance is equal to $R_1 + R_2$. These resistors are said to be connected in *series*. (*Hint:* Make use of the fact that the same current flows through both resistors and add together the potential differences across each.)

11. Prove that when two resistors R_1 and R_2 are connected together at both terminals so that the current in any circuit of which they are a part is divided between them, their combined resistance is equal to $R_1R_2/(R_1 + R_2)$. These resistors are said to be connected in *parallel*. (*Hint:* Make use of the fact that the same potential difference exists across each resistor and add together the currents that flow through each.)

12. A copper wire 100 ft long is to have a maximum resistance of 25 ohms. What is the smallest wire diameter that can be used?

13. A copper rod 1 m long and 1 cm in diameter is drawn out into a wire 1 mm in diameter. (a) What is the length of the wire? (b) Compare the resistance of the rod with the resistance of the wire.

14. Aluminum wires are sometimes used to transmit electric power instead of copper wires. What is the ratio between the weights of an aluminum and a copper wire of the same length whose cross-sectional areas are such that they have the same resistance? (Table 7–3 gives the densities of aluminum and copper.)

15. A 25-μf capacitor is connected to a source of potential difference of 1000 volts. What is the resulting charge on the capacitor?

16. A 10-μf capacitor can withstand a maximum potential difference of 4000 volts. What is the maximum energy it can store?

17. How much work must be done to charge a 12-μf capacitor until the potential difference between its plates is 250 volts?

18. A parallel-plate capacitor of capacitance C is given the charge Q and then disconnected from the circuit. How much work is required to pull the plates of this capacitor to twice their original separation?

19. The plates of a parallel-plate capacitor are 50 cm² in area and 1 mm apart. (a) What is its capacitance? (b) When the capacitor is connected to a 45-volt battery, what is the charge on either plate? (c) What is the energy of the charged capacitor?

20. The space between the plates of the capacitor of the previous problem is filled with sulfur. Answer questions (a), (b), and (c) for this case.

21. The capacitance of a parallel-plate capacitor is increased from 8 μf to 50 μf when a sheet of glass is inserted between its plates. What is the dielectric constant of the glass?

22. A sheet of mica whose dielectric constant is 5 is placed between the plates of a charged, isolated parallel-plate capacitor. How is the potential difference across the capacitor affected? How is the charge on the capacitor affected?

23. A capacitor with air between its plates is connected to a battery and each of its plates receives a charge of 10^{-4} coul. While still connected to the battery the capacitor is immersed in oil, and a further charge of 10^{-4} coul is added to each plate. What is the dielectric constant of the oil?

24. A reusable flash bulb requires an energy of 100 j for its discharge. A 450-volt battery is used to charge a capacitor for this purpose. The resistance of the charging circuit is 15 ohms. (a) What is the required capacitance? (b) What is the time constant of the circuit?

25. A 5-μf capacitor is connected across a 1000-volt battery with wires whose resistance is a total of 5000 ohms. (a) What is the time constant of this circuit? (b) What is the initial current that flows when the battery is connected? (c) How long would it take to charge the capacitor if this current remained constant?

26. The electric field intensity near the earth's surface is about 100 volts/m. How much electrical energy is stored in the lowest kilometer of the atmosphere?

27. Dry air is an insulator provided the electric field intensity in it does not exceed about 3×10^6 volts/m. What energy density does this correspond to?

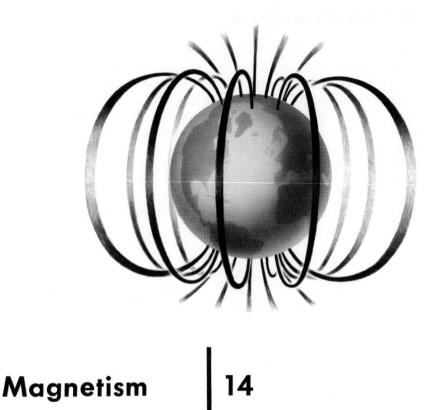

Magnetism | 14

Electric charges in motion are able to exert forces of a kind quite different from those they can exert while at rest. For instance, if we place a current-carrying wire parallel with and near another current-carrying wire, with their currents in the same direction, we find that they attract each other; if the currents are in opposite directions, the force is repulsive. These observations cannot be accounted for on the basis of Coulomb's law. The forces that come into being whenever electric currents are brought near one another are called *magnetic* forces. Of course, the word "magnetic" suggests ordinary magnets and their familiar attraction for iron objects, but, as we shall see, this is but one aspect of the whole subject of magnetism.

247

14–1 Oersted's experiment

While magnetic phenomena have been known for thousands of years, a knowledge of their connection with electric currents is only about 150 years old. The word "magnetism" comes from Magnesia, the name of a region in ancient Asia Minor where naturally magnetic pieces of ferric oxide (Fe_3O_4), called *lodestones*, are found. The ancients knew that lodestones attract iron objects and can even magnetize such objects so that they, in turn, attract other pieces of iron. It is an easy matter to show that static electric charges and stationary magnets have no attraction for one another, and, as a perhaps unreliable legend relates, it was in the course of a demonstration that electric currents and magnets also do not interact that Hans Oersted (1777–1851) discovered the contrary to be true.

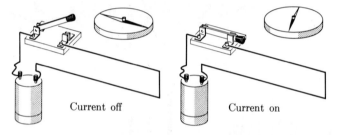

Current off   Current on

FIG. 14–1. Oersted's experiment.

Oersted used an experimental arrangement like that shown in Fig. 14–1. When a wire is placed parallel to the compass needle and the current turned on, the needle rotates until it is almost perpendicular to the wire. When the wire is initially perpendicular to the compass needle, there is no deflection when the current is turned on. The two conclusions that can be drawn are, first, *an electric current somehow exerts a twisting force on a magnet near it,* and second, *the magnitude of the force depends upon the relative orientation of the current and the magnet.*

When studying magnetism, as with gravitation and electricity, it is useful to think in terms of *fields* rather than specific forces. A current alters the properties of space in its vicinity such that a piece of iron there experiences a force; it is the interaction between the magnetic field of the current and the piece of iron that leads to the force, rather than the current and iron acting upon one another directly. Once we have determined the magnetic field at any point in space, we can readily compute its effect on anything we may choose to place there. Otherwise, if there are a

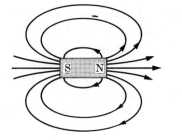

F‌IG. 14–2. The pattern of magnetic lines of force around a bar magnet.

number of sources of magnetic field present, we would have a difficult calculation to make each time. Even more important than this practical consideration is the fact that a magnetic field may possess both energy and momentum, and therefore must be accepted as a distinct entity.

As before, it is convenient to visualize a magnetic field with the help of lines of force. The directions of magnetic lines of force are those in which a suitable test object (for instance a compass needle) aligns itself at each point in the field, and their spacing is smaller the stronger the field. If iron filings are sprinkled on a card held over a bar magnet, the pattern of Fig. 14–2 results; to verify that this pattern actually represents the magnetic field of the magnet, we can place a compass needle in various locations and compare its direction and the force with which it is held in that direction with the pattern of iron filings. (Lines of force do not, as we have said before, exist as such in nature. They are no more than aids to our imagination.)

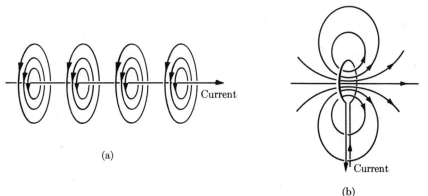

F‌IG. 14–3. The patterns of magnetic lines of force around (a) a straight current-carrying wire and around (b) a current-carrying wire loop. The lines of force in (a) are concentric circles. The lines of force in (b) are the same as those around a very short bar magnet; a coil consisting of several such loops in line has a magnetic field exactly like that shown in **Fig 14–2**.

The patterns of lines of force surrounding a straight current-carrying wire and a wire loop are shown in Fig. 14–3. We note that the magnetic field around a current loop is the same as that around a bar magnet. This might lead us to surmise that permanent magnets possess their magnetic properties by virtue of minute subatomic currents, which is, in fact, very close to the truth. This is the reason that there is no such thing as an isolated magnetic pole, even though there are isolated electric charges.

14–2 Magnetic induction

The relationship between an electric current and the magnetic field it gives rise to is considerably more complicated than that between a mass and its gravitational field or that between a charge and its electric field. We should keep in mind, however, that electric and magnetic fields are merely different manifestations of a single *electromagnetic* field, with the differences between them arising from the relative motion of the charges that constitute a current and the observer.

In our study of electricity, we defined the electric field intensity **E** at a point in space as

$$\mathbf{E} = \frac{\mathbf{F}}{q},$$

the ratio between the force **F** an electric field exerts on a test charge q at that point and the magnitude of q. The corresponding quantity that describes a magnetic field is called its *magnetic induction,* the symbol for which is **B**. One way of defining the magnetic induction at a point is in terms of the magnetic force **F** at that point which the field exerts on a small length of wire that carries an electric current (Fig. 14–4). If the wire is Δl long and has the current i flowing through it, we define the

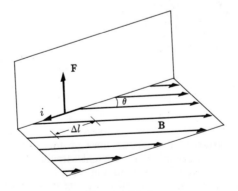

FIGURE 14–4

magnitude of **B** as

$$B = \frac{F}{i \, \Delta l \sin \theta},$$ (14-1)

where θ is the angle between the direction of i and that of **B**. (The direction of **B** can be empirically determined with the help of iron filings, for instance.) The presence of θ in Eq. (14-1) arises from the vector nature of electric current, which flows in a specific direction at a particular point in a conductor; the sine of θ is specified as a result of experiment. No such angular reference appears in the definition of **E** since electric charge, unlike current, is a scalar quantity without directional character.

From Eq. (14-1) we see that the unit of magnetic induction is the n/amp·m. This unit has been designated the weber/m²:

$$1 \frac{\text{weber}}{\text{m}^2} = 1 \frac{\text{n}}{\text{amp·m}}.$$

The aptness of the weber/m² follows from the description of magnetic field in terms of lines of force, which are sometimes called simply *magnetic flux*. The unit of magnetic flux is the *weber*. The stronger the field, the closer together the lines of force and the greater the density of flux, so it is appropriate to describe magnetic induction as a flux density and use the weber/m² as the unit.

Some representative values of magnetic induction may help in acquiring a feeling for the magnitude of the weber/m². The magnetic induction within our galaxy is believed to average about 10^{-9} weber/m²; the magnetic induction of the earth's magnetic field at sea level is about 3×10^{-5} weber/m²; the magnetic induction near a strong permanent magnet is about 1 weber/m²; the magnetic induction produced at the nucleus of a hydrogen atom by the electron circling it is about 14 webers/m²; and the most powerful magnetic fields achieved in the laboratory have magnetic inductions in the neighborhood of 100 webers/m².

14–3 Magnetic field of a current

Having defined magnetic induction **B** in terms of its effects, our next task is to find a means of calculating the magnetic induction produced by a current. To accomplish this we first perform experiments to determine how the magnetic force **F** between two wires varies with their directions, currents, and separation. Then we substitute the empirical formula for **F** in the defining equation for magnetic induction, just as we substituted Coulomb's law into the defining equation for electric field intensity in Chapter 12. The result is called the *Biot-Savart law*.

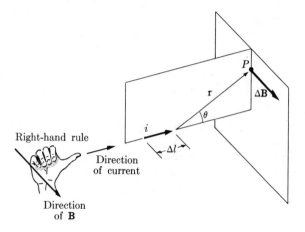

FIGURE 14–5

Let us consider a small length of wire Δl long that carries a current i, as in Fig. 14–5. The magnetic induction ΔB due to this current element at a point P a distance r away from the wire is

$$\Delta B = \frac{\mu}{4\pi}\ \frac{i\,\Delta l \sin\theta}{r^2}, \qquad\qquad (14\text{–}2)$$

where θ here is the angle between the direction of the current and that of r. Equation (14–2) is the *Biot-Savart law,* and it can be used for determining the magnetic induction around any current. The value of the constant μ depends upon the medium in which the magnetic field exists, and is referred to as the *permeability* of the medium. Its value in empty space, denoted μ_0, is

$$\mu_0 = 12.6 \times 10^{-7}\ \frac{\text{weber}}{\text{amp·m}}\ ;$$

its value in air is very close to this. The direction of ΔB is, as shown in Fig. 14–5, perpendicular to the plane formed by Δl and r. It is easy to figure out the direction with the help of the *right-hand rule:* grasp the wire with the right hand so that the thumb points in the direction of the current; then the curled fingers of that hand point in the direction of ΔB.

To apply the Biot-Savart law in an actual situation, we must compute the values of ΔB at some point P for each of the tiny lengths of wire Δl that make up the real wire, and then add up the results to find the total field. It is necessary to take the directions of each ΔB into account, since they all may not be parallel to one another and therefore will cancel out in part.

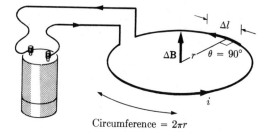

Circumference $= 2\pi r$

FIG. 14–6. The magnetic induction at the center of a current-carrying wire loop is perpendicular to the plane of the loop. Figure 14–3(b) shows the configuration of the entire magnetic field around the loop.

Let us use the Biot-Savart law to find the magnetic induction at the center of a circular current loop (Fig. 14–6). This is perhaps the simplest example of its application. Because we are interested in the value of **B** at the center of the loop, the distance r is constant and the angle θ is always 90°. Hence, since

$$\sin 90° = 1,$$

we have for the magnitude of $\Delta \mathbf{B}$

$$\Delta B = \frac{\mu i \, \Delta l}{4\pi r^2}.$$

Now we sum up all the $\Delta \mathbf{B}$ values around the loop. Since the factors i, $\mu/4\pi$, and r^2 are the same for all of them, this addition is the same as replacing Δl with the circumference of the loop, $2\pi r$. The total induction B at the center of the loop is therefore

$$B = \frac{\mu i \times 2\pi r}{4\pi r^2} = \frac{\mu i}{2r}. \tag{14–3}$$

The direction of **B** is perpendicular to the plane of the loop, as indicated in Fig. 14–6. If the loop consists of more than one turn of wire, the inductions of each individual loop add up algebraically; if there are n turns, then,

$$B = \frac{\mu n i}{2r}.$$

In the event there are so many turns that the coil is long compared with its diameter, the field within it is uniform (except near the ends) and has the flux density

$$B = \mu n i.$$

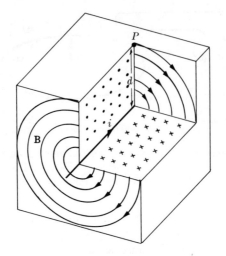

FIG. 14–7. The magnetic field around a long, straight, current-carrying wire consists of lines of force in the form of concentric circles.

A similar calculation in the case of the magnetic field around a long, straight wire is considerably more difficult, since each small section of the wire is both at a different distance from a specific point P and is at a different angle with respect to a line joining it with P. The result is

$$B = \frac{\mu}{2\pi} \frac{i}{d}, \qquad (14–4)$$

where d is the distance from the wire to the point P (Fig. 14–7). The lines of force of **B** are in the form of concentric circles about the wire; their direction is that in which the fingers of the right hand point when the thumb is placed in the direction of the current.

Let us use Eq. (14–4) to determine the magnetic induction 1 cm (0.01 m) from a wire carrying a current of 1 amp. We have

$$B = \frac{\mu}{2\pi} \frac{i}{d} = \frac{12.6 \times 10^{-7} \text{ weber/m}}{2\pi \times 0.01 \text{ m}}$$

$$= 2 \times 10^{-5} \frac{\text{weber}}{\text{m}^2},$$

a little less than the magnetic induction of the earth's magnetic field. This is the reason that great care is taken aboard ships to keep current-carrying wires away from compasses.

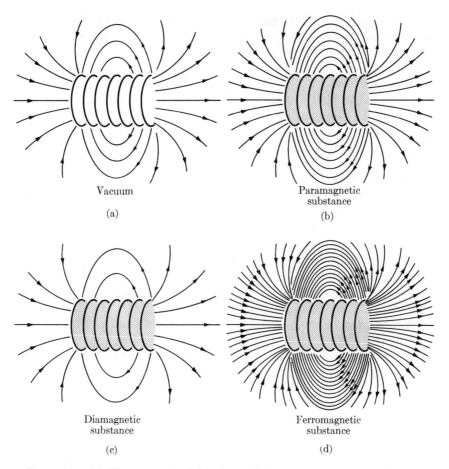

FIG. 14–8. (a) The magnetic field of a coil that carries an electric current. (b) When a rod of a paramagnetic substance is inside the coil, the field is slightly more intense. (c) When a rod of a diamagnetic substance is inside the coil, the field is slightly less intense. (d) When a rod of a ferromagnetic substance is inside the coil, the field is very much more intense. Paramagnetism, diamagnetism, and ferromagnetism have their explanations in the magnetic properties of atoms.

14–4 Magnetic properties of matter

The magnetic field of a coil of wire that carries a current, shown in Fig. 14–8(a), is similar to that of a single current loop, and we can calculate the magnetic induction in its vicinity. When we insert rods of various different materials into the coil, however, we find that the magnetic field, while retaining the same configuration, is slightly less intense, slightly more intense, or very much more intense, depending upon the nature of the material, as shown in Figs. 14–8(b), (c), and (d). These effects have an adequate explanation in terms of the internal structure of the atoms of the material.

Those substances that lead to a slight increase in B are called *paramagnetic*. Their effects are interpreted as resulting from the tendency of their permanent atomic current loops, each of which behaves like a tiny bar magnet, to partially align themselves parallel to the external field, thereby enhancing it. In several substances, notably iron, nickel, and cobalt, the alignment is especially marked, and the total magnetic induction may be many times greater than that produced by the current alone (Fig. 14–9). Substances of the latter kind are called *ferromagnetic*. The atomic current loops in ferromagnetic substances remain aligned even after the original magnetic field of external origin is removed, making possible *permanent magnets,* which are really assemblies of aligned permanent current loops. When a permanent magnet is heated to a sufficiently high temperature (760°C in the case of an iron magnet), the atoms acquire enough kinetic energy to overcome the interatomic forces that hold them in alignment. The current loops then become randomly oriented, and the "permanent magnet" returns to its original status as a metal bar.

Diamagnetic substances, on the other hand, lead to a decrease in B. The magnetic fields of any currents that may be present within each

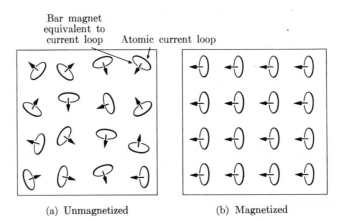

Bar magnet
equivalent to
current loop   Atomic current loop

(a) Unmagnetized   (b) Magnetized

Fig. 14–9. In an unmagnetized substance, the atomic current loops, each of which behaves like a tiny bar magnet, are randomly oriented. When a ferromagnetic substance is placed in a magnetic field, its atomic current loops align themselves with that field. The alignment remains even after the original field is removed, which makes permanent magnets possible.

of their atoms cancel one another out, so that each atom normally has no magnetic properties. In the presence of an external magnetic field, though, the atomic currents are so affected that the cancellation is no longer complete, and they now act as if they are tiny bar magnets pointed *opposite* to the original magnetic field, thereby reducing its magnitude.

Paramagnetic and ferromagnetic substances are attracted by magnetic fields, while diamagnetic ones are repelled.

14–5 Charged particles

From the definition of electric current as a flow of charge, we would expect moving charged particles to be affected by magnetic fields in a manner similar to that of currents. According to Eq. (14–1), the force on a wire of length Δl carrying a current i when it is in a magnetic field of induction B is

$$F = i\,\Delta l B \sin\theta, \tag{14–5}$$

where θ is the angle between the direction of the current and that of **B**. What we must do in order to determine the magnetic force on a moving charged particle is replace the i and Δl in Eq. (14–5), which refer to electric current, with the appropriate quantities for moving charges.

We shall consider a particle of charge q and velocity v, as in Fig. 14–10. This particle requires the time

$$t = \frac{\Delta l}{v}$$

in order to travel the distance Δl. During this time it behaves as though

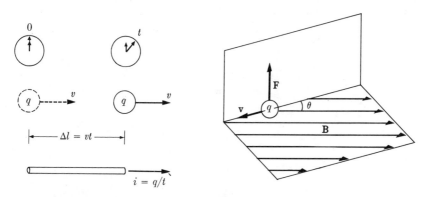

Fig. 14–10. The force on a charge q moving with the velocity v in a magnetic field is the same as that on a wire Δl long carrying the current i, where $i\,\Delta l = qv$.

it is a current of magnitude

$$i = \frac{q}{t}$$

in Δl. Hence we can substitute

$$i = \frac{q}{t} \quad \text{and} \quad \Delta l = vt$$

in Eq. (14–5), with the result that the force on a charged particle in a magnetic field is

$$F = i\,\Delta l\,B\,\sin\theta = \left(\frac{q}{t}\right)(vt)B\,\sin\theta = qvB\,\sin\theta, \qquad (14\text{–}6)$$

where θ is now the angle between the direction of the particle's motion and that of $\mathbf{B}$. The direction of $\mathbf{F}$ is perpendicular to both $\mathbf{v}$ and $\mathbf{B}$, as shown in Fig. 14–10. Because the force exerted by a constant magnetic field on a moving charged particle has no component parallel to the particle's velocity, no work is done on it by the field.

In the above derivation we have assumed that an individual moving charge behaves magnetically in the same way as a continuous current. This is a reasonable assumption, but it nevertheless must be verified experimentally. One method is to check Eq. (14–6) itself, which has been done with electrons, protons, and charged atoms and molecules. An easier alternative is a direct comparison of the magnetic fields of a current and of a series of moving charges. This was first accomplished by the American physicist Henry Rowland (1848–1901), whose experiment is shown schematically in Fig. 14–11. In both cases discrete moving charges and currents in conductors were found to have identical magnetic behavior, justifying our setting $i\,\Delta l$ equal to qv.

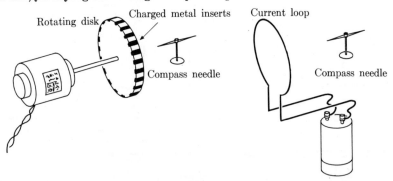

Fig. 14–11. Rowland's experiment. The magnetic effects of a rotating disk made of an insulating material whose periphery contains charged metal inserts are identical with those of a current loop.

14–6 The cyclotron

Suppose we have a positively charged particle traveling in a uniform magnetic field at right angles to **B**, as in Fig. 14–12. Since $\theta = 90°$ and $\sin 90° = 1$, the particle will experience the force

$$F = qvB. \tag{14–7}$$

This force is directed perpendicular to the direction of the particle's motion, and it therefore moves in the circular path shown. To find the radius R of the circle, we note that the magnetic force qvB provides the particle with the centripetal force mv^2/R that keeps it moving in a circle. Equating the magnetic and centripetal forces, then, we have

$$F_{\text{magnetic}} = F_{\text{centripetal}},$$

$$qvB = \frac{mv^2}{R},$$

$$R = \frac{mv}{qB}. \tag{14–8}$$

The radius of a charged particle's orbit in a uniform magnetic field is directly proportional to its momentum and inversely proportional to its charge and to the induction of the field. Because the force on a charged particle moving in a magnetic field is perpendicular to its direction of motion, the force does no work on it. Hence it keeps the same speed and energy it had when it entered the field, even though it has been deflected by the magnetic field.

The cyclotron, a device for accelerating charged particles, is based upon Eq. (14–8). The time T required for a charged particle in a magnetic

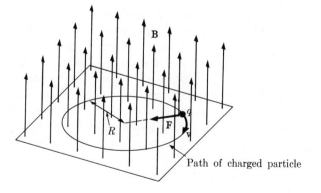

Path of charged particle

FIGURE 14–12

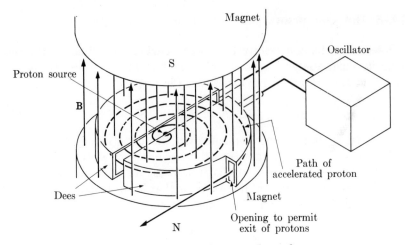

Fig. 14–13. Schematic diagram of a cyclotron.

field to make a complete revolution, since the distance it moves in that time is the circumference $2\pi R$ of its circular path, is

$$T = \frac{\text{distance}}{\text{speed}} = \frac{2\pi R}{v} = \frac{2\pi mv}{vqB} = \frac{2\pi m}{qB},$$

which does *not* depend upon the speed of the particle or the radius of its orbit! The particle goes around the circle in a period of time that varies only with its mass and charge and with the magnetic induction.

Figure 14–13 is a sketch of a cyclotron. The magnetic field is directed upward as shown. The *dees** are hollow copper boxes that are connected to a source of alternating potential (called an *oscillator*), so that their polarity changes sign regularly. When a charged particle, for instance a proton, is injected into the space between the dees from a suitable source, it is attracted by the dee that is then negative since its own charge is positive. Within the dee the magnetic field compels the proton to travel in a semicircle, as in Fig. 14–13. When it comes out at the other side, if the alternating current has the proper frequency, the opposite dee will be negative and the proton will be accelerated across the gap between the dees. Then it circulates within the second dee and again receives acceleration when it emerges, and so on. Ultimately the proton has sufficient energy to leave the cyclotron through the opening shown. The principle of the cyclotron, then, is to use a relatively small electric field to accelerate charged particles by causing them to be acted upon by the field repeatedly. If the period of the oscillator is exactly equal to

---

* So called because they are shaped like the letter "D."

the period of the protons in the magnetic field, they will always be attracted to the opposite dee when they reach the gap between the dees even though their speed (and the radius of their orbit) is greater each time they arrive there.

A typical cyclotron might have a magnetic induction of 1.5 weber/m². If it is accelerating protons, then, their period is

$$T = \frac{2\pi m}{qB}$$

$$= \frac{2\pi \times 1.7 \times 10^{-27}\ \mathrm{kg}}{1.6 \times 10^{-19}\ \mathrm{coul} \times 1.5\ \mathrm{weber/m^2}}$$

$$= 4.5 \times 10^{-8}\ \mathrm{sec}$$

since the charge of the proton is 1.6×10^{-19} coul and its mass is 1.7×10^{-27} kg. This means that the oscillator should have a frequency of

$$f = \frac{1}{T} = 2.2 \times 10^{7}\ \mathrm{sec}^{-1},$$

which is 22 million cycles/sec, a perfectly feasible operating frequency. If the potential difference between the dees is 100,000 volts, which is about as high as it can conveniently be made, each proton receives 10^5 ev of energy each time it passes through the gap. In the course of 100 passages, corresponding to 50 complete revolutions, the proton therefore receives a total of 10^7 ev (10 million electron volts, often abbreviated simply 10 Mev). To accelerate a proton directly to an energy of 10 Mev requires that a potential difference of 10^7 volts be established, which is impossible for practical reasons.

What is the speed of a 10-Mev proton? As we learned in Chapter 12,

$$1\ \mathrm{ev} = 1.6 \times 10^{-19}\ \mathrm{j},$$

and so such a proton has a kinetic energy of

$$\mathrm{KE} = 10^7\ \mathrm{ev} \times 1.6 \times 10^{-19}\ \mathrm{j/ev} = 1.6 \times 10^{-12}\ \mathrm{j}.$$

Hence

$$\mathrm{KE} = \tfrac{1}{2}mv^2,$$

$$v = \sqrt{\frac{2\ \mathrm{KE}}{m}} = \sqrt{\frac{2 \times 1.6 \times 10^{-12}\ \mathrm{j}}{1.7 \times 10^{-27}\ \mathrm{kg}}} = 4.4 \times 10^{7}\ \frac{\mathrm{m}}{\mathrm{sec}}.$$

At speeds comparable to that of light, 3×10^8 m/sec, the mass of a body increases with increasing speed (see Chapter 18). Under these circumstances the period of revolution of a particle in a cyclotron is no longer a constant, independent of its speed, and it will no longer be accelerated since it will not reach the gap between the dees only when the electric field there is favorable. The energy limit that this relativistic effect imposes upon cyclotrons is several tens of Mev. More sophisticated accelerators have been designed which are able to cope with the mass changes attendant upon high particle speeds, and machines are being constructed or are under design that will generate particles with energies of as much as 100,000 Mev (100 Bev).

IMPORTANT TERMS

The forces that act between electric currents are called *magnetic* forces.

Lines of force are helpful in visualizing a magnetic field. The directions of magnetic lines of force are those in which a test object (for instance, a compass needle) aligns itself at each point in the field, and their spacing is smaller the stronger the field.

The quantity describing a magnetic field that corresponds to electric field intensity **E** in the case of an electric field is the *magnetic induction* **B**. The unit of **B** is the weber/m². The force a magnetic field of induction **B** exerts on a wire Δl long that carries the current i is $F = i \, \Delta l \, B \sin \theta$, where θ is the angle between the direction of the current and that of **B**.

The *Biot-Savart law* is a formula that may be used to determine the magnetic induction around any electric current. According to the *right-hand rule*, when a current-carrying wire is grasped with the right hand so that the thumb points in the direction of the current, the curled fingers of that hand then point in the direction of the magnetic field.

When different substances are inserted in a current-carrying wire coil, the magnetic induction in its vicinity changes. Those substances that lead to a slight increase in B are called *paramagnetic*, those that lead to a great increase in B are called *ferromagnetic*, and those that lead to a slight decrease in B are called *diamagnetic*. A *permanent magnet* is an object composed of a ferromagnetic material whose atomic current loops have been aligned by an external current.

A *cyclotron* is a device for accelerating charged particles to high speeds by subjecting them repeatedly to the action of a relatively small electric field. A magnetic field is used to keep the particles in circular orbits in which their periods of revolution are all the same, making possible the above process.

PROBLEMS

1. Discuss the similarities and differences between Coulomb's law and the Biot-Savart law.

2. A current-carrying wire is in a magnetic field. (a) What angle should the wire make with **B** for the force on it to be zero? (b) What should the angle be for the force to be a maximum?

3. A wire 1 m long is perpendicular to a magnetic field of induction 5×10^{-2} weber/m^2. What is the force on the wire when it carries a current of 2 amp?

4. A certain size of copper wire has a mass of 5 kg per 100 m. A length of this wire is placed horizontally in a 1 weber/m^2 magnetic field whose lines of force are horizontal and perpendicular to the direction of the wire. What should the current in the wire be for it to be supported against gravity by the magnetic force on it?

5. What should the current be in a wire loop 1 m in diameter if the magnetic induction at the center of the loop is to be 1 weber/m^2?

6. A 50-turn circular coil has a radius of 10 cm. What is the magnetic induction at the center of the coil when the current in it is 0.7 amp?

7. According to a particular model of the hydrogen atom, its electron circles the nucleus 7×10^{15} times per second in an orbit 5×10^{-11} m in radius. What is the magnetic induction at the nucleus due to the electron's motion?

8. What should the current be in a long, straight wire if the magnetic induction 10 cm from it is to be 2×10^{-2} weber/m^2?

9. Use diagrams to show why paramagnetic substances are attracted by magnets while diamagnetic substances are repelled.

10. Why is a piece of iron attracted by *either* pole of a magnet?

11. An electron in a television picture tube travels at a speed of 3×10^7 m/sec and is acted upon both by gravity and by the earth's magnetic field. Which of these exerts the greater force on the electron?

12. A charge of $+10^{-6}$ coul is moving with a speed of 5×10^2 m/sec along a path parallel to a long, straight wire and 0.1 m from it. The wire carries a current of 2 amp in the same direction as that of the charge. What is the magnitude and direction of the force on the charge?

13. A long, straight wire carries a current of 100 amp. (a) What is the force on an electron traveling parallel to the wire, in the opposite direction to the current, at a speed of 10^7 m/sec when it is 10 cm from the wire? (b) Find the force on the electron under the above circumstances when it is traveling perpendicularly toward the wire.

14. (a) Prove, starting from Eqs. (9–1) and (9–4), that the force per unit length between two parallel wires d apart, each carrying the current i, is $\mu i^2/2\pi d$. (b) In what direction is the force when the currents are in the same direction? (c) In what direction is the force when the currents are in opposite directions?

15. A particle of charge q is moving with a speed v perpendicular to an electric field of intensity E. What should be the magnitude and direction of a magnetic field in the same region that will exactly cancel out the effects of the electric field on the moving charge?

16. An electron moving through an electric field of intensity 500 volts/m and a magnetic field of induction 0.1 weber/m² experiences no force. The two fields and the electron's direction of motion are all mutually perpendicular. What is the speed of the electron?

17. What is the radius of the path of a 4×10^4 ev electron in a magnetic field of 0.02 weber/m²?

18. Compute the radius of curvature in the earth's magnetic field at sea level, assuming $B = 3 \times 10^{-5}$ weber/m², of (a) a proton whose speed is 2×10^7 m/sec, and (b) an electron of the same speed.

19. An *alpha particle* (which is actually the nucleus of a helium atom) has a mass approximately four times that of a proton and a charge twice that of a proton. If alpha particles are being accelerated in the cyclotron described in the text, (a) what should the frequency of the oscillator be? (b) How much energy will the alpha particle have after 50 revolutions? (c) What will its speed be after 50 revolutions?

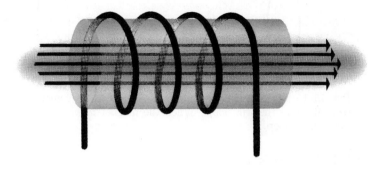

Electromagnetic Induction

15

We have seen that, by causing a current to flow, an electric field is able to produce a magnetic field. Is there any way in which a magnetic field can produce an electric field? This problem was unsuccessfully tackled by many of the early workers in electricity and magnetism. Finally, in 1831, Michael Faraday in England and Joseph Henry in the United States independently discovered the phenomenon of electromagnetic induction, which we shall consider in this chapter together with several related topics.

265

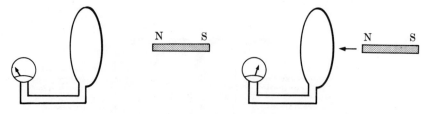

Fig. 15–1. Electromagnetic induction. When the magnet is stationary, no current flows in the wire loop. When the magnet moves, however, a current is induced in the loop.

15–1 Faraday's law

When a wire loop surrounds a stationary magnet, no current flows in the loop. What Faraday and Henry discovered was, in essence, that *moving* the magnet causes a current to flow in the loop (Fig. 15–1). It does not matter whether the magnet or the loop is actually being moved; the origin of the current lies in the *relative motion* between a conductor and a magnetic field. This phenomenon is known as *electromagnetic induction*.

Faraday was soon able to generalize the above observation by employing the notion of lines of force:

An electric field is produced in a conductor whenever magnetic lines of force move across it.

It is this electric field that leads to the current observed whenever there is relative motion between a conductor and a magnetic field. In fact, it is not even necessary for there to be actual motion of either a wire or a source of magnetic field, because a magnetic field that changes in strength has associated with it moving lines of force. Figure 15–2 shows how a coil of wire induces a current in another coil when the current in it is turned on or off. At (a) the switch connecting loop (1) with the battery is open, and no current flows in either loop. At (b) the switch has just been closed, and the expanding lines of force resulting from the increasing current in loop (1) cut across loop (2), inducing a current in it. The current in loop (2) flows in the *opposite* direction to that in loop (1) for a reason we shall discuss shortly. At (c) a constant current flows in loop (1), and since the flux through it is constant, no current is induced in loop (2). At (d) the switch has just been opened, and the contracting lines of force resulting from the decreasing current in loop (1) cut across loop (2), inducing a current in it. The current in loop (2) flows now in the *same* direction as that in loop (1). At (e) the current in loop (1) has disappeared, and no current flows in either loop.

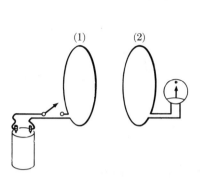

(1) (2)

(a)

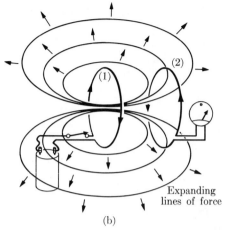

(2)

(1)

(2)

Expanding
lines of force

(b)

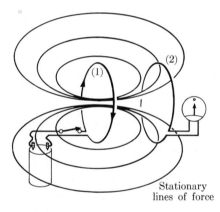

(2)

(1)

(2)

Stationary
lines of force

(c)

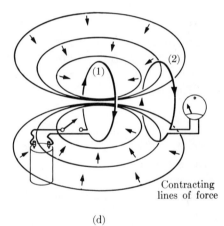

(2)

(1)

(2)

Contracting
lines of force

(d)

(1) (2)

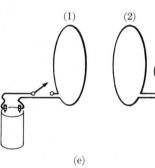

(e)

FIGURE 15–2

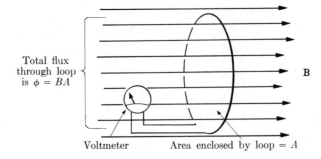

Total flux through loop is $\phi = BA$

B

Voltmeter   Area enclosed by loop $= A$

FIG. 15–3. Electromagnetic induction.

Electromagnetic induction obeys a very simple quantitative relationship. If the area of a wire loop is A and it is perpendicular to a magnetic field of induction B, as in Fig. 15–3, the *total magnetic flux* ϕ through the loop is defined as

$$\phi = BA. \tag{15-1}$$

Since the unit of B is the weber/m² and that of A is m², the unit of flux is the weber. Faraday found that the potential difference V between the ends of such a wire loop is *equal to the rate of change of the flux through it*. That is,

$$V = -\frac{\Delta\phi}{\Delta t}, \tag{15-2}$$

where $\Delta\phi$ is the change in the flux ϕ that takes place during a period of time Δt. The values of the magnetic induction B and the loop area A are, in themselves, irrelevant; only the speed with which either or both of them changes is important. Equation (15–2) is called Faraday's law of electromagnetic induction.

When the magnetic flux through a closed wire loop changes, the current that flows is said to result from an induced *electromotive force* in the loop. This electromotive force (abbreviated emf) is equal to the induced potential difference V_i of Eq. (15–2) that would be found between the ends of an identical *open* wire loop. The notion of emf is useful because it is meaningless to speak of the potential difference around a closed circuit; the emf in a closed circuit is the potential difference that would be found between the ends of the circuit if it were cut anywhere. Often the potential difference across the terminals of an unconnected battery is called the emf of the battery, since it is this potential difference that produces a current when the battery is connected in a circuit. The actual potential difference measured across the terminals of a battery in a circuit is slightly less than its emf because of the internal resistance of the battery itself to the flow of current through it; the difference is seldom important.

Electromagnetic induction is of immense practical importance since it is the means whereby nearly all the world's electric power is generated. In a *dynamo* a coil of wire is rotated in a magnetic field so that the flux through the coil changes constantly. The resulting potential difference across the ends of the coil causes a current to flow in an external circuit, and this current can be transmitted by a suitable system of wires for long distances from its origin.

15–2 Lenz's law

The minus sign in Eq. (15–2) is a consequence of the law of conservation of energy. If the sign of V were the same as that of $\Delta\phi/\Delta t$, the induced electric current would be in such a direction that its own magnetic field would *add* to that of the external field B; this additional changing field would then augment the existing rate of change of the flux ϕ, and more and more current would flow even if the external contribution to ϕ were to stay constant! Since energy must be supplied at all times to maintain a current, owing to the continual energy loss in the form of heat in every conductor in which there is a current, removing the source of energy (here the changing flux of external origin) must cause the current to cease. Therefore *the direction of the induced current must be such that its own magnetic field opposes the changes in flux that are inducing it,* a conclusion known as Lenz's law.

Lenz's law is illustrated in Fig. 15–2, where, at B, the induced current in loop (2) flows opposite to that in loop (1), since the increasing current in (1) is what is ultimately responsible for the current in (2). Similarly, at D, the induced current in loop (2) flows now in the same direction as that in loop (1), since it is the decreasing current in (1) that leads to the current in (2).

15–3 The betatron

An interesting application of electromagnetic induction is the betatron. In a betatron (Fig. 15–4) electrons are accelerated to high speeds through the action of an increasing magnetic field. As the magnetic field B increases, the associated electric field contributes to the energy of an electron moving in a circular path of area A by an amount per revolution equal to its increase in energy when it moves through a potential difference of

$$V = -\frac{\Delta\phi}{\Delta t} = -A\,\frac{\Delta B}{\Delta t}.$$

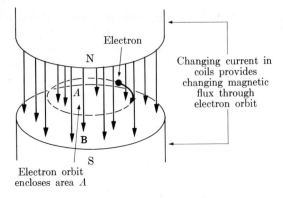

FIG. 15–4. Schematic diagram of a betatron.

That is, we can regard a circular path of this kind exactly as though it is a loop of wire insofar as electromagnetic induction is concerned. As the electron goes faster and faster, gaining the energy qV in each revolution, it will require a greater magnetic induction **B** if it is to stay in the same orbit; it is not difficult to arrange matters so that the very increase in **B** that accelerates the electron in the first place exactly keeps pace with its increasing speed, thereby maintaining a constant orbit radius.

A typical betatron that accelerates electrons to 100 Mev might have an orbit radius of 1 meter and a magnetic field changing at the rate of 100 webers/m²·sec. Hence the orbit area is

$$A = \pi R^2 = 3.1 \text{ m}^2,$$

and the potential difference per revolution is

$$V = -A\,\frac{\Delta B}{\Delta t} = -3.1 \text{ m}^2 \times 100\,\frac{\text{webers}}{\text{m}^2\cdot\text{sec}} = 310 \text{ volts.}$$

An electron in this betatron acquires 310 ev each time it makes a complete circle; it must make over 323,000 revolutions before it has an energy of 100 Mev! The relativistic limitation on particle speed that affects the cyclotron does not affect the betatron, since the electrons in the latter are accelerated continuously and do not have to make each turn in precisely the same period of time.

15–4 Motion of a wire in a magnetic field

When a wire moves through a magnetic field **B** with a velocity **v** which is not parallel to **B**, a potential difference V_i is induced between the ends of the wire. There are two quite different ways of calculating V_i which

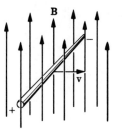

F<small>IG</small>. 15–5. When a wire is moved through a magnetic field, the potential difference induced between its ends causes equal and opposite electric charges to appear at them.

lead to the same result, and it is worth while examining both in order to verify the essential unity of electromagnetic phenomena. Let us assume, for convenience, that the wire, its direction of motion, and the flux density **B** are all perpendicular to one another, as in Fig. 15–5.

We shall first analyze the problem on a microscopic basis in terms of the forces that act on the electrons able to move freely within the wire. According to Eq. (14–7) the force on a charge $+q$ when it moves with the velocity **v** perpendicular to a magnetic field **B** has the magnitude

$$F = qvB .$$

The direction of this force is along the wire, as in Fig. 15–6. As the wire moves through the magnetic field, electrons $(q = -e)$ move opposite to the direction shown, so that one end of the wire becomes negatively charged and the other end positively charged. These charges build up until the electric forces they exert on the remaining electrons in the wire are just enough to balance the forces qvB on each of them due to the motion through the magnetic field. When this balance is reached there is no net force on the electrons, and no further charge accumulates at the ends of the wire. If E is the electric field intensity in the wire, the condition for balance is

$$Eq = qvB, \qquad E = vB.$$

An electric field E acting along the length L of the wire leads to a potential difference of

$$V_i = EL$$

between its ends, and so, substituting vB for E, we find that

$$V_i = BLv . \qquad (15\text{–}3)$$

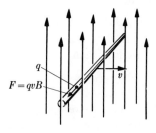

F<small>IGURE</small> 15–6

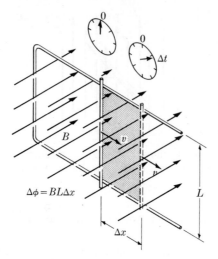

FIGURE 15-7

The alternative approach is a macroscopic one based upon Faraday's law of induction. Let us suppose that the moving wire slides across the legs of a U-shaped metal frame as in Fig. 15–7, so that the frame completes the circuit. At the instant in which the moving wire is the distance x from the closed end of the frame, the flux enclosed is

$$\phi = BA \; = \; BLx.$$

As the wire moves, the enclosed flux changes at the rate

$$\frac{\Delta\phi}{\Delta t} = BL \frac{\Delta x}{\Delta t}.$$

Since the speed of the wire here is

$$v = - \frac{\Delta x}{\Delta t},$$

and the induced emf is

$$V_i = - \frac{\Delta\phi}{\Delta t},$$

we find that

$$V_i = BLv,$$

just as before.

As an example, let us compute the potential difference between the wing tips of a jet airplane induced by its motion through the earth's magnetic field. If the total wing span of the airplane is 40 meters and its speed is 300 m/sec in a region where the vertical component of the flux density of the earth's field is 3×10^{-5} w/m², we have

$$V_i = BLv = 3 \times 10^{-5} \frac{w}{m^2} \times 40 \text{ m} \times 300 \frac{m}{\text{sec}} = 0.36 \text{ volt.}$$

In the general case, when **B** is not perpendicular to **v**, V_i is equal to $BLv \sin\theta$, where $B \sin\theta$ is the component of **B** perpendicular to **v**.

15–5 Inductance

An electromotive force is induced in a wire loop or other conducting circuit whenever the magnetic flux enclosed by it undergoes a change. The changes we have thus far considered originated, like the flux itself, *outside* the circuit, either in a permanent magnet or in an electric current. We now come to the interesting situation of an isolated current-carrying circuit whose own current changes, thereby leading to a change in the magnetic flux it encloses. A change in flux is a change in flux, however it comes about, and the result is a *self-induced emf* that, by Lenz's law, is such as to tend to oppose the change that causes it. The response of a circuit to a drop in current is a self-induced emf in the same direction as the current, tending to reestablish it, while a rise in current results in a self-induced emf opposite to the current, tending to keep it at its former value.

The magnitude of the self-induced emf depends upon the rate of change $\Delta\phi/\Delta t$ of the flux through the circuit. Since we assume that the geometry of the circuit is fixed, ϕ is proportional to the current i in the circuit, and $\Delta\phi/\Delta t$ is proportional to $\Delta i/\Delta t$, the rate of change of i. We can express this proportionality as

$$V_i = -L\frac{\Delta i}{\Delta t}, \tag{15-4}$$

where L, the constant of proportionality, is called the *inductance* of the circuit. As in Eq. (15–2) the minus sign is an expression of Lenz's law. The unit of inductance is the *henry*, abbreviated h, where

$$1 \text{ h} = 1 \frac{\text{volt-sec}}{\text{amp}}.$$

The inductance of a circuit depends upon its geometry and upon the proximity of magnetic materials (Fig. 15–8). A coil consisting of many turns has a much greater inductance than a single loop both because the emf's induced in the turns add together and because the more turns there are, the more flux there is to undergo change when the current through them changes. A core in a coil changes its inductance by changing the

High inductance   Low inductance

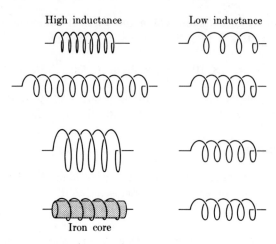

Iron core

FIGURE 15–8

flux through the coil (Fig. 14–8). Thus a diamagnetic core decreases the inductance slightly, a paramagnetic core increases it slightly, and a ferromagnetic core increases it considerably—as much as a factor of 10,000 in certain cases.

The inductance of a coil whose length l is greater than its diameter is

$$L = \mu_0 n^2 l A , \qquad (15\text{–}5)$$

where n is the number of turns per unit length and A is the cross-sectional area of the coil. A typical coil might be 0.1 meter long, might have a cross-sectional area of 2×10^{-3} m², and might be wound with 10^3 turns/m. Its inductance is

$$L = \mu_0 n^2 l A$$

$$= 12.6 \times 10^{-7} \, \frac{\text{w}}{\text{amp-m}} \times (10^3/\text{m})^2 \times 0.1 \, \text{m} \times 2 \times 10^{-3} \, \text{m}^2$$

$$= 2.52 \times 10^{-4} \, \text{h} .$$

Evidently the henry, like the farad, may be too large a unit for convenience. Inductances are accordingly often expressed in *millihenries* (mh) or *microhenries* (μh), where

$$1 \, \text{mh} = 10^{-3} \, \text{h},$$

$$1 \, \mu\text{h} = 10^{-6} \, \text{h} .$$

The inductance of the above coil would be referred to as 0.252 mh.

15–6 Growth and decay of a current

When a coil is connected to a battery, the current in the circuit does not rise instantly to its ultimate value $i = V/R$, where V is the emf of the battery and R is the resistance in the circuit. When the switch in Fig. 15–9 is closed, the current starts to grow, and, as a result, the induced emf $V_i = -L(\Delta i/\Delta t)$ comes into being in the opposite direction to the battery emf V. The net emf acting to establish current in the circuit is therefore $V - L(\Delta i/\Delta t)$, and the current reaches its ultimate value of V/R in a gradual manner.

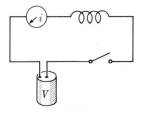

FIGURE 15–9

The graph in Fig. 15–10 shows how i varies with time when a current is being established in a circuit containing inductance. A mathematical analysis shows that, after a time interval of L/R, the current reaches 63% of its final value V/R. The time L/R is therefore a convenient measure of how rapidly a current rises in a circuit containing inductance, and, like RC in a circuit containing capacitance, is called the *time constant* of the circuit. To verify that L/R has the dimensions of time, we note that the dimensions of L and R correspond to

$$L = \frac{Vt}{i}, \qquad R = \frac{V}{i}, \qquad \text{and so} \qquad \frac{L}{R} = \frac{Vt/i}{V/i} = t.$$

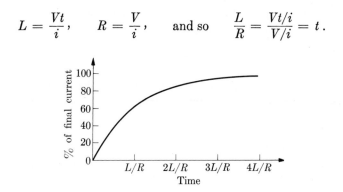

FIG. 15–10. The growth of current in a circuit containing inductance and resistance.

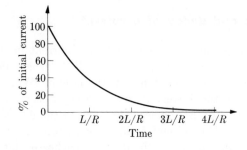

Fɪɢ. 15–11. The decay of current in a circuit containing inductance and resistance when the battery is short-circuited.

The right-hand part of Fig. 15–11 shows the gradual drop in i when the battery is short-circuited by a wire; here the induced emf tends to maintain the existing current. The behavior illustrated in Figs. 15–10 and 15–11 represents no more than the operation of Lenz's law.

15–7 Magnetic potential energy

A magnetic field, like an electric field, contains potential energy. Let us determine the amount of energy W stored in a coil of inductance L when a steady current i flows through it. This energy is equal to the work that had to be done against the self-induced emf V_i in order to establish the current starting from $i = 0$. If we suppose that the current rises at a uniform rate from $i = 0$ to $i = i$ in a time interval Δt, the average current $\bar{\imath}$ during Δt is

$$\bar{\imath} = \frac{i_{\text{final}} + i_{\text{initial}}}{2} = \frac{i + 0}{2} = \frac{1}{2}i.$$

Thus the total charge Q that passed through the coil while the current was building up to its final value i is

$$Q = \bar{\imath}\,\Delta t = \tfrac{1}{2}i\,\Delta t.$$

The self-induced emf V_i against which the charge Q had to be sent has the magnitude

$$V_i = L\,\frac{i}{\Delta t}$$

during the establishment of the current. The total work done against V_i is therefore

$$W = QV_i = \frac{1}{2}i\,\Delta t \times L\,\frac{i}{\Delta t} = \frac{1}{2}Li^2. \tag{15–6}$$

A coil of inductance L has an energy of $\frac{1}{2}Li^2$ stored in it when a current i is present in it. When the potential difference across the coil that is responsible for the current is removed, the energy $\frac{1}{2}Li^2$ is what powers the self-induced emf that retards the drop in current. The gradual rise of current in a circuit containing inductance may be thought of as the result of the initial absorption of $\frac{1}{2}Li^2$ of potential energy by the circuit, and its gradual drop may be thought of as the result of the restoration of the potential energy to the circuit.

A coil whose inductance is 12 mh and whose resistance is 5 ohms has a potential difference of 20 volts applied across it. The time constant of this circuit is

$$\frac{L}{R} = \frac{12 \times 10^{-3}\,\text{h}}{5\ \text{ohms}} = 2.4 \times 10^{-3}\ \text{sec}.$$

The final current will be

$$i = \frac{V}{R} = \frac{20\ \text{volts}}{5\ \text{ohms}} = 4\ \text{amp},$$

and so the energy stored in the coil will be

$$W = \tfrac{1}{2}Li^2 = \tfrac{1}{2} \times 12 \times 10^{-3}\,\text{h} \times (4\ \text{amp})^2 = 0.096\ \text{j}$$

when this current flows through it.

As we said earlier, the potential energy stored in a current-carrying coil resides in the magnetic field of the coil. Let us compute the energy density in a magnetic field in a manner similar to that which we employed in computing the energy density in an electric field. The inductance of a coil of length l and cross-sectional area A which has n turns/m is

$$L = \mu_0 n^2 l A,$$

and its potential energy when it carries the current i is

$$W = \tfrac{1}{2}Li^2 = \tfrac{1}{2}\mu_0 n^2 l A i^2. \tag{15–7}$$

The magnetic field within a coil whose length is large relative to its diameter has a flux density of

$$B = \mu_0 n i,$$

according to Section 14–3. Hence we can express W in terms of B within the coil instead of in terms of i by substituting

$$i = \frac{B}{\mu_0 n}$$

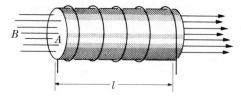

FIGURE 15-12

in Eq. (15-7). This yields

$$W = \frac{\frac{1}{2}B^2}{\mu_0} \, lA \, .$$ (15-8)

We note that lA, the product of the length of the coil and its cross-sectional area, is the volume it encloses (Fig. 15-12). If we define the energy density w of a magnetic field as the magnetic potential energy per unit volume associated with it, we have here

$$w = \frac{\text{total energy of magnetic field}}{\text{volume occupied by magnetic field}}$$

$$= \frac{W}{lA} = \frac{\frac{1}{2}B^2}{\mu_0} \, .$$ (15-9)

The energy density of a magnetic field is proportional to the square of its flux density B. Equation (15-9) is a perfectly general result even though we have used the specific case of a current-carrying coil to derive it. It is interesting to see the resemblance between Eq. (15-9) and the corresponding formula

$$W = \frac{1}{2}\epsilon_0 E^2$$

for the energy density in an electric field.

IMPORTANT TERMS

Electromagnetic induction refers to the production of an electric field in a conductor whenever magnetic lines of force move across it. The potential difference induced between the ends of a wire loop of area A that is perpendicular to a magnetic field of induction B is

$$V = -\frac{\Delta\phi}{\Delta t}, \qquad V = -h\frac{\Delta\phi}{\Delta t}$$

where $\Delta\phi$ is the change in the *magnetic flux* $\phi = BA$ that occurs in the time interval Δt. *Lenz's law* states that the direction of the induced current must be such that its own magnetic field opposes the changes in flux that are inducing it.

A *betatron* is a device for accelerating electrons to high speeds through the action of an increasing magnetic field. The field simultaneously keeps pace with the electron speed to maintain a constant orbit radius.

The *inductance L* of a circuit is the ratio between the negative of the self-induced emf V_i due to a changing current in it and the rate of change $\Delta i/\Delta t$ of the current. The unit of inductance is the *henry*, which is equal to 1 volt·sec/amp.

The *time constant* of a circuit containing inductance and capacitance is a measure of how rapidly a current can be established or can disappear in it.

The *energy density* of a magnetic field is the magnetic energy per unit volume associated with it. The energy density of a magnetic field in free space of flux density **B** is $\frac{1}{2}B^2/\mu_0$.

PROBLEMS

1. A rectangular wire loop 5 cm × 10 cm in size is perpendicular to a magnetic field of induction 10^{-3} weber/m². (a) What is the flux through the loop? (b) If the magnetic field drops to zero in 3 sec, what is the potential difference induced between the ends of the loop during that period?

2. One end of a bar magnet is thrust into a coil, and the induced current is clockwise as seen from the front of the coil. (a) Was the end of the magnet its north or south pole? (b) What will be the direction of the induced current when the magnet is withdrawn?

3. A square wire loop 10 cm on a side is situated so that its plane is perpendicular to a magnetic field. The resistance of the loop is 5 ohms. How rapidly should the magnetic induction change if a current of 2 amps is to flow in the loop?

4. A wire loop is 5 cm in diameter and is situated so that its plane is perpendicular to a magnetic field. How rapidly should the magnetic induction change if a potential difference of 1 volt is to appear across the ends of the loop?

5. Explain why, when a wire loop is rotated in a magnetic field about a diameter perpendicular to the direction of the field, the current induced in the loop reverses itself twice in each full rotation.

6. Why is it easy to turn the shaft of a dynamo when it is not connected to an outside circuit, but much harder when such a connection is made?

7. *Magnetic storms* occur when swarms of charged particles from the sun enter the earth's magnetic field. Explain why such storms always produce a decrease in the magnetic field at sea level.

8. In what ways does a betatron resemble a cyclotron? In what ways does it differ from a cyclotron?

9. In a particular betatron the direction in which the electrons rotate is clockwise as seen from above. What is the direction of the magnetic field?

10. What is the induced emf in a 15-h coil when the current through it is changing at the rate of 4 amp/sec?

11. The magnetic flux through a 20-turn coil drops from 0.3 weber to 0 in 1 sec. What is the induced potential difference across the ends of the coil?

12. A 2-h coil carries a current of 0.5 amp. (a) How much energy is stored in it? (b) In how much time should the current drop to 0 if an emf of 100 volts is to be induced in it?

13. An automobile has a speed of 30 m/sec on a road where the vertical component of the earth's magnetic field is 8×10^{-5} weber/m². What is the potential difference between the ends of its axles, which are 2 m long?

14. Why does an electric motor require more current when it is turned on than when it is running continuously?

15. What is the inductance of a coil whose resistance is 14 ohms and whose time constant is 0.1 sec?

16. A 50-mh coil with a resistance of 20 ohms is connected to a 90-volt battery. (a) What is the time constant of the circuit? (b) How much energy is stored in the magnetic field of the coil when the current has reached its final value?

17. What is the inductance of a coil 40 cm long and 4 cm in diameter that has 1000 turns of wire?

18. A coil 20 cm long and 3 cm in diameter is tightly wound with copper wire 1 mm in diameter. What is its time constant?

19. The strongest magnetic fields that have been produced in the laboratory have had flux densities of about 10^2 webers/m². (a) How much energy is contained in a 10^3 cm³ volume of such a field? (b) What intensity would an electric field require to have the same energy density?

20. What is the energy density of the earth's magnetic field at a location where its flux density is 6×10^{-5} weber/m²?

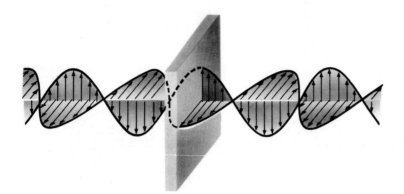

Electromagnetic Waves | 16

Among the most noteworthy achievements of 19th-century science was the realization that light consists of electromagnetic waves. Electromagnetic waves themselves were predicted by James Clerk Maxwell in 1864 on the basis of his theory of electric and magnetic fields and their interrelation. Maxwell noted that the speed of such waves in vacuum, as obtained from his calculations, was, within the limits of experimental uncertainty, identical with the speed of light in vacuum, and he surmised that, since both were also transverse waves, they were the same phenomenon. This surmise has since been verified in every detail, and Maxwell's electromagnetic theory of light today occupies a niche in the history of physics alongside Newton's theory of universal gravitation and Einstein's theory of relativity. In this chapter we shall examine some characteristic properties of electromagnetic waves and how they are manifested in the behavior of light.

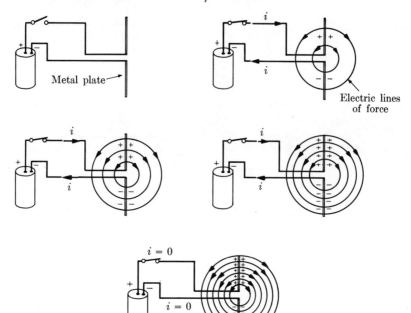

FIGURE 16–1

16–1 Maxwell's hypothesis

As we learned in the previous chapter, a changing magnetic field produces an electric field. This effect is known as electromagnetic induction, and the consequent electric field is present regardless of whether the magnetic field variations occur in free space or in a material medium. There is also a converse process, in which a changing electric field produces a magnetic field—even in empty space, where electric currents cannot flow. No simple experiment can directly demonstrate the latter effect, and it was first proposed by Maxwell on the basis of somewhat indirect reasoning.

When Maxwell expressed mathematically what was known in his time about electricity and magnetism in terms of Faraday's intuitive ideas of fields of force, he found that the resulting equations became particularly simple and symmetrical if he assumed that a magnetic field arises from a varying electric field as well as from an actual electric current. This assumption is actually less arbitrary than it may seem. Suppose that we attach two metal plates to the opposite terminals of a battery, as in Fig. 16–1. At first currents flow in the wires, and electric charge of opposite signs accumulates on the plates. Ultimately enough charge is present for the potential difference between the plates to equal that between the terminals of the battery, and the currents cease. While charge builds up on the plates, the electric field around them changes accordingly: the larger the current, the greater the rate of change of the field, and the smaller the current, the less the rate of change of the field. The behavior of the current is thus mirrored in the *rate of change* of the electric field surrounding the termination points of the current. By asserting that a varying electric field produces a magnetic field, then, Maxwell was no more than extending the correspondence between electric currents and varying electric fields to include magnetic effects. That such an extension also simplified the equations of electricity and magnetism encouraged Maxwell to take this notion seriously, because fundamental physical principles usually (though not always) may be expressed in simple form.

It is not surprising that electromagnetic induction became known many years before its converse was suspected. Even a slight electric field causes a current to flow in a conductor, and, if the resistance of the conductor is low enough, the current may be sufficiently large to detect despite the feebleness of the field itself. However, electric current has no magnetic counterpart because single magnetic poles do not exist; the opposite poles of a magnet cannot be separated from one another the way opposite electric charges can. Weak magnetic fields are therefore hard to measure no matter where they occur, unlike weak electric fields, and those due to changing electric fields are seldom strong. Maxwell's notion that an electric field that varies with time gives rise to a magnetic field accordingly did not originate in an observation, but instead developed from an intuitive feeling for order in the natural world.

One of the tenets of the scientific method of inquiry is that all hypotheses must be capable of being experimentally verified, directly or indirectly, if they are to mean anything at all. Maxwell's next step was to seek a phenomenon that was a unique consequence of his hypothesis.

16–2 Electromagnetic waves

It follows from electromagnetic induction that whenever there is a change in a magnetic field, an electric field is produced, and it follows from Maxwell's hypothesis that whenever there is a change in an electric field, a magnetic field is produced. Evidently it is impossible to have either effect occur alone. The electric field that arises from a change in a magnetic field is in itself a change in the pre-existing electric field (which might have had any original value, including zero), and therefore

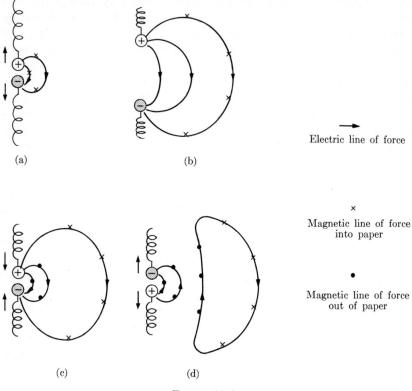

(a)

(b)

→

Electric line of force

×

Magnetic line of force
into paper

•

Magnetic line of force
out of paper

(c)

(d)

FIGURE 16–2

causes another magnetic field. The latter magnetic field, too, represents a change, and from the change an electric field is in turn produced. The process continues indefinitely, with a definite coupling between the fluctuating electric and magnetic fields. On the basis of his hypothesis, together with the other principles of electricity and magnetism, Maxwell was able to develop a detailed picture of how these field fluctuations travel through space.

The first idea that emerged from Maxwell's analysis was that the field fluctuations spread out in space from an initial disturbance in the same manner that waves spread out from a disturbance in a body of water; hence the name *electromagnetic waves* to describe them. If we throw a stone into a pond or otherwise alter the state of the water surface at some point, oscillations occur in which energy is continually interchanged between the kinetic energy of moving water and the potential energy of water higher than its normal level. These oscillations begin where the stone lands, and spread out as waves across the surface of the pond. The wave speed depends upon the properties of the pond water, varying with temperature, impurity content, and so on, but it is independent of the wave amplitude. As we saw in Chapter 8, this is typical wave behavior.

When electromagnetic waves spread out from an electric or magnetic disturbance, their energy is constantly being interchanged between the fluctuating electric field and the fluctuating magnetic field of the waves. Let us suppose that we have a periodic source of electromagnetic waves, say a pair of electric charges connected by a vibrating spring as in Fig. 16–2. In part (a) of the figure the charges are moving apart; the electric lines of force surrounding the charges are as shown, and the magnetic lines of force produced by the motion of the charges, which are concentric circles perpendicular to the paper, are indicated by crosses when their direction is into the paper and by circles when their direction is out of the paper. In (b) the charges have reached the limit of their motion and have stopped, so that they cease to produce a magnetic field. The outer magnetic lines of force do not disappear instantly because of the finite speed at which changes in electric and magnetic fields travel. If the charges remain at rest permanently, the entire magnetic field would in time vanish. However, as in (c), the charges now move toward each other, generating a magnetic field whose lines of force have the opposite direction to those created in (a). The electric lines of force have the same direction they had initially. In (d) the charges have passed each other and now are moving apart; the electric lines of force near them accordingly reverse their direction, but the magnetic lines of force are in the same direction as those produced in (c).

Owing to this sequence of changes in the fields, the outermost electric and magnetic lines of force respectively form into closed loops. These loops of ˚force, which lie in perpendicular planes, are divorced from the oscillating charges that gave rise to them and continue moving outward, constituting an electromagnetic wave. As the charges continue oscillating back and forth, further associated loops of electric and magnetic lines of force are emitted, forming an expanding pattern of loops.

Figure 16–3 shows the configuration of the electric and magnetic fields that spread outward from a pair of oscillating charges. The actual fields are present in three dimensions, so that the magnetic lines of force form loops in planes perpendicular to the line joining the charges. There are three significant things we should note about these electromagnetic waves, which are as follows:

(1) The variations occur simultaneously in both fields (except close to the oscillating charges), so that the electric and magnetic fields have maxima and minima at the same times and in the same places.

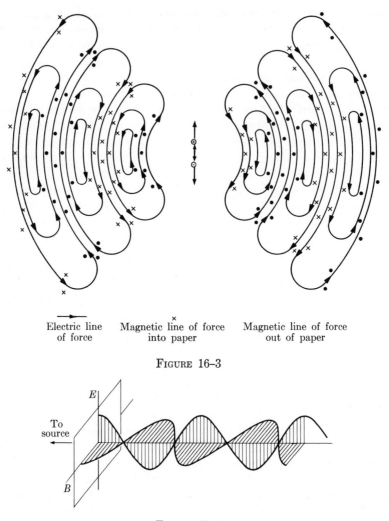

	×	
→	Magnetic line of force	Magnetic line of force
Electric line	into paper	out of paper
of force		

FIGURE 16–3

FIGURE 16–4

(2) The directions of the electric and magnetic fields are perpendicular to each other and to the direction in which the waves are moving. Light waves are therefore transverse.

(3) The speed of the waves depends only upon the electric and magnetic properties of the medium they travel in, and not upon the amplitudes of the field variations.

Figure 16–4 is an attempt at portraying (1) and (2) of the above in terms of superimposed graphs showing the electric and magnetic field intensities and directions in an electromagnetic wave at a particular instant. The actual configuration of the fields in space is shown in Fig. 16–3. It is worth keeping in mind that, unlike the other types of wave we have considered—water waves, sound waves, waves in a stretched string—nothing material moves in the path of an electromagnetic wave. The only changes are in field intensities.

Maxwell's theory of electromagnetic waves showed that their speed c in vacuum depends solely upon ϵ_0 and μ_0, the permittivity and permeability of free space. Maxwell found that the speed c is given by

$$c = \frac{1}{\sqrt{\epsilon_0 \mu_0}} \tag{16–1}$$

$$= \frac{1}{\sqrt{8.85 \times 10^{-12} \text{ coul}^2/\text{n·m}^2 \times 1.26 \times 10^{-6} \text{ w/amp·m}}}$$

$$= 3.00 \times 10^8 \text{ m/sec},$$

the same speed that had been experimentally measured for light waves in free space! The correspondence was too great to be accidental, and, as further evidence became known, the electromagnetic nature of light found universal acceptance.

While the correct value of c is

$$c = 2.998 \times 10^8 \text{ m/sec}$$

to four significant figures, we shall use the value

$$c = 3 \times 10^8 \text{ m/sec}$$

in what follows.

16–3 Varieties of electromagnetic waves

Light is not the only example of an electromagnetic wave. While all electromagnetic waves share certain basic properties, other features of their behavior depend upon their frequencies. Light waves themselves span a brief frequency interval, from 3×10^{14} cycles/sec for red light to 8×10^{14} cycles/sec for blue light. Electromagnetic waves with frequencies between these limits are the only ones that the eye responds to, and specialized instruments of various kinds are required to detect waves with higher and lower frequencies. Figure 16–5 shows the electromagnetic wave *spectrum* from the low frequencies used in radio communication to the high frequencies found in x-rays and gamma rays (about which we shall have more to say in later chapters). The wavelengths corresponding to the various frequencies are also shown.

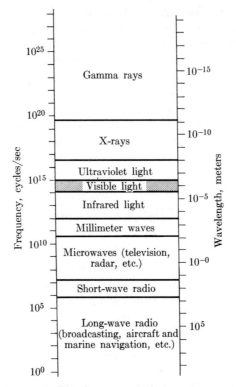

Fɪɢ. 16–5. The electromagnetic wave spectrum.

As we recall, the product of the frequency f of a wave and its wavelength λ is just the wave speed, here c. Hence, given the wavelength or frequency of a particular electromagnetic wave, we can immediately find the complementary quantity. The wavelength of red light, for instance, is

$$\lambda = \frac{c}{f},$$

$$\lambda = \frac{3 \times 10^8 \text{ m/sec}}{4 \times 10^{14} \text{ cycles/sec}}$$

$$\lambda = 7.5 \times 10^{-7} \text{ m},$$

which is less than 1/10,000 of a millimeter. By contrast, a 1 megacycle/sec radio wave has a wavelength of 300 meters. (1 kilocycle = 10^3 cycles; 1 megacycle = 10^3 kilocycles = 10^6 cycles.)*

---

* The abbreviations for megacycle and kilocycle are mc and kc.

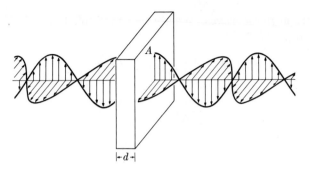

FIGURE 16–6

16–4 The Poynting vector

Associated with every electric field of intensity E is an energy density $\frac{1}{2}\epsilon_0 E^2$ and with every magnetic field of flux density B is an energy density $\frac{1}{2}B^2/\mu_0$. Electromagnetic waves therefore contain energy, and their motion involves the flow of energy. The *Poynting vector* **S** describes this flow of energy. The direction of **S** is the direction of the electromagnetic wave in question, and its magnitude S is equal to the rate at which energy is being transported by the wave per unit cross-sectional area. We can derive a formula for S in terms of the simple situation of Fig. 16–6, which shows an electromagnetic wave passing through an imaginary box of area A and thickness d. If E and B are the values of the electric field intensity and magnetic flux density of the wave in the box at any instant, the total energy density w in the box at that instant is

$$w = \frac{1}{2}\,\epsilon_0 E^2 + \frac{1}{2}\,\frac{B^2}{\mu_0}. \qquad (16\text{–}2)$$

The total energy W in the box is

$$W = wV,$$

where V is the volume of the box. Since $V = Ad$, we have

$$W = \frac{1}{2}\left(\epsilon_0 E^2 + \frac{B^2}{\mu_0}\right)Ad. \qquad (16\text{–}3)$$

One of Maxwell's theoretical findings was that the magnitudes of the electric field intensity and of the magnetic flux density in an electromagnetic wave are proportional to each other, with the velocity of light c as the proportionality constant. That is,

$$E = cB \qquad (16\text{–}4)$$

in all electromagnetic waves. We are therefore able to rewrite Eq. (16–3) in the form

$$W = \frac{1}{2}\left(\epsilon_0 cEB + \frac{EB}{\mu_0 c}\right) Ad$$

$$= \frac{1}{2}\left(\epsilon_0 c + \frac{1}{\mu_0 c}\right) EBAd. \tag{16–5}$$

Because

$$c = \frac{1}{\sqrt{\epsilon_0 \mu_0}}, \tag{16–1}$$

we see that

$$\epsilon_0 c = \frac{1}{\mu_0 c},$$

and so

$$\left(\epsilon_0 c + \frac{1}{\mu_0 c}\right) = \frac{2}{\mu_0 c}.$$

Therefore Eq. (16–5) becomes

$$W = \frac{EBAd}{\mu_0 c}. \tag{16–6}$$

The amount of energy W contained in the box at any instant is traveling at the speed c, and in the time $\Delta t = d/c$ it will all have passed out of the box. Thus the rate at which energy passes through the box is

$$P = \frac{W}{\Delta t} = \frac{Wc}{d} = \frac{EBA}{\mu_0}.$$

The magnitude S of the Poynting vector is the rate of energy flow per unit cross-sectional area. Here the area of the box is A, and we have

$$S = \frac{P}{A} = \frac{EB}{\mu_0}. \tag{16–7}$$

Since **S** represents the flow of energy per unit area, it is expressed in watts/m².

In an actual electromagnetic wave both E and B vary in magnitude as shown in Fig. 16–4. While the average values of E and B individually are evidently zero, since each is negative as much as it is positive, the average value of the product EB is *not* zero. When E is negative B is negative, and when E is positive B is positive; their product is therefore always a positive quantity. The average value of EB, which we write

$\overline{EB}$, is related to the maximum values of E and B by the formula

$$\overline{EB} = \tfrac{1}{2}E_{\max}B_{\max}. \tag{16-8}$$

Hence the average magnitude $\overline{S}$ of the Poynting vector is

$$\overline{S} = \frac{E_{\max}B_{\max}}{2\mu_0}. \tag{16-9}$$

It is also true that the average values of E^2 and B^2 are related to their maximum values by

$$\overline{E}^2 = \tfrac{1}{2}E_{\max}^2, \qquad \overline{B}^2 = \tfrac{1}{2}B_{\max}^2. \tag{16-10}$$

We can therefore express the average magnitude of the Poynting vector in the alternative forms

$$\overline{S} = \frac{cB_{\max}^2}{2\mu_0} = \frac{E_{\max}^2}{2c\mu_0} \tag{16-11}$$

in view of Eq. (16-4).

Electromagnetic radiation from the sun arrives at the earth at the rate of about 1400 watts/m². A single electromagnetic wave of the kind we have been discussing (as distinct from the complex of different individual waves that comes from the sun) whose Poynting vector has this magnitude has a maximum electric field intensity of

$$E_{\max} = \sqrt{2c\mu_0\overline{S}}$$

$$= \sqrt{2 \times 3 \times 10^8 \text{ m/sec} \times 1.26 \times 10^{-6} \text{ w/amp·m} \times 1400 \text{ watts/m}}$$

$$= 1.03 \times 10^3 \text{ volts/m}.$$

The corresponding maximum magnetic flux density is

$$B_{\max} = \frac{E_{\max}}{c} = \frac{1.03 \times 10^3 \text{ volts/m}}{3 \times 10^8 \text{ m/sec}}$$

$$= 3.43 \times 10^{-6} \frac{\text{w}}{\text{m}^2}.$$

16–5 Radiation pressure

Electromagnetic waves transport momentum as well as energy. The simplest method for demonstrating this makes use of the mass-energy relationship

$$E = mc^2.$$

A portion of an electromagnetic wave whose energy content is E is equivalent to a mass of

$$m = \frac{E}{c^2}.$$

Since the wave travels at the speed c, the momentum of this portion is

$$mc = \frac{E}{c}.$$

Hence the *momentum density* (momentum per unit volume) of an electromagnetic wave whose energy density is w is w/c, and the rate of flow of momentum per unit cross-sectional area of the wave is $\mathbf{S}/c$, where $\mathbf{S}$ is the Poynting vector of the wave.

The most interesting consequence of momentum transport by electromagnetic waves is the phenomenon of *radiation pressure*, the pressure these waves exert on any surface they fall upon. We recall from Chapter 10 that the force a stream of particles exerts on a wall is equal to the rate of change of their momenta upon striking the wall. An electromagnetic wave whose Poynting vector has the magnitude S loses the momentum S/c per unit area per unit time when it is absorbed by a surface, and so the force it exerts upon the wall is S/c per unit area. Since pressure is force per unit area, the pressure p of the wave is

$$p = \frac{S}{c} \text{ (absorbed)} \tag{16–12}$$

as it is absorbed. If the wave is totally reflected, its change in momentum per unit area per unit time is doubled to $2S/c$ (since momentum is a vector quantity), and the pressure it exerts is also doubled to

$$p = \frac{2S}{c} \text{ (reflected)} \cdot \tag{16–13}$$

As we noted in the previous section, the approximate magnitude of the Poynting vector for solar radiation at the earth is $S = 1400 \text{ watts/m}^2$. If this radiation is totally absorbed, the resulting pressure on the absorber is

$$p = \frac{S}{c} = \frac{1400 \text{ watts/m}^2}{3 \times 10^8 \text{ m/sec}} = 4.7 \times 10^{-6} \text{ n/m}^2 \cdot$$

The seemingly trivial pressure of solar radiation is responsible for the deflection of comet tails so that they always point away from the sun; it has even been seriously suggested as a means of interplanetary propulsion through the use of large reflecting "sails" on space ships.

IMPORTANT TERMS

Maxwell's hypothesis was that a changing electric field produces a magnetic field.

Electromagnetic waves consist of coupled electric and magnetic field oscillations. They exist as a consequence of electromagnetic induction, which states that whenever there is a change in a magnetic field, an electric field is produced, and of Maxwell's hypothesis, which states that whenever there is a change in an electric field, a magnetic field is produced. Radio waves, light waves, x-rays, and gamma rays are all electromagnetic waves differing only in their frequency.

The *Poynting vector* **S** describes the flow of energy in an electromagnetic wave. Its direction is that of the wave, and its magnitude is equal to the rate at which energy is being transported by the wave per unit cross-sectional area. The *momentum density* of an electromagnetic wave is the ratio between the Poynting vector of the wave and the speed of light. When electromagnetic waves strike a surface, they exert *radiation pressure* on it because of their momenta.

PROBLEMS

1. A certain radio station transmits at a frequency of 880 kc/sec. What is the wavelength of these waves?

2. A small marine radar operates at a wavelength of 3.2 cm and a power of 7 kilowatts per pulse. (a) What is the frequency of the radar waves? (b) How much reaction force is exerted on the antenna when each pulse is emitted?

3. The intensity of solar radiation varies inversely with the square of the distance from the sun. The earth is an average of 1.5×10^8 km from the sun. How close to the sun is it necessary to be for its radiation pressure to be 1 n/m^2?

4. A radio wave is found to exert a pressure of 10^{-8} n/m^2 on a reflecting surface. (a) How much power per unit area does the wave transport? (b) What are the maximum magnitudes of its electric intensity and magnetic flux density?

5. For good reception a radio wave should have a maximum electric field intensity of at least 10^{-4} volt/m when it arrives at the receiving antenna. (a) What is the maximum flux density of the magnetic field of such a wave? (b) What is the magnitude of the Poynting vector of such a wave? (c) What radiation pressure does it exert when it is absorbed?

6. The entire 1000-watt luminous output of a searchlight is focused on a spot 2 m in diameter. If half of the light is absorbed and half reflected, how much pressure does it exert?

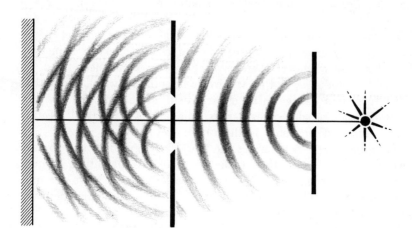

Light | 17

Light was recognized as a wave phenomenon well before its electro-magnetic character became known. The problem of the nature of light is an old one: Newton felt sure that light consists of a stream of tiny particles, while his contemporary, Christian Huygens (1629–1695), thought it a succession of waves. Neither man offered any hypothesis as to what kind of particle or wave was involved. There are a number of ways of resolving the question, and we shall consider several in the remainder of this chapter. The first one we shall take up is based upon the refraction of light.

17-1 Wave nature of light

It is a matter of experience that a beam of light passing obliquely from one medium to another, say from air to water, is deflected at the surface between the two media. The usual way of describing this effect involves the angles made by the beam with the *normal* to the surface, which is a line drawn perpendicular to the surface at the point where the beam crosses it. When a light beam goes from air into water along the normal, it simply continues along the same path (Fig. 17–1a), but when it enters the water at any other angle, it is bent toward the normal (Fig. 17–1b). The paths are reversible; that is, a light beam emerging from the water is bent *away* from the normal as it enters the air (Fig. 17–1c). The bending of a light beam when it passes from one medium to another is called *refraction.*

Let us first examine how the particle theory of light explains refraction. The velocity of an oblique light beam may be thought of as being composed of two components, one parallel to the normal and one parallel to the surface (Fig. 17–2). Because it does not make any difference from which side of the normal the light particles arrive, passing through the

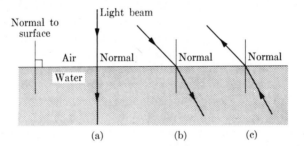

FIG. 17–1. Refraction.

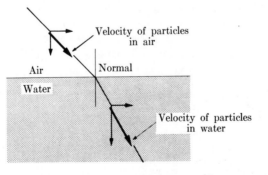

FIG. 17–2. Refraction according to the particle theory of light.

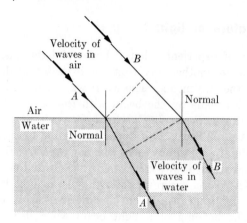

F<small>IG</small>. 17–3. Refraction according to the wave theory of light.

surface cannot affect that part of their motion parallel to the surface. However, it is plausible to suppose that a force may be exerted on the light particles as they enter the water, thereby increasing their velocity components parallel to the normal. Then, since the beam speed is greater in the downward direction in water than in air, while remaining the same parallel to the surface, the net effect of passing into the water is a bending of the beam toward the normal. The speed of light particles in water is therefore *greater* than in air, according to this theory.

The wave theory of light reaches the opposite conclusion. Let us examine two parallel light rays approaching the water from the air, as in Fig. 17–3. The ray on the left, A, reaches the water first; if its speed in the water is less than it was in air, this ray travels a shorter distance in the water than the ray on the right, B, does in air while B is approaching the water surface. After B enters the water, the two again proceed parallel to each other, an experimental fact, but now in a direction which makes a smaller angle than before with the normal. Since this is what is actually observed, we conclude that the speed of light waves in water is *less* than in air if the wave theory is correct.

There is thus a clear experimental criterion for deciding whether light is a wave or a particle phenomenon: is its speed in a medium at whose surface it is bent toward the normal less than or greater than it is in the medium from which it came? While conceptually clear, the experiment is not an easy one to perform, and it was not until the middle of the 19th century that the speeds of light in air and in water were actually compared. The finding was that the speed in air is greater, strong evidence for the wave nature of light.

TABLE 17-1

Indexes of refraction

Substance	n
Air	1.0003
Benzene	1.50
Carbon disulfide	1.63
Diamond	2.42
Ethyl alcohol	1.36
Glass, crown	1.52
Glass, flint	1.63
Glycerin	1.47
Ice	1.31
Quartz	1.46
Water	1.34

17-2 Dispersion

The ratio between the speed of light c in free space and its speed v in a particular medium is called the *index of refraction* of the medium, the symbol for which is n. That is,

$$n = \frac{c}{v}.$$ (17-1)

Table 17-1 is a list of the values of n for a number of substances. The greater the index of refraction, the greater the extent to which a light beam is deflected upon entering or leaving that medium. Diamonds owe their sparkle and brilliance to their very high index of refraction.

Generally the index of refraction of a medium depends to some extent upon the frequency of the light involved, with the highest frequencies having the highest values of n. In ordinary glass the index of refraction

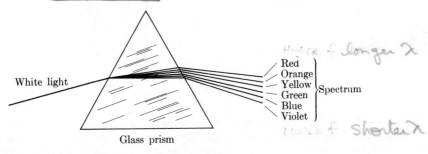

FIG. 17-4. Dispersion by a prism.

for violet light is about one per cent greater than that for red light, for example. Since a different index of refraction means a different degree of deflection when a light beam enters or leaves a medium, a beam containing more than one frequency is split into a corresponding number of different beams when it is refracted. This effect, called *dispersion*, is illustrated in Fig. 17–4, which shows the result of directing a beam of white light at one face of a glass prism. It is separated into beams of various colors, from which we conclude that white light is actually a mixture of light of these different colors. The band of colors that emerges from the prism is known as a *spectrum*.

17–3 Diffraction

Waves are able to bend around the edge of an obstacles in their path, a property called *diffraction*. We have all heard sounds that originated around the corner of a building from where we were standing, for example. These sound waves cannot have traveled in a straight line from their source to our ears, and refraction cannot account for their behavior. Water waves, too, diffract, as the simple experiment illustrated in Fig. 17–5 shows. A beam of particles behaves quite differently: bullets fired from another side of a building cannot possibly reach us by bending around the corner (Fig. 17–6), though we can hear the sound of the gun. Because sharp-looking shadows result when a light beam is partially obstructed, Newton felt sure that light must be corpuscular in nature. However, diffraction is pronounced only when the dimensions of an obstacle or opening in an obstacle are comparable to the wavelength of the waves striking them. A typical audible sound wave might have a

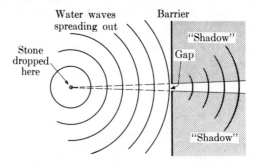

Fig. 17–5. The diffraction of water waves at a gap in a barrier. Note that the waves on the far side of the gap spread out into the geometrical "shadow" of the gap's edges, though with reduced amplitude. In the absence of diffraction, no waves would be present in the shadow region. The diffracted waves spread out as thought they originated at the gap.

Fig. 17-6. While sound waves diffract around corners, particles do not. The man on the right can hear the machine gun being fired around the corner, but the bullets from it cannot reach him.

wavelength of perhaps 1 meter, and a typical ocean wave might have a wavelength of perhaps 100 meters; it is therefore not hard to observe diffraction effects with these types of wave. Visible light, however, contains wavelengths smaller than 10^{-6} meter, making it more of a feat to exhibit the diffraction of light.

The first convincing diffraction experiment with light waves was performed in 1801 by Thomas Young, though the fact that the edges of

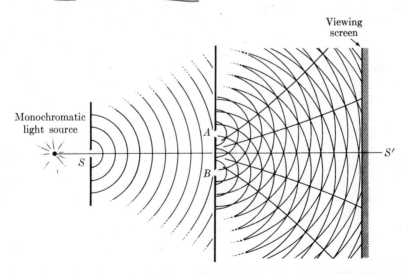

Fig. 17-7. Young's double-slit experiment. In the absence of diffraction the viewing screen would be completely dark, since no light ray can reach it from the source along a straight path.

shadows are not perfectly sharp had been observed much earlier. Young used an arrangement similar to that shown in Fig. 17–7: a source of monochromatic light (that is, light consisting of only a single wavelength) is placed behind a narrow slit S in an opaque screen, and another screen having two similar narrow slits A and B close together is placed on the other side. Light from S passes through both A and B and then to the viewing screen. If light were not a wave phenomenon, we would expect to find the viewing screen completely dark, since no light ray can reach it from the source along a straight path. Instead, owing to diffraction, the entire screen is illuminated! Even the point S', separated from S by the opaque barrier between the slits A and B, turns out to be bright rather than dark.

17–4 Interference

An interesting feature of Young's experiment, and even stronger evidence for the wave nature of light, is the striated appearance of the viewing screen (Fig. 17–8). The diffracted light does not evenly illuminate the screen, but instead produces a regular pattern of alternate light and dark lines. To understand the origin of these lines we must first discuss the *principle of superposition,* one of the fundamental laws governing wave behavior.

To illustrate the principle of superposition, let us imagine two trains of parallel water waves of the same amplitude meeting each other head on, as in Fig. 17–9. What happens to the surface of the water as the various waves pass by? The answer is that the surface responds to *both* sets of waves, with each set acting as though the other did not exist. If two crests simultaneously pass a given point (Fig. 17–9a), the water level there rises to a height equal to the sum of the individual heights of each crest. Similarly, if two troughs simultaneously pass the point (Fig. 17–9b), the water level there falls to a depth equal to the sum of the individual depths of each trough. In the event that a crest belonging to one wave train meets a trough belonging to the other (Fig. 17–9c), since the amplitude of both is the same, there is an exact cancellation and the water level neither rises nor falls.

The principle of superposition is a statement of the above behavior, and applies to all wave motions, sound and light as well as water waves. The principle states that

When two or more waves of the same nature travel past a given point at the same time, the amplitude at the point is the sum of the instantaneous amplitudes of the individual waves.

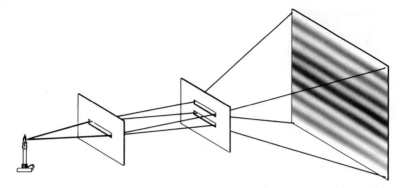

FIG. 17–8. The illumination of the screen in Young's double-slit experiment is not uniform, but consists of alternate bright and dark lines.

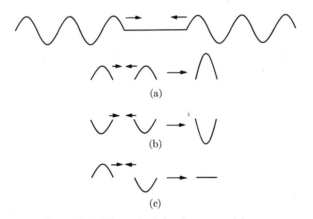

(a)

(b)

(c)

FIG. 17–9. The principle of superposition.

In other words, each wave train proceeds as though the others did not exist. Should the waves come together in such a way that crest meets crest and trough meets trough, the resulting composite wave has an amplitude greater than that of either of the original waves. When this occurs the waves are said to *interfere constructively* with one another. Should the waves come together in such a way that crest meets trough and trough meets crest, the composite wave has an amplitude less than that of either of the original waves. When this occurs the waves are said to *destructively interfere* with one another (Fig. 17–10).

Waves can interfere with one another even if their wavelengths differ. In this event sometimes the interference is constructive and at other times it is destructive. An example of interference involving different

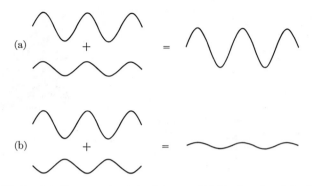

FIG. 17-10. (a) Constructive interference. (b) Destructive interference.

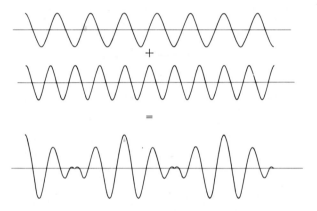

FIG. 17-11. Diagram illustrating the origin of beats.

wavelengths occurs in sound waves. If two tuning forks (or other sources of single-frequency sound waves) whose frequencies are slightly different are struck at the same time, the sound that we hear fluctuates in intensity. At one instant we hear a loud tone, then virtual silence, then the loud tone again, then virtual silence, and so on. The origin of this behavior is shown schematically in Fig. 17-11; the loud tones occur when the waves from the two forks interfere constructively, thus reinforcing one another, and the quiet periods occur when the waves interfere destructively, thus partially or wholly canceling one another out. These regular pulsations are called *beats*. Observation and calculation both yield the result that the number of beats per second equals the difference between the frequencies of the two sound sources: if tuning forks are used whose frequencies are, say 440 and 445 cycles/sec, five beats per second will be heard.

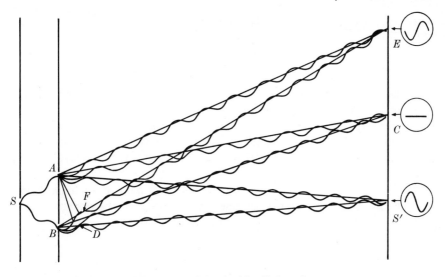

FIG. 17–12. The origin of the double-slit interference pattern.

17–5 The double slit

The interference of light waves readily accounts for the pattern of alternate light and dark lines observed on the screen in Young's double-slit experiment. We see in Fig. 17–12 that light waves from slits A and B travel identical distances in reaching the central position S' on the screen, and consequently they interfere constructively there. The reason for having the light reaching A and B come from the same source S is now evident. Owing to this arrangement a light wave leaving A is always in exactly the same part of its cycle as that leaving B at the same time, so that when a crest leaves A a crest also leaves B, when a trough leaves A a trough also leaves B, and so on. When waves from the two slits meet at S' they therefore reinforce one another to produce a bright line. If different sources of light were used to illuminate A and B, the light waves leaving A and B would be independent of one another, and when they meet at S' they would usually be in different parts of their cycles. No interference patterns of light and dark lines would result, although diffraction at the slits would cause the screen to be weakly illuminated. Interference phenomena are only possible when a single original light beam is split into two or more parts which travel along different paths and later recombine.

Let us examine what happens at the position C on the screen located to one side of S'. The distance BC is longer than the distance AC by the amount BD, as shown in Fig. 17–12. Suppose that BD equals exactly half a wavelength of the light being used, that is, that

$$BD = \tfrac{1}{2}\lambda.$$

When a crest from A reaches C, this difference in path length means that a trough from B arrives there at the same time, since $\tfrac{1}{2}\lambda$ separates a crest and a trough in the same wave. The two cancel each other out, the light intensity at C is zero, and a dark line results on the screen there. At S' the equality of path length gives rise to constructive interference; at C the difference of $\tfrac{1}{2}\lambda$ in path length gives rise to destructive interference.

If we go past C on the screen we will come to a point E such that the distance BE is greater by exactly one wavelength than the distance AE. That is, the difference BF between BE and AE is

$$BF = \lambda.$$

Consequently, when a crest from A reaches E, a crest from B also arrives there, although the latter crest left B earlier than that from A owing to the longer path it had to cover. Because $BF = \lambda$, waves arriving at E from both slits are always in the same part of their cycles, and they constructively interfere to produce a bright line at E.

By continuing the same analysis, we find that the alternate light and dark lines actually observed on the screen correspond respectively to locations where constructive and destructive interference occurs. Waves reaching the screen from A and B along paths that are equal or differ by a whole number of wavelengths (λ, 2λ, 3λ, and so on) reinforce, while those whose paths differ by an odd number of half wavelengths ($\tfrac{1}{2}\lambda$, $\tfrac{3}{2}\lambda$, $\tfrac{5}{2}\lambda$, and so on) cancel. At intermediate locations on the screen the interference is only partial, so that the light intensity on the screen varies gradually between the bright and dark lines.

The double-slit experiment demonstrates that both diffraction and interference occur in light. Since both are phenomena characteristic of waves, this experiment is further evidence for the wave nature of light.

17–6 The wavelength of light

Interference patterns like those produced by the double slit permit us to determine the wavelength of the light used. To illustrate the procedure, we shall consider the double-slit experiment quantitatively. Figure 17–13 is a diagram of the experiment: d is the separation between the slits, L is the distance from the slits to the screen, and y is the distance from the central point S' on the screen to the point Q whose illumination we

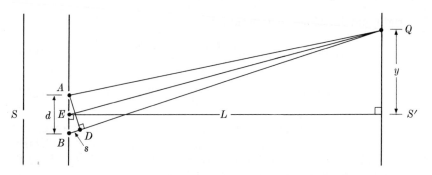

Fɪɢ. 17–13. A diagram of the double-slit experiment.

are observing. Waves traveling from slit B to Q must travel s farther than those from slit A. As we have seen, when the path difference s is

$$s = 0, \quad \lambda, \quad 2\lambda, \quad 3\lambda, \ldots, \tag{17–2}$$

where λ is the wavelength of the light from the source, waves from A and B arrive at Q in the same stage of their cycles and reinforce one another to produce a bright line there. When the path difference s is

$$s = \tfrac{1}{2}\lambda, \quad \tfrac{3}{2}\lambda, \quad \tfrac{5}{2}\lambda, \ldots, \tag{17–3}$$

on the other hand, waves from A and B arrive at Q in the opposite stages of their cycles and cancel one another out to produce a dark line there. The triangles ABD and QES' are similar, since each is a right triangle and two sides of each are perpendicular to two sides of the other. Corresponding sides of similar triangles are proportional, and so

$$\frac{s}{y} = \frac{d}{EQ}. \tag{17–4}$$

In an actual experiment, y is very much smaller than L (their ratio is usually less than 1:1000; in Fig. 17–13 y is exaggerated for clarity), which means that EQ is almost exactly equal to L. At the expense of introducing a negligible error, we may substitute L for EQ in Eq. (17–4), with the result that

$$\frac{s}{y} = \frac{d}{L},$$

$$s = \frac{dy}{L}. \tag{17–5}$$

Combining Eqs. (17–2) and (17–3) with Eq. (17–5) yields the conditions for bright and dark lines to occur at Q:

$$\text{Bright lines} \quad y = 0, \quad \frac{L\lambda}{d}, \quad \frac{2L\lambda}{d}, \quad \frac{3L\lambda}{d}, \cdots, \qquad (17\text{--}6)$$

$$\text{Dark lines} \quad y = \frac{L\lambda}{2d}, \quad \frac{3L\lambda}{2d}, \quad \frac{5L\lambda}{2d}, \cdots. \qquad (17\text{--}7)$$

According to Eq. (17–6), there is a bright line when $y = 0$, corresponding to the center of the screen S', a bright line on either side of S' a distance $L\lambda/d$ from it, another bright line on either side a distance $2L\lambda/d$ from S', and so on (Fig. 17–8). Similarly, Eq. (17–7) states that there is a dark line on either side of S' a distance $L\lambda/2d$ from it, another dark line on either side a distance $3L\lambda/2d$ from it, and so on.

In an actual experiment, L and d are known initially, and the distance y from the center of the screen to any particular light or dark line can be measured. The wavelength of the light may then be computed by using the equation corresponding to the particular line. As an example, let us consider monochromatic yellow light from a sodium vapor lamp illuminating two narrow slits 1 mm apart. The viewing screen is 1 meter from the slits, and we find that the distance from the central bright line to the bright line nearest it is 0.589 mm. Hence

$$y = \frac{L\lambda}{d},$$

$$\lambda = \frac{yd}{L} = \frac{5.89 \times 10^{-4} \text{ m} \times 10^{-3} \text{ m}}{1 \text{ m}} = 5.89 \times 10^{-7} \text{ m}.$$

The wavelength of the light from the lamp is 5.89×10^{-7} meter. Since the wavelength and frequency of a wave are related to its speed by the formula

$$c = f\lambda,$$

the frequency corresponding to the above wavelength is

$$f = \frac{c}{\lambda} = \frac{3 \times 10^8 \text{ m/sec}}{5.89 \times 10^{-7} \text{ m}} = 5.09 \times 10^{14} \text{ sec}^{-1}.$$

17–7 Polarization

While all the experimental evidence we have examined thus far supports the view that light is a wave phenomenon, none of it tells us whether light waves are longitudinal or transverse. These classes of waves were discussed in Chapter 8, and we recall that longitudinal waves are chacterized by motion of the particles of the medium parallel to the direction in

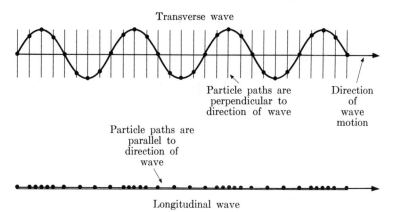

FIG. 17–14. Transverse and longitudinal waves.

which the waves travel, while transverse waves are characterized by motion of the particles of the medium perpendicular to the direction in which the waves travel (Fig. 17–14). Sound waves are longitudinal, for example, and the waves that move down a stretched string are transverse. Refraction, diffraction, and interference occur in the propagation of both longitudinal and transverse waves, and we require further experimental information in order to ascertain in which of the latter categories light belongs.

A definitive distinction between longitudinal and transverse waves may be made based upon their capacities to be *polarized*. A polarized beam of transverse waves is one whose vibrations occur in only a single direction perpendicular to the direction in which the beam travels (Fig. 17–15), so that the entire wave motion is confined to a plane called the *plane of polarization*. When many different directions of polarization are present in a beam of transverse waves, vibrations occur equally often in all directions perpendicular to the direction of motion, and the beam is then said to be *unpolarized*. Since the vibrations constituting longitudinal waves can only take place in one direction, namely that in which the waves travel, longitudinal waves cannot be polarized. By establishing whether or not light can be polarized, then, we have a means for determining which type of wave it consists of.

How can we find out whether a particular wave phenomenon is polarizable or not? To clarify this question we shall consider the behavior of transverse waves in a stretched string. If the string passes through a tiny hole in a fence, as in Fig. 17–16(a), waves traveling down the string are stopped since the string cannot vibrate there at all. When the hole is replaced by a vertical slot, as in Fig. 17–16(b), waves whose vibrations

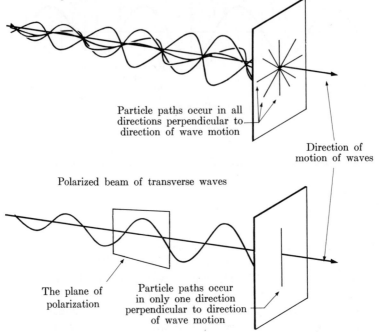

FIG. 17–15. An unpolarized and a polarized beam of transverse waves.

are vertical can get through the fence although waves with vibrations in all other directions cannot. In a situation in which several waves vibrating in different directions simultaneously move down the string, the slot stops all but vertical vibrations: an initially unpolarized series of waves has become polarized. Hence we can devise a very effective method for determining whether a particular series of waves can be polarized or not. In the case of the stretched string, what we do is erect another fence a short distance from the first. If the slot in the new fence is also vertical, as in Fig. 17–16(c), those waves that can get through the first fence can also get through the second. If the slot in the new fence is horizontal, however, it will stop all waves that reach it from the first fence (Fig. 17–16d). Should longitudinal rather than transverse waves, say in a spring, go through the fences (Fig. 17–16e and f), it is perfectly possible that their amplitudes might decrease in passing through each slot, but the relative alignments of the slots would not matter, while it is the critical factor in the case of transverse waves.

The above chain of reasoning made it possible for the transverse nature of light waves to be established in the last century. A number

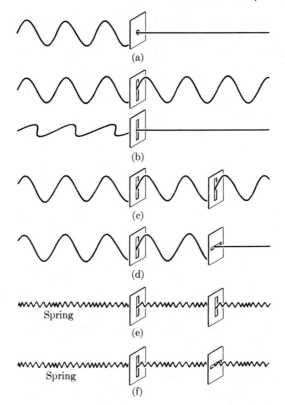

FIG. 17-16. Mechanical analogies of fundamental polarization phenomena.

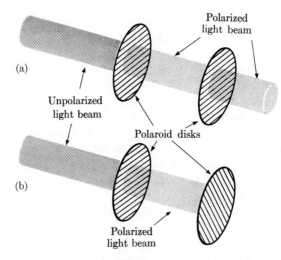

FIG. 17-17. Diagram showing the transverse nature of light waves.

of substances, for instance quartz, calcite, and tourmaline, are known which have the property of transmitting light in only a single plane of polarization. When a beam of unpolarized light is incident upon such a substance, only those of its waves whose planes of polarization are parallel to a particular plane in the substance emerge from the other side. The remainder of the waves are absorbed or deflected. *Polaroid* is an artificially made polarizing material in wide use. It consists of two sheets of plastic with a thin layer of needlelike quinine iodosulfate crystals between them; the crystals are aligned by means of a strong electric field, and the resulting clear material transmits only light with a single plane of polarization. To exhibit the transverse nature of light waves, we first place two Polaroid disks in line so that their axes of polarization are parallel, Fig. 17–17(a), and note that all light passing through one disk also passes through the other. Then we turn one disk until its axis of polarization is perpendicular to that of the other, Fig. 17–17(b), and note that all light passing through one disk is now *stopped* by the other.

Just what is it whose vibrations are aligned in a beam of polarized light? As we said earlier in this chapter, light waves actually consist of oscillating electric and magnetic fields perpendicular to each other. Because it is the electric fields of light waves whose interactions with matter produce nearly all common optical effects, the plane of polarization of a light wave is considered to be that in which both the direction of its electric field and its direction of motion lie (the plane of the paper in Fig. 16–4). Even though nothing material moves during the passage of a light wave, then, it is possible to establish its transverse nature and identify its plane of polarization.

IMPORTANT TERMS

The bending of a light beam when passing from one medium to another is called *refraction.* The quantity that governs the degree to which a light beam will be deflected in entering a medium is its *index of refraction,* defined as the ratio between the speed of light in free space and its value in the medium. *Dispersion* refers to the splitting up of a beam of light containing different frequencies by passage through a substance whose index of refraction varies with frequency.

The ability of waves to bend around the edges of obstacles in their paths is called *diffraction.*

The *principle of superposition* states that when two or more waves of the same nature travel past a given point at the same time, the amplitude at the point is the sum of the amplitudes of the individual waves. The interaction of different waves is called *interference: constructive inter-*

ference results when the resulting composite wave has an amplitude greater than that of either of the original waves, and *destructive inter-ference* results when the resulting composite wave has an amplitude less than that of either of the original waves.

A *polarized* beam of transverse waves is one whose vibrations occur in only a single direction perpendicular to the direction in which the beam travels, so that the entire wave motion is confined to a plane called the *plane of polarization*. An *unpolarized* beam of transverse waves is one whose vibrations occur equally often in all directions perpendicular to the direction of motion.

PROBLEMS

1. What types of waves can be transmitted through (a) solids, (b) liquids, (c) gases, (d) empty space?

2. What is the index of refraction of a substance in which the speed of light is 2.3×10^8 m/sec?

3. Using the indexes of refraction given in Table 17–1 and taking the speed of light in free space as 3×10^8 m/sec, find the speed of light in (a) air, (b) diamond, (c) crown glass, (d) water.

4. Does the frequency or the wavelength of a light wave change when it goes from air into glass? Explain your answer.

5. When a straight stick is placed in water at an oblique angle to its surface, the part under the surface appears as though bent upward. Explain this effect with the help of a diagram.

6. *Fermat's principle* states that light travels between two points in such a manner that the transit time is a minimum. How is the phenomenon of refraction in accord with Fermat's principle?

7. The duration of daylight is extended to a small extent owing to the refraction of sunlight by the earth's atmosphere. Use a diagram to illustrate this effect.

8. The waves used to carry television signals cannot reach receivers beyond the visual horizon of their transmitting antennas, while ordinary radio waves readily travel beyond the visual horizon of their transmitting antennas. Can you think of a reason for this contradictory behavior?

9. Flint glass and carbon disulfide have almost the same index of refraction. How does this explain the fact that a flint glass rod immersed in carbon disulfide is nearly invisible?

10. Why is a beam of white light not dispersed into its component colors when it passes perpendicularly through a pane of glass?

11. Explain why a cut diamond held in white light shows flashes of color. What would happen if it were held in red light?

12. Radio waves diffract pronouncedly around buildings, while light waves, which are also electromagnetic waves, do not. Why?

13. A man has two tuning forks, one marked "256 cycles/sec" and the other of unknown frequency. He strikes them simultaneously, and hears 10 beats per second. "Aha," he says, "the other tuning fork has a frequency of 266 cycles/sec." What is wrong with his conclusion?

14. In Young's double-slit experiment, which effects are due to diffraction and which to interference?

15. What do diffraction and interference have in common? How do they differ?

16. In Young's experiment, how does the spacing of the slits affect the pattern of bright and dark lines?

17. Light of unknown wavelength is used to illuminate two parallel slits 1 mm apart. Adjacent bright lines on the interference pattern that results on a screen 1.5 m away are 0.75 mm apart. What is the wavelength of the light?

18. Two parallel slits 0.12 mm apart are illuminated by light of wavelength 6×10^{-7} m. A viewing screen is 1.5 m from the slits. (a) How far from the central bright line is the next bright line? (b) How far is the first dark line? (c) How far is the fifth dark line?

19. Two parallel slits 0.1 mm apart are illuminated by light of wavelength 5.46×10^{-7} m. A viewing screen is 0.8 m from the slits. (a) How far from the central bright line is the second bright line? (b) How far is the third bright line? (c) How far is the third dark line?

20. Two parallel slits 0.2 mm apart are illuminated by light of two wavelengths, 5×10^{-7} m and 6×10^{-7} m. A viewing screen is 2 m away from the slits. How far are the bright lines of one wavelength from the bright lines of the other?

21. Two parallel slits are illuminated by light of two wavelengths, one of which is 5.8×10^{-7} m. On a viewing screen an unknown distance from the slits the fourth dark line of the light of the known wavelength coincides with the fifth bright line of the light of the unknown wavelength. Find the unknown wavelength.

22. Which of the following can occur in (a) transverse waves and (b) longitudinal waves? Refraction, dispersion, interference, diffraction, polarization.

23. Explain the peculiar appearance of a distant light source when seen through a piece of finely woven cloth.

24. How can a single sheet of Polaroid be used to show that sky light is partially polarized?

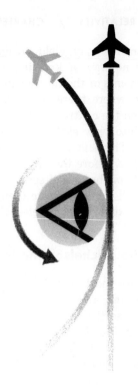

Relativity | 18

The great French mathematician and physicist J. L. Lagrange (1736–1813) once wrote that "Newton was the greatest genius who ever lived, and the most fortunate, for there cannot be more than once a system of the world to establish." Though Newton has lost none of his glory since this was written, our own century has seen the basis for the second part of Lagrange's tribute disappear. Not only have relativity and quantum theory forced major revisions in our ideas of the "system of the world," but they have also shown the folly in assuming that we have complete understanding of any major aspect of the physical universe. Today's measurements may yield no discrepancy with a particular theory, but we have no way of knowing what the more precise or more searching measurements of the future will have to say. Science is not like mountain climbing, where there is a definite goal to reach and either success or failure; on the contrary, it is a progressive endeavor, with further heights coming into view each time we think we have, at long last, achieved a summit. There is no better illustration of this than the situation of physics just before the turn of the last century, when the belief was current that this branch of science presented no mysteries that could not be solved with the knowledge then in hand. We shall learn in this chapter and the next just how baseless this belief really was.

18–1 The Michelson-Morley experiment

When we speak of light waves traveling through space, precisely what do we mean? If space is empty, there is nothing to vibrate back and forth, and so the concept of a wave moving through empty space is not easy to accept. In the 19th century, physicists avoided the problem by assuming that the universe is pervaded with a curious fluid called *ether* whose only property is an ability to support electromagnetic waves. This seems a harmless enough idea, but it contains an implication that led directly to its downfall and thereby to the theory of relativity. Let us see what this implication is through a simple analogy.

Figure 18–1 is a sketch of a river flowing with the speed v. The river's width is D. Two identical boats start out from one bank of the river with the same speed V; boat A is to cross the river to a point on the other bank directly opposite the starting place and then to return, while boat B is to proceed downstream for the distance D and then to return to the starting place. How much time will each round trip take?

We shall first consider boat A. If A heads directly across the river (as in Fig. 18–2a), the current will carry it downstream from its destina-

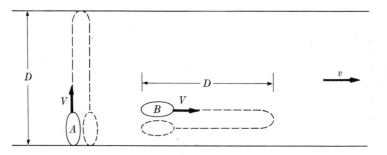

FIGURE 18–1

tion on the other bank opposite where it started, and so it must head somewhat upstream (Fig. 18–2b) in order to compensate for the current. When it does this, it will have an upstream component of its velocity equal to exactly $-v$ in order to cancel out the downstream current v, leaving the component V' as its net speed across the river. From Fig. 18–2 we see that these speeds are related by

$$V^2 = V'^2 + v^2,$$

so that the actual speed with which it crosses the river is

$$V' = \sqrt{V^2 - v^2} = V\sqrt{1 - (v^2/V^2)}.$$

Hence the time for the crossing is the distance D divided by the speed V', and, since the reverse crossing involves exactly the same amount of time, the total round-trip time t_A is twice D/V', or

$$t_A = \frac{2D/V}{\sqrt{1 - (v^2/V^2)}}. \tag{18–1}$$

Boat B is in a different situation. As it proceeds downstream, its speed relative to the shore is its own speed V *plus* the speed v of the river

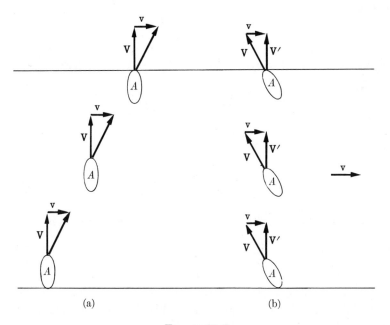

(a)   (b)

FIGURE 18–2

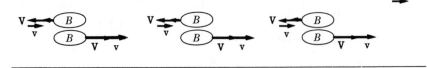

FIGURE 18–3

(Fig. 18–3), and it travels the distance D downstream in the time

$$\frac{D}{V + v}.$$

On the return trip, however, B's speed relative to the shore is its own speed V *minus* the speed v of the river, which now opposes its motion, and it requires the longer time

$$\frac{D}{V - v}$$

to travel upstream the distance D. The total round trip time t_B is the sum of these two, or

$$t_B = \frac{D}{V + v} + \frac{D}{V - v}.$$

Using the common denominator $(V + v)(V - v)$ for both terms,

$$t_B = \frac{D(V - v) + D(V + v)}{(V + v)(V - v)}$$

$$= \frac{2DV}{V^2 - v^2}$$

$$= \frac{2D/V}{1 - (v^2/V^2)}, \tag{18–2}$$

which is longer than t_A. The ratio between the times t_A and t_B is

$$\frac{t_A}{t_B} = \sqrt{1 - (v^2/V^2)}. \tag{18–3}$$

By sending out a pair of boats in this way, then, a measurement of t_A/t_B together with a knowledge of the boats' speed V enables us to calculate the speed v of the river current.

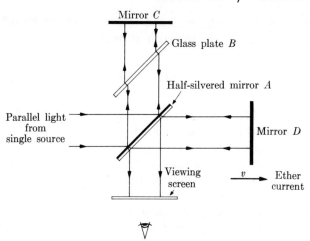

Fig. 18–4. Schematic diagram of the apparatus used by Michelson and Morley to measure the speed of the ether current past the earth.

Now let us restate the above problem in terms of the passage of light waves through the ether. Instead of a current of water relative to the shore we shall have a current of ether relative to the earth as the earth moves around the sun within the conjectured universal sea of ether, and instead of a pair of boats we shall have a pair of light beams from a single source. Figure 18–4 is a schematic representation of the apparatus used in the famous experiment performed by the American physicists Michelson and Morley in 1887. As shown in the figure, parallel light strikes the half-silvered mirror A and is split into two beams, one proceeding in the original direction to the right and the other reflected upward. The former beam is reflected by the mirror D and returns to A, where some of it is further reflected to a viewing screen. The upward beam is reflected by the mirror C and also returns to A, where part of it is transmitted to reach the viewing screen. The distances from A to C and from A to D are the same, and a glass plate B is inserted in the path of the upward beam, so that each beam passes through identical thicknesses of air and glass between its entry into the apparatus and its arrival at the viewing screen. The presence of an "ether current" as shown would result in the different beams requiring different transit times to go from the source to the viewing screen. Instead of arriving at exactly the same stage in their cycles, the light waves of the two beams would be slightly—but definitely—out of step with each other, and hence would interfere destructively to a corresponding extent.

Michelson and Morley expected to find a value for t_A/t_B of 0.999999995, corresponding to the substitution in Eq. (18–3) of the speed of light c

(186,000 mi/sec) for the boats' speed V and of the earth's orbital speed (averaging 18.5 mi/sec) for the river current v, namely

$$\frac{t_A}{t_B} = \sqrt{1 - (v^2/c^2)}. \tag{18–4}$$

The results of the Michelson-Morley experiment were startling: although their apparatus was capable of measuring the minute time difference expected, *none was found.* When the experiment was performed at different seasons and in different locations, and when different experiments were tried for the same purpose, the conclusions were always identical—no motion through the ether could be detected. To give a somewhat fanciful analogy, it was as if our boats made their round trips in a river bed running down the side of a mountain and returned at the same time: if there had been water in the river bed, it would have been in the form of a strong current; if there had been a strong current, the travel times would have differed; they did not differ, hence the river bed must have been empty. (For this analogy to hold we assume the boats were also equipped with wheels, permitting them to travel with equal facility on dry land or in water.)

18–2 Frames of reference

The immediate conclusion we can draw from the failure of the Michelson-Morley experiment to detect any effect of the ether on the motion of light waves is that the speed of light in free space is a constant everywhere, completely independent of any relative motion of the ether. As a corollary we must discard the notion of the ether entirely; the sole property of the ether is its ability to support electromagnetic waves, something it cannot be doing if such waves are not affected by its motion past source and observer.

The absence of an ether has a profound consequence that is not evident from our discussion thus far. Let us consider for a moment what we mean by so elementary a notion as velocity. When we say that a body is moving we imply the existence of some sort of frame of reference (coordinate system) in which its velocity **v** is to be measured. Suppose that a car traveling down a road goes past a slower car. There is a thick fog, so neither driver can see anything but the road and the other car. To the driver in the fast car, the other car appears to be moving backwards. The driver in the slow car, while observing that the fast car is going past him, nevertheless is sure that he is moving in the same direction as it is.

Who is correct? The modern answer is that both are correct in terms of their own frames of reference, with each driver determining his own motion relative to the ground while determining that of the other car relative to himself. Of course, we tend to think offhand that the driver of the slow car is "correct" because the earth is the frame of reference for our daily activities. But the earth itself is rotating on its axis, revolving about the sun, and traveling with the sun as the sun revolves about the center of our galaxy and partakes with the galaxy of the latter's motion through space. How then can we separate "apparent" from "true" motions, and establish the "truth" about what is happening in the universe?

At one time the ether looked like the solution to the problem of finding the "true" frame of reference of the universe. Newton's laws were supposed to hold exactly when phenomena are viewed from a frame of reference stationary relative to the ether. Now there would be no ambiguity in physical laws, all of which could be referred to the ether. However, as the Michelson-Morley experiment revealed, there is *no* ether, and therefore no way of specifying any universal frame of reference. In other words, if we observe something changing its position with respect to us, we have no way (even in principle) of knowing whether *it* is moving or *we* are moving. If we were alone in the universe we would have no way of finding out whether we are moving or not. All motion is relative to the observer; there is no such thing as absolute motion.

18–3 The special theory of relativity

The theory of relativity is a largely successful attempt to work out the physical consequences of the absence of a universal frame of reference. The special theory of relativity, published in 1905 by Albert Einstein, confines itself to problems involving the motion of frames of reference at constant velocity (that is, both constant speed and constant direction) with respect to one another; the general theory of relativity, published 10 years later by Einstein, deals with problems involving frames of reference accelerated with respect to one another. The special theory has had an enormous impact on all of physics, and since it is also simpler mathematically than the general theory, we shall consider it exclusively.

Einstein developed the special theory of relativity from two postulates. The first states that

The laws of physics may be expressed in the same set of equations for all frames of reference moving at constant velocity with respect to one another.

This postulate follows directly from the absence of a universal frame of

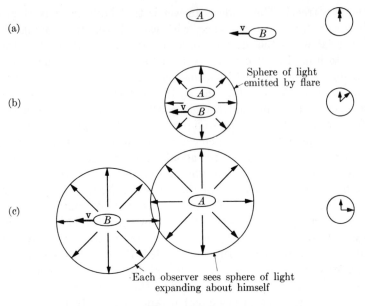

FIGURE 18–5

reference. If the laws of physics took on different forms for different observers in relative motion, they could infer from these differences which of them were "stationary" in space and which were "moving"; but this distinction does not exist in nature, since there is no universal frame of reference, and the above postulate is an expression of this fact.

The second postulate states that

The speed of light in free space has the same value for all observers, regardless of their state of motion.

This postulate expresses the result of the Michelson-Morley experiment.

At first glance these postulates do not seem so very extraordinary, a misleading impression indeed, as we can show with a simple example. Suppose that we have the two boats A and B once more, this time in a large lake (Fig. 18–5a). Boat A is stationary and boat B is drifting at the constant velocity v. There is a fog present, and the observers in each boat have no idea which boat is moving. At the instant that B is abreast of A (Fig. 18–5b), somebody in one of the boats fires a flare. That light from the flare spreads out uniformly in all directions, something we could verify if our visual responses were fast enough by watching the light scattered by the fog. An observer on either boat must see a sphere of light expanding with _him_ at its center (Fig. 18–5c), according to the postulates of

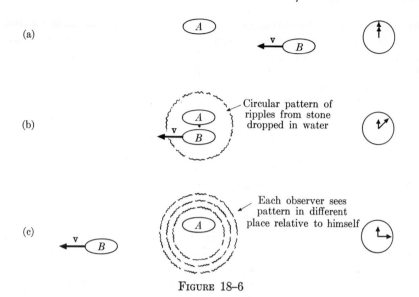

FIGURE 18–6

special relativity, even though one of them is changing his position with respect to the point where the flare went off. But it is impossible for the observers to tell which of them is undergoing such a change in position, since (owing in this case to the fog) there is no frame of reference other than each boat itself, and so, since the speed of light is the same for both of them, they must both see the identical phenomenon. And, of course, the fog is irrelevant; nothing would be changed in its absence except our ability to visualize what is going on.

What is so remarkable about the situation of Fig. 18–5? Let us consider a more familiar analog. It is a clear day and, instead of firing a flare, somebody drops a stone in the water. A circular pattern of ripples spreads out, as in Fig. 18–6, *which appears different* to observers on each boat. Merely by observing whether or not they are at the center of the pattern of ripples, the observers can tell whether they are stationary with respect to the water or not. Here the water behaves as a frame of reference, and an observer on a boat moving through the water will measure a ripple speed with respect to him different from that measured by an observer on a stationary boat. It is important for us to note that motion and waves *in water* are entirely different from motion and waves *in space;* water is in itself a frame of reference while space is not, and wave speeds in water vary with the observer's motion while wave speeds in space do not.

There is only one way of accounting for the fact that observers in the two boats observe identical expanding spheres of light, and that is to assume that the coordinate system of each observer, as seen by the other, is affected by their relative motion. When this idea is worked out, using no more than the ordinary laws of physics and Einstein's postulates, it is found that a host of peculiar effects arise—and all of them, no matter how strange to us, have been confirmed by experiment.

Among the first of these effects is the conclusion that measurements of space and time made by an observer on distances and events in a frame of reference moving with respect to him depend upon the ratio between the relative speed v and the speed of light c. If we find that the length of a rocket is L_0 when it is on its launching pad, we will find its length L when it moves with the speed v to be

$$L = L_0\sqrt{1 - (v^2/c^2)}. \tag{18-5}$$

The length of an object in motion with respect to an observer is measured by the observer to be shorter than when it is at rest with respect to him. This shortening works both ways; to a man in a rocket, measurements indicate that objects on the earth behind him are shorter than they were when he was on the ground by the amount specified in Eq. (18–5). The length of an object is a maximum when measured in a reference frame in which it is stationary, and its length is less when measured in a reference frame in which it is moving. [The ratio between L and L_0 in Eq. (18–5) is the same as that in Eq. (18–4), which referred to the times of travel of the two light beams, so that we can interpret the result of the Michelson-Morley experiment, if we wish, as evidence for the contraction of the length of their apparatus in the direction of the earth's motion.]

The relativistic length contraction is negligible for ordinary speeds, but is an important effect at speeds close to the speed of light. A speed of 1000 mi/sec seems enormous to us, yet it results in a shortening in the direction of motion to only

$$\frac{L}{L_0} = \sqrt{1 - (v^2/c^2)} = \sqrt{1 - \frac{(1000 \text{ mi/sec})^2}{(186{,}000 \text{ mi/sec})^2}}$$

$$= 0.999985 = 99.9985\%$$

of the length at rest. On the other hand, something traveling at nine-tenths the speed of light is shortened to

$$\frac{L}{L_0} = \sqrt{1 - \frac{(0.9c)^2}{c^2}} = 0.436 = 43.6\%$$

of its length at rest, a significant change.

Time intervals, too, are affected by relative motion. A clock moving with respect to an observer appears to tick less rapidly than it does when it is at rest with respect to him; that is, if we observe the length of time some event requires in a frame of reference in motion relative to us, our clock would indicate a longer time interval than a clock in the moving frame. If t_0 is the interval measured by a clock in the moving frame of reference and t the interval we measure,

$$t = \frac{t_0}{\sqrt{1 - (v^2/c^2)}} . \tag{18-6}$$

A good illustration of both the *time dilation* of Eq. (18–6) and the length contraction of Eq. (18–5) occurs in the decay of unstable particles called mu-mesons, whose properties we shall discuss in greater detail in Chapter 25. For the moment what concerns us is the fact that a mu-meson decays into an electron an average of 2×10^{-6} sec after it comes into being. Now mu-mesons are created in the atmosphere thousands of meters above the ground by fast cosmic-ray particles incident upon the earth from space, and reach sea level in profusion. Their speeds are observed to be about 2.994×10^8 m/sec, or 0.998 of the velocity of light c. But in $t_0 = 2 \times 10^{-6}$ sec, the meson lifetime, they can travel a distance of only

$$h = vt_0 = 2.994 \times 10^8 \frac{\text{m}}{\text{sec}} \times 2 \times 10^{-6} \text{sec} = 599 \text{ m},$$

while they are actually created at elevations 10 or more times greater than this.

We can resolve the meson paradox by using the results of the special theory of relativity. Let us first do this from the frame of reference of the meson, where its lifetime is 2×10^{-6} sec. While the meson lifetime is unaffected by the motion, its distance to the ground appears shortened by the factor,

$$\frac{h}{h_0} = \sqrt{1 - (v^2/c^2)}.$$

That is, while we, on the ground, measure the altitude of the meson as h_0, the meson sees it as h. If we let h be 599 meters, the maximum distance the meson can go in its own frame of reference at the speed $0.998c$ before decaying, we find that the corresponding distance h_0 in *our* reference frame is

$$h_0 = \frac{h}{\sqrt{1 - (v^2/c^2)}} = \frac{599 \text{ m}}{\sqrt{1 - [(0.998c)^2/c^2]}}$$

$$= \frac{599 \text{ m}}{\sqrt{1 - 0.996}} = \frac{599 \text{ m}}{0.063} = 9510 \text{ m}.$$

Hence, despite its brief life span, the meson is able to reach the ground from quite respectable altitudes.

Now let us apply the special theory of relativity from the point of view of an observer on the ground. On the ground we find the meson altitude to be h_0, but its lifetime is no longer $t_0 = 2 \times 10^{-6}$ sec; its apparent lifetime t *to us* has been extended owing to the relative motion to the value

$$t = \frac{t_0}{\sqrt{1 - (v^2/c^2)}} = \frac{2 \times 10^{-6} \text{ sec}}{\sqrt{1 - [(0.998c)^2/c^2]}}$$

$$= \frac{2 \times 10^{-6} \text{ sec}}{0.063} = 31.7 \times 10^{-6} \text{ sec.}$$

almost 16 times longer than when it is at rest with respect to us. In 31.7×10^{-6} sec a meson whose speed is $0.998c$ can travel a distance

$$h_0 = vt = 2.994 \times 10^8 \frac{\text{m}}{\text{sec}} \times 31.7 \times 10^{-6} \text{ sec}$$

$$= 9510 \text{ m,}$$

the same distance we found previously. The two points of view give consistent results.

18–4 The relativity of mass

Another conclusion from the theory of relativity is that mass is no more independent of motion than space and time are. As seen from the earth, a rocket ship in flight is shorter than its twin still on the ground, its clocks tick more slowly, and, in addition, its mass is greater. If the *rest mass* of an object (that is, its mass as measured by somebody stationary with respect to it) is m_0, the mass m that will be measured by somebody moving with the speed v with respect to it is

$$m = \frac{m_0}{\sqrt{1 - (v^2/c^2)}}. \tag{18–7}$$

Since the denominator of this equation is always less than one, an object will always appear more massive when in relative motion than when at rest. This mass increase is reciprocal; to a measuring device on the rocket ship in flight its twin ship on the ground also appears to have a mass m greater than its own mass m_0.

Relativistic mass increases are significant only at speeds approaching that of light. At a speed one-tenth that of light, the mass increase amounts to only 0.5%, but this increase is over 100% at a speed of nine-tenths that of light. Only atomic particles such as electrons, protons, mesons, and so on have sufficiently high speeds for relativistic effects to take place, and in dealing with such particles, as we have seen, these effects must be carefully taken into account. Historically, the first confirmation of Eq. (18–7) was the discovery of Bucherer in 1908 that the ratio e/m of the electron's charge to its mass was smaller for fast electrons than for slow ones; this equation, like the others of special relativity, has been verified so many times that it is now accepted as one of the basic formulas of physics.

Equation (18–7) has something interesting to say about the greatest speed an object can have. The closer v approaches c, the closer v^2/c^2 approaches one, and the closer $\sqrt{1 - (v^2/c^2)}$ approaches zero. As the denominator of Eq. (18–7) becomes smaller, the mass m becomes larger, so that if the relative speed v actually were equal to the speed of light, the object's mass would be infinite. The concept of an infinite mass anywhere in the universe is, of course, nonsense on many counts; it would have required an infinite force to have accelerated it to the speed of light, its length in the direction of motion would be zero by Eq. (18–5) so that its volume would be zero, and it would exert an infinite gravitational force on all other bodies in the universe. Hence we interpret Eq. (18–7) to mean that no material body can ever equal or exceed the speed of light.

We should keep in mind that c in the formulas of relativity refers to the speed of light in free space, 3×10^8 m/sec. In all material media, such as water, glass, and air, light travels more slowly than this, and atomic particles are capable of moving with higher speeds in such media than the speed of light *in them,* though never faster than the speed of light in free space. When an electrically charged particle moves through a substance at a speed exceeding that of light in the substance, a cone of light waves is emitted in a process roughly similar to that in which a ship produces a bow wave as it moves through the water faster than water waves do. The emission of light waves under these circumstances constitutes the *Cerenkov effect.*

18–5 Mass and energy

The most famous of the results Einstein obtained in 1905 from the postulates of special relativity is the fact that mass is a form of energy. We may remember that the conversion factor between heat and energy, called the mechanical equivalent of heat, is 4.186 j/cal; Einstein found the conversion factor between mass and energy to be c^2, the square of the speed of light, or 9×10^{16} j/kg! In the form of an equation,

$$E = mc^2. \tag{18-8}$$

Equation (18–8) states that the total energy (in joules) of a body of mass m (in kilograms) is the product of m and the square of the speed of light (in meters per second).

Notice that E is the *total* energy of the body to an observer who measures the body's mass as m. If the body is at rest relative to the observer, its mass is just its rest mass m_0 and the energy

$$E_0 = m_0 c^2 \tag{18-9}$$

is called the *rest energy* of the body. Thus a 1-kg mass is equivalent to 9×10^{16} joules. Even a minute bit of matter can yield an immense amount of energy. In fact, the conversion of matter into energy is the energy source of the sun and stars as well as that of atomic and hydrogen bombs, of chemical reactions, and, indeed, of *all* processes in which energy is liberated.

If the E of Eq. (18–8) is the total energy of a moving body and the E_0 of Eq. (18–9) is its rest energy, the difference between them must be the body's kinetic energy. Hence

$$\text{KE} = E - E_0 = mc^2 - m_0 c^2.$$

Substituting for the mass m from Eq. (18–7),

$$\text{KE} = \frac{m_0 c^2}{\sqrt{1 - (v^2/c^2)}} - m_0 c^2. \tag{18-10}$$

When the relative speed v is small compared with c, we expect that the formula for kinetic energy should become just the $\frac{1}{2}m_0 v^2$ we are familiar with and which has been verified by experiment at low speeds. Let us see whether this is true. In general, the binomial theorem of algebra tells us that if some quantity x is much smaller than 1,

$$(1 + x)^n = 1 + nx.$$

Here, with

$$x = \frac{-v^2}{c^2}, \quad \text{and} \quad n = -\tfrac{1}{2}$$

[since $(1/\sqrt{Y}) = Y^{-1/2}$], we find that Eq. (18–10) indeed becomes

$$\bullet \quad \text{KE} = \left(1 + \frac{\frac{1}{2}v^2}{c^2}\right) m_0 c^2 - m_0 c^2 = \tfrac{1}{2}m_0 v^2.$$

In the foregoing calculation, relativity has met an important test; it has yielded exactly the same results as those of ordinary mechanics at low speeds, where we know by experience that the latter is perfectly valid. It is proper, then, to think of the principles of mechanics that we developed in Chapters 1 through 6 as no more than approximations to the correct relativistic principles of mechanics, valid approximations in nearly all common situations but requiring modification under certain circumstances.

IMPORTANT TERMS

The *Michelson-Morley experiment* showed that the *ether,* a fluid medium supposedly necessary for the propagation of electromagnetic waves, does not exist. The absence of an ether means that there is no universal frame of reference in which measurements can be made, and hence there can be no distinction made between "true" and "apparent" motions.

The *special theory of relativity* is a formulation of the consequences of the absence of a universal frame of reference. It has two postulates: (1) the laws of physics may be expressed in the same set of equations for all frames of reference moving at constant velocity with respect to one another; (2) the speed of light in free space has the same value for all observers, regardless of their state of motion.

The term *relativistic length contraction* refers to the fact that the measured length of an object decreases when it is moving relative to an observer. The term *relativistic time dilation* refers to the fact that clocks moving with respect to an observer appear to tick less rapidly than they do to an observer traveling with the clock. The term *relativity of mass* refers to the fact that a body in motion relative to an observer is measured by him to have a greater mass than when it is at rest relative to him.

The *total energy* of a body is mc^2, the product of its mass and the square of the speed of light. The *rest energy* of a body is m_0c^2, the product of its mass when at rest and the square of the speed of light; the difference between the total energy of a moving body and its rest energy is equal to its kinetic energy.

PROBLEMS

1. Can an observer in a windowless laboratory in principle determine if the earth is (a) moving through space with a uniform velocity, (b) moving through space with a uniform linear acceleration, (c) rotating on its axis?
2. The hypothetical speed of the earth through the ether is its orbital speed of 3×10^4 m/sec. (a) If light takes precisely 3×10^{-7} sec to travel

through the Michelson-Morley apparatus in the direction parallel to this motion, how long will it take to travel through it in the direction perpendicular to this motion? (b) If light of wavelength 6×10^{-7} m is used, what fraction of a wavelength does the difference in the above times represent?

3. Two observers, A on earth and B in a rocket ship whose speed is 2×10^8 m/sec, both set their watches to 1:00 when the ship is abreast of the earth. (a) When A's watch reads 1:30 he looks at B's watch through a telescope. What does it read? (b) When B's watch reads 1:30 he looks at A's watch through a telescope. What does it read?

4. A certain process requires 10^{-6} sec to occur in an atom at rest in the laboratory. How much time will this process require to an observer in the laboratory when the atom is moving at a speed of 5×10^7 m/sec?

5. How fast would a rocket ship have to go for each year on the ship to correspond to two years on the earth?

6. What is the relativistic contraction of an airplane 100 ft long when it moves directly away from an observer at 400 mi/hr?

7. How fast would a rocket ship have to go relative to an observer for its length to be contracted to 99% of its length when at rest?

8. The specific heat of lead is 0.03 kcal/kg·°C. If a bar of lead has a mass of 100 kg at 20°C, how much mass will it gain when heated to 300°C?

9. How much mass is lost by 1 kg of water at 0°C when it turns to ice at 0°C?

10. Compute the rest energy of a proton in joules and in electron volts.

11. Dynamite liberates 1.3×10^3 kcal/kg when it explodes. What percentage of its total energy content is this?

12. Typical chemical reactions absorb or release energy at the rate of several electron volts per molecular change. What change in mass is associated with the absorption of 1 ev of energy? With the release of 1 ev of energy?

13. What speed must a body have if its mass is to double?

14. (a) How much energy is required to double the speed of an electron whose initial speed is 7×10^7 m/sec? (b) How much energy is required to double the speed of an electron whose initial speed is 1.4×10^8 m/sec?

15. In Chapter 14 it was stated that the relativistic increase in mass of moving protons limits the ability of a cyclotron to accelerate them. (a) At what speed will a proton's mass exceed its rest mass by 1%? (b) What kinetic energy in electron volts does this speed correspond to?

16. An electron is accelerated through a potential difference of 10^5 volts. (a) What is its speed? (b) What is its mass? (c) Compare the answers to (a) and (b) with the values that classical physics would predict.

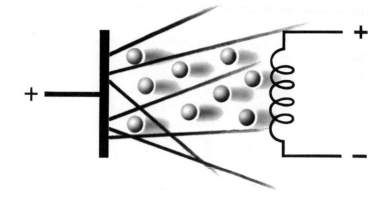

Quanta | 19

The formulation of the theory of relativity and that of the quantum theory of light, both of which took place early in this century, profoundly altered our ways of thinking about the physical world. In the previous chapter we examined some of the remarkable consequences of relativity, consequences so alien to our intuition that our first feeling must be one of astonishment. Now we come to the realm of quanta, which will be no less remarkable or astonishing. But these subjective terms are really not justified, because they are merely based upon the limits to our imagination that are imposed by our experience; if we were in a world where we were about the same size as an electron, relativistic and quantum phenomena would be familiar (though then most macroscopic phenomena would not). We shall find that the ideas presented in this chapter are not only intrinsically of great interest, but are also essential if we are to understand the material on the atom and the nucleus that follows.

19–1 The photoelectric effect

Toward the end of the 19th century a number of experiments were performed that revealed the emission of electrons from a metal surface when light (particularly ultraviolet light) is shined on it (Fig. 19–1). This phenomenon is known as the *photoelectric effect*. It is not, at first glance, anything to surprise us, for light waves carry energy, and some of the energy absorbed by the metal may somehow concentrate on individual electrons and reappear as kinetic energy. Upon closer inspection of the data, however, we find that the photoelectric effect can hardly be explained in so straightforward a manner.

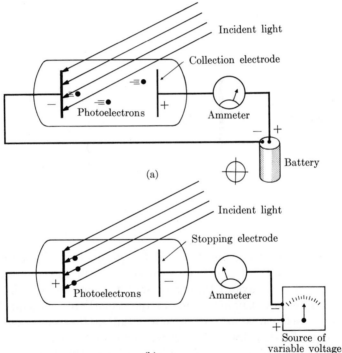

Fig. 19–1. (a) A method of detecting the photoelectric effect. The photoelectrons ejected from the metal plate being irradiated are attracted to the positive collection electrode at the other end of the tube, and the current that results is measured with the ammeter. (b) A method of detecting the maximum energy of the photoelectrons. As the stopping electrode is made more negative, the slower photoelectrons are repelled before they can reach it. Finally a voltage will be reached at which no photoelectrons whatever are received at the stopping electrode, as indicated by the current dropping to zero, and this voltage corresponds to the maximum photoelectron energy.

The first peculiarity of the photoelectric effect is that the energy of the emitted electrons (which are called *photoelectrons*) does not depend upon the intensity of the light. A bright light yields more electrons than a dim one, but their average energy remains the same. Furthermore, even when the metal surface is faintly illuminated, the photoelectrons leave the surface immediately. This behavior contradicts the electromagnetic theory of light, which predicts that the energy of photoelectrons should depend upon the intensity of the light beam responsible for them and that, if the beam has a very low intensity, a certain period of time (perhaps even days) must elapse on the average before any individual electrons accumulate enough energy to leave the metal.

Still more puzzling is the fact that the photoelectron energy depends upon the *frequency* of the light employed. At frequencies below a certain critical one (which is characteristic of the particular metal), no electrons whatever are emitted. Above this threshold frequency, the photoelectrons have a range of energies from zero to a certain maximum value, and *this maximum energy increases with increasing frequency*. High frequencies result in high maximum photoelectron energies, low frequencies in low maximum photoelectron energies. Thus a faint blue light produces electrons with more energy than those produced by a bright red light, although the latter yields a greater number of them.

Figure 19–2 is a graph of maximum photoelectron energy plotted against the frequency f of the incident light in a particular experiment.

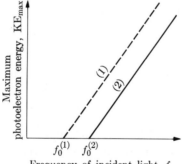

FIG. 19–2. The variation of maximum photoelectron energy with the frequency of the incident light for two target metals. No photoelectrons are emitted for frequencies less than $f_0^{(1)}$ in the case of metal (1) and less than $f_0^{(2)}$ in the case of metal (2). In both cases, however, the angle between the experimental line and either axis is the same. Hence we may write the equation of the lines as $KE_{max} = h(f - f_0)$, where h has the same value in all cases but f_0, the minimum frequency required for photoelectric emission to occur, depends upon the nature of the target metal.

It is clear that the relationship between maximum photoelectron energy, KE_{max}, and light frequency is a simple one, which we can express in equation form as

$$KE_{max} = h(f - f_0) = hf - hf_0, \qquad (19\text{-}1)$$

where f_0 is the threshold frequency below which no photoemission occurs and h is a constant. Significantly, the value of h is the same in *all* cases, although f_0 varies with the particular metal being illuminated.

19-2 Quantum theory of light

Aware that the electromagnetic theory of light, despite its notable success in accounting for other light phenomena, failed to explain the photoelectric effect, Albert Einstein in 1905 sought some other way of understanding it. He found what he wanted in a strange assumption that Max Planck, a German physicist, had had to make some years earlier in order to derive a formula that agreed with the experimental results on the radiation given off by an object so hot that it is luminous (a poker thrust in a fire, for instance). Planck found that the accepted physical laws of the time predicted the observed characteristics of this radiation *provided* that the radiation is considered as though being emitted in little bursts of energy, rather than continuously. These bursts of energy are called quanta. Planck showed that the energy E of each quantum had to be related to the light frequency f by

$$E = hf, \qquad (19\text{-}2)$$

where h is a constant (known today as Planck's constant) whose value is

$$h = 6.63 \times 10^{-34}\,\text{j·sec}.$$

While the energy radiated by a heated object must be assumed to come off intermittently, in order for theory and experiment to agree, Planck held to the conventional view that it nevertheless propagates through space continuously.

Einstein saw that Planck's idea could be used to interpret the photoelectric effect if light not only is emitted a quantum at a time but also travels as quanta. Then the h of Eq. (19-1) is the same as the h of (19-2), and the significance of the former equation becomes clear when it is rewritten

$$hf = KE_{max} + hf_0. \qquad (19\text{-}3)$$

What this equation states is simply that

Quantum energy = maximum electron energy
+ energy required to eject an electron.

The reason for a threshold frequency is clear: it corresponds to the energy required to dislodge an electron from the metal surface. (There must be such a minimum energy, or electrons would leave metals all the time.) And, of course, there are several plausible reasons why not all photoelectrons have the same maximum energy. For instance, not all of the quantum energy hf may be transferred to a single electron, and an electron may lose some of its initial energy in collisions within the metal before it actually emerges from the surface.

Let us apply Eq. (19-3) to a specific problem. The threshold frequency for copper is $1.1 \times 10^{15} \sec^{-1}$. When ultraviolet light of frequency $1.5 \times 10^{15} \sec^{-1}$ is shined on a copper surface, what is the maximum energy of the photoelectrons? We find that

$$\text{KE}_{\max} = hf - hf_0 = h(f - f_0)$$

$$= 6.63 \times 10^{-34} \text{ j·sec} \times (1.5 - 1.1) \times 10^{15} \sec^{-1}$$

$$= 2.7 \times 10^{-19} \text{ j.}$$

Since $1 \text{ ev} = 1.6 \times 10^{-19}$ joule, the maximum photoelectron energy is

$$\frac{2.7 \times 10^{-19} \text{ j}}{1.6 \times 10^{-19} \text{ j/coul}} = 1.7 \text{ ev.}$$

Einstein's notion that light travels as a series of little packets of energy (sometimes referred to as quanta, sometimes as *photons*) is in complete contradiction with the wave theory of light (Fig. 19-3). And the wave theory, as we know, has some powerful observational evidence on its side. There is no other way of explaining interference and diffraction effects, for example. According to the wave theory, light spreads out from a source in a manner analogous to the spreading out of ripples on the surface of a lake when a stone is dropped into it, with the energy of the light distributed continuously throughout the wave pattern. According to the quantum theory, light spreads out from a source as a succession of localized packets of energy, each sufficiently small to permit its being absorbed by a single electron. Yet, despite the particle picture of light that it presents, the quantum theory requires a knowledge of the light frequency f, a wave quantity, in order to determine the energy of each quantum.

On the other hand, the quantum theory of light is able to explain the photoelectric effect. It predicts that the maximum photoelectron energy

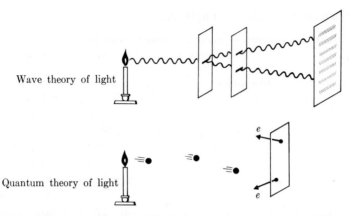

Wave theory of light

Quantum theory of light

FIG. 19–3. The wave theory of light is necessary to explain diffraction and interference phenomena, which the quantum theory cannot explain. The quantum theory of light is necessary to explain the photoelectric effect, which the wave theory cannot explain.

should depend upon the frequency of the incident light and not upon its intensity, precisely the opposite of what the wave theory suggests, and it is able to explain why even the feeblest light can lead to the immediate emission of photoelectrons. The wave theory can give no reason why there should be a threshold frequency below which no photoelectrons are observed, no matter how strong the light beam, something that follows naturally from the quantum theory.

Which theory is correct? The history of physics is filled with examples of physical ideas that required revision or even replacement when new empirical data conflicted with them, but this is the first occasion in which two completely different theories are both required to explain a single physical phenomenon. In thinking about this, it is important for us to note that, in a particular situation, light behaves *either* as though it has a wave nature *or* a particle nature. While light sometimes assumes one guise and sometimes the other, there is no physical process in which both are exhibited. The same light beam can diffract around an obstacle and then impinge on a metal surface to eject photoelectrons, but these two processes occur separately. The electromagnetic theory of light and the quantum theory of light complement each other; each by itself is "correct" in certain experiments, and there are no relevant experiments which neither can explain. Light must be thought of as a phenomenon that sometimes manifests itself as particles and sometimes as waves; while we cannot visualize its "true nature," these complementary theories of light are able to account for its behavior, and we have no choice but to accept them both.

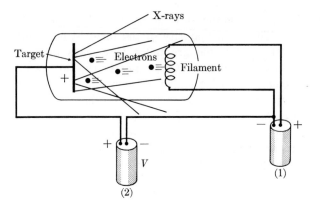

Fɪɢ. 19–4. Diagram of an x-ray tube.

19–3 X-rays

If photons of light can give up their energy to electrons, can the kinetic energy of moving electrons be converted into photons? The answer is that such a transformation is not only possible, but had in fact been discovered (though not understood) prior to the work of Planck and Einstein. In 1895 Roentgen found that a mysterious, highly penetrating radiation is emitted when high-speed electrons impinge on matter. The x-rays (so called because their nature was then unknown) caused phosphorescent substances to glow, exposed photographic plates, traveled in straight lines, and were not affected by electric or magnetic fields. The more energetic the electrons, the more penetrating the x-rays, and the greater the number of electrons, the greater the density of the resulting x-ray beam.

After over 10 years of study, it was finally established that x-rays exhibit, under certain circumstances, both interference and polarization effects, leading to the conclusion that they are electromagnetic waves. From the interference experiments their frequencies were found to be very high, above those in ultraviolet light.

Figure 19–4 is a diagram of an x-ray tube. Battery (1) sends a current through the filament, heating it until it emits electrons. These electrons are then accelerated toward a metallic target by the potential difference V provided by battery (2). The tube is evacuated to permit the electrons to reach the target unimpeded. The impact of the electrons causes the evolution of x-rays from the target.

What is the physical process involved in the production of x-rays? It is known that charged particles emit electromagnetic waves whenever they are accelerated, and so we may reasonably identify x-rays as the radiation accompanying the slowing down of fast electrons when they strike a metal. The great majority of the incident electrons, to be sure, lose their kinetic energy too gradually for x-rays to be evolved, and merely act to heat the target. (Consequently the targets in x-ray tubes are made of metals with high melting points, and a means for cooling the target is often provided.) A few electrons, however, lose much or all of their energy in single collisions with target atoms, and this is the energy that appears as x-rays. In other words, we may regard x-ray production as an inverse photoelectric effect. The highest frequency f_{max} found in the x-rays emitted from a particular tube should therefore correspond to a quantum energy of hf_{max}, where hf_{max} equals the kinetic energy of the electrons. That is,

$$KE = hf_{max}.$$

If we denote the charge of an electron by e, its kinetic energy after being accelerated through a potential difference V is

$$KE = eV;$$

hence we expect that

$$hf_{max} = eV \qquad\qquad (19\text{–}4)$$

in the operation of an x-ray tube.

A conventional x-ray machine might have an operating potential of 50,000 volts. To find the highest frequency present in its radiation, we use Eq. (19–4):

$$f_{max} = \frac{eV}{h} = \frac{1.6 \times 10^{-19} \text{ coul} \times 5 \times 10^4 \text{ volts}}{6.6 \times 10^{-34} \text{ j·sec}}$$

$$= 1.2 \times 10^{19} \text{ sec}^{-1}.$$

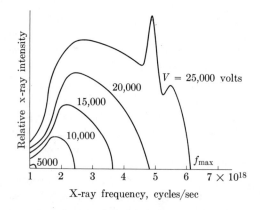

FIGURE 19–5

Figure 19–5 shows the x-ray spectra that result when a molybdenum target is bombarded by electrons at several different accelerating voltages. The proportionality between f_{max} and V predicted by Eq. (19–4) agrees with the measurements, which provides confirmation for the quantum theory of light. The curves show that all frequencies below f_{max} are emitted, as we expect, and that their intensities vary in a regular manner. In addition, the highest-voltage curve possesses curious, sharp "spikes" that indicate the enhanced production of certain frequencies. These spikes originate in rearrangements of the electron structures of the target atoms that are induced by the bombarding electrons. A number of specific energies may be involved in these rearrangements, depending upon the particular target material, so that the presence of discrete x-ray frequencies has a straightforward explanation.

19–4 The Compton effect

In the photoelectric effect, a light photon gives up most or all of its energy to an electron, leading us to attribute particle properties to the photon. How far can we carry the analogy? For instance, can we analyze the case of a photon making a glancing collision with an electron just as if both were billiard balls? Let us see what we can expect in such a collision if, indeed, photons behave like particles. Figure 19–6 shows the collision; an x-ray photon strikes an electron, which is initially at rest, and is deflected to one side, while the electron starts to move. As a result of the impact, the photon loses an amount of energy equal to the

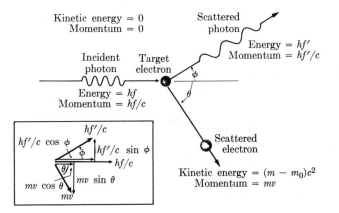

Fig. 19–6. The Compton effect. A vector diagram showing the various momenta and their projections parallel and perpendicular to the original photon direction is included to help in understanding the origin of Eqs. (19–7) and (19–8).

kinetic energy $(m - m_0)c^2$ gained by the electron. (The relativistic formula for kinetic energy is the proper one to use here.) If the original frequency associated with the photon is f, its loss of energy means that its subsequent frequency will be the lower value f', where

Change in photon energy = electron kinetic energy

$$hf - hf' = (m - m_0)c^2. \qquad (19\text{--}5)$$

A moving electron has momentum mv as well as energy, so a photon must have momentum if it is to transfer any to an electron. The momentum of a photon, if it behaves like a particle, should be

Photon momentum $= m_p c$

since it travels with the speed of light c. While a photon has no rest mass, we can take its energy hf as equal to the energy $m_p c^2$ of a particle whose relativistic mass is m_p. Hence

$$hf = m_p c^2, \qquad m_p = \frac{hf}{c^2},$$

and the photon momentum is

$$m_p c = \frac{hf}{c}. \qquad (19\text{--}6)$$

Momentum is a vector quantity, possessing direction as well as magnitude, and we must separately conserve momentum in the original direction of the photon and in a direction perpendicular to it in the plane of the electron and scattered photon. The initial momentum of the photon is hf/c, its final momentum is hf'/c, and the initial and final momenta of the electron are respectively zero and mv. In the original photon direction, as Fig. 19–6 indicates,

Initial momentum = final momentum

$$\frac{hf}{c} + 0 = \frac{hf'}{c} \cos \phi + mv \cos \theta, \qquad (19\text{--}7)$$

and perpendicular to this direction,

Initial momentum = final momentum

$$0 = \frac{hf'}{c} \sin \phi - mv \sin \theta. \qquad (19\text{--}8)$$

When Eqs. (19–5), (19–7), and (19–8) are solved, the new photon frequency f' is found to be related to the original frequency f and to the angle ϕ between the initial photon direction and the scattered photon direction by the formula

$$\frac{1}{f'} = \frac{1}{f} + \frac{h}{m_0 c^2}\,(1 - \cos\phi), \qquad (19\text{–}9)$$

where m_0 is the rest mass of the electron. This formula can readily be checked by experiment; a beam of x-rays of a single, known frequency f is directed at some target, and the frequencies of the scattered x-rays measured at various angles. According to Eq. (19–9) the scattered x-rays should have lower frequencies than those in the original beam, with the change in frequency greatest for the greatest scattering angle. Experiments of this kind were performed by A. H. Compton in the early 1920's, and he was led to derive Eq. (19–9) in order to account for the results. Theory and experiment agree, and the phenomenon they refer to has become known as the *Compton effect*.

19–5 Matter waves

As we have seen, electromagnetic waves under certain circumstances have properties indistinguishable from those of particles. It requires no greater stretch of the imagination to speculate whether what we normally think of as particles might not have wave properties, too. This speculation was first made in 1924 by Louis de Broglie. He started with the formula for the momentum of a photon,

$$\text{Momentum} = \frac{hf}{c},$$

which can be expressed in terms of wavelength λ as

$$\text{Momentum} = \frac{h}{\lambda}$$

since

$$\lambda f = c.$$

Hence, for a photon,

$$\lambda = \frac{h}{\text{momentum}}.$$

De Broglie suggested that this equation for wavelength is a perfectly general one, applying to material particles as well as to photons. In the case of particles,

$$\text{Momentum} = mv,$$

and so the *de Broglie wavelength* of a particle is

$$\lambda = \frac{h}{mv}. \tag{19-10}$$

The higher the momentum of the particle, the shorter its wavelength. $m_0/\sqrt{1 - (v^2/c^2)}$ must be used for m in Eq. (19–10).

How can de Broglie's hypothesis be verified? Perhaps the most striking exhibition of wave behavior is a diffraction pattern, which depends upon the ability of waves both to bend around obstacles and to interfere constructively and destructively with one another. Several years after de Broglie's work, Davisson and Germer in the United States and G. P. Thomson in England independently demonstrated that streams of electrons are diffracted when they are scattered from crystals whose atoms have appropriate spacing. The diffraction patterns they observed were in complete accord with the electron wavelength predicted by Eq. (19–10). As is true for electromagnetic waves, the wave and particle aspects of moving bodies never appear together in the same experiment, so we cannot determine which is the "correct" description. All we can say is that at certain times we must think of a moving body as a particle (in the sense of Newton's laws of motion) and at other times as a wave; whatever the ultimate nature of matter, its manifestations to us exhibit this inescapable duality.

Let us calculate the de Broglie wavelengths of two moving objects, an automobile whose speed is 60 mi/hr and an electron whose speed is 10^7 m/sec. The automobile might weigh 3200 lb, corresponding to a mass of 100 slugs, and 60 mi/hr = 88 ft/sec; hence,

$$\lambda = \frac{h}{mv} = \frac{6.62 \times 10^{-34}\,\text{j·sec} \times 0.738\,\text{ft·lb/j}}{100\,\text{slugs} \times 88\,\text{ft/sec}} = 5.5 \times 10^{-38}\,\text{ft.}$$

The wavelength of the automobile is so small relative to its dimensions that we expect no wave aspects in its behavior. For the electron, however,

$$\lambda = \frac{h}{mv} = \frac{6.62 \times 10^{-34}\,\text{j·sec}}{9.1 \times 10^{-31}\,\text{kg} \times 10^7\,\text{m/sec}} = 7.3 \times 10^{-9}\,\text{m,}$$

which is of the order of magnitude of atomic dimensions. It is not surprising, then, that electrons with velocities of about 10^7 m/sec exhibit diffraction effects when incident upon crystals.

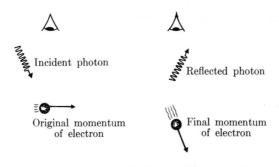

Incident photon

Reflected photon

Original momentum
of electron

Final momentum
of electron

FIG. 19–7. An example of the uncertainty principle: it is impossible to accurately determine the position and momentum of a body at the same time.

19–6 The uncertainty principle

The wave nature of particles and the particle nature of waves have a rather unexpected concomitant. Suppose that we wish to measure the position and momentum of something at a particular instant. To do so, we must touch it with something else that will carry the required information back to us; that is, we must poke it with a stick, shine light on it, or perform some similar act. Suppose we are looking at an electron with light of wavelength λ, as in Fig. 19–7. Each photon of this light has the momentum h/λ. When one of these photons bounces off the electron (which must occur if we are to "see" it), the electron's original momentum will be changed. The exact amount of the change Δmv cannot be predicted, but it will be of the same order of magnitude as the photon momentum h/λ. Hence

$$\Delta mv \approx h/\lambda; \qquad (19\text{--}11)$$

the *larger* the wavelength, the smaller the uncertainty in momentum. Because light is a wave phenomenon, we cannot expect to determine the electron's position with perfect accuracy even with the best of instruments; a reasonable estimate of the irreducible uncertainty Δx in the measurement might be one wavelength. That is

$$\Delta x \approx \lambda; \qquad (19\text{--}12)$$

the *smaller* the wavelength, the smaller the uncertainty in position. Hence if we use light of short wavelength to increase the accuracy of our position measurement, there will be a corresponding decrease in the accuracy of our momentum measurement, while light of long wavelength will yield an accurate momentum but an inaccurate position.

Combining Eqs. (19–11) and (19–12) gives

$$\Delta x \, \Delta mv \approx h, \qquad\qquad (19\text{--}13)$$

which is one form of the *uncertainty principle* first obtained by Werner Heisenberg in 1927. It states that the product of the uncertainty Δx in the position of a body at some instant and the uncertainty Δmv in its momentum at the same instant is approximately equal to Planck's constant. In other words, we cannot ever hope to simultaneously measure both position and momentum with perfect accuracy. The uncertainty is not in our instruments but in nature. On a macroscopic scale, since h is so minute, the limitation imposed upon measurements by the uncertainty principle is negligible, but on a microscopic scale the uncertainty principle dominates many phenomena.

A significant application of the uncertainty principle is to the question of whether electrons are present within atomic nuclei. As we shall learn in Chapter 23, nuclei are roughly 10^{-15} meter across. If an electron is to be confined within a nucleus, the uncertainty in its position cannot exceed 10^{-15} meter. The consequent inherent uncertainty in the electron's momentum is

$$\Delta(mv) \approx \frac{h}{\Delta x} \approx \frac{6.62 \times 10^{-34} \, \text{j·sec}}{10^{-15} \, \text{m}}$$

$$\approx 6.62 \times 10^{-19} \, \frac{\text{kg·m}}{\text{sec}} \, .$$

This momentum uncertainty corresponds to an energy uncertainty of over 10^9 ev, which is a billion electron volts! But there is no evidence that electrons found in atoms ever have much more than 1/1000 of this huge amount of energy, and we therefore conclude that electrons are not present inside atomic nuclei.

Another form of the uncertainty principle is easily obtained. Suppose that we are measuring the energy E emitted in some process during the period of time Δt. If the energy is in the form of electromagnetic waves, the time Δt limits the accuracy with which we can measure the wave frequency f. Because the uncertainty in the number of waves we count in a given wave train is of the order of one wave, and

$$\text{Frequency} = \frac{\text{number of waves}}{\text{time interval}},$$

the uncertainty in frequency Δf of our determination is

$$\Delta f \approx \frac{1}{\Delta t} \, .$$

The corresponding uncertainty in energy ΔE is

$$\Delta E \approx h \, \Delta f,$$

and so

$$\Delta E \approx \frac{h}{\Delta t}$$

or

$$\Delta E \, \Delta t \approx h. \tag{19-14}$$

Equation (19–14) states that the product of the uncertainty in a measured amount of energy and the time available for the measurement is approximately equal to Planck's constant.

19–7 Causality

Until now we have taken for granted the notion of causality, namely that a given cause always produces a specific effect. The "laws" of physics that we have studied are all cause-and-effect relationships. For example, $\mathbf{F} = m\mathbf{a}$, Newton's second law of motion, states that the acceleration a particle experiences as a result of the application of a force is invariably proportional to the magnitude of the force and inversely proportional to the particle's mass. But, because of the uncertainty principle, it is impossible ever to perform experiments of infinite accuracy, and therefore impossible ever to establish that $\mathbf{F}$ is invariably equal to precisely $m\mathbf{a}$.

How seriously can we take relationships whose truth cannot possibly be verified? The answer is not as clear-cut as we would prefer. While causal relations break down when applied to individual elementary particles, and hence cannot be regarded as expressions of *truth*, they may nevertheless be regarded as expressions of *probability*. If sufficiently accurate measurements could be made of the responses of electrons to applied forces, we would expect to find that $\mathbf{F} = m\mathbf{a}$ is obeyed *on the average*, but that the individual data would show deviations from this behavior. The ordinary material bodies we encounter in our lives are aggregates of such vast numbers of particles that their average behavior is all that we perceive. Hence causality is a reasonably reliable guide to macroscopic events, but we should not expect it to hold in applications remote from our direct experience.

IMPORTANT TERMS

The *photoelectric effect* is the emission of electrons from a metal surface when light is shined on it.

The *quantum theory of light* states that light travels in tiny bursts of energy called *quanta* or *photons*. If the frequency of the light is f, each burst has the energy hf, where h is known as *Planck's constant*. The quantum theory of light is required to account for the photoelectric effect.

X-rays are high-frequency electromagnetic waves emitted when fast electrons impinge on matter.

The *Compton effect* is the scattering of an x-ray photon by an electron, with the photon changing frequency in the process.

A moving body behaves as though it has a wave nature. The waves representing such a particle are called *de Broglie waves*.

The *uncertainty principle* states that it is impossible to simultaneously measure the position and momentum of a body with perfect accuracy. The product of the uncertainties in the position and momentum of a body is approximately equal to Planck's constant h.

PROBLEMS

1. If light transfers energy by means of separate quanta, why do we not perceive a faint light as a series of tiny flashes?

2. Yellow light has a wavelength of 6×10^{-7} m. How many photons are emitted per second by a yellow lamp radiating at a power of 10 watts? (1 watt = 1 j/sec.)

3. A radio transmitter operates at a frequency of 880 kc/sec (880×10^3 cycles/sec) and a power of 10,000 watts. How many photons per second does it emit?

4. What is the lowest-frequency light that will cause the emission of photoelectrons from a surface whose nature is such that 1.9 ev is required to eject an electron?

5. Photoelectrons are emitted with a maximum speed of 7×10^5 m/sec from a surface when light of frequency 8×10^{14} cycles/sec is shined on it. What is the threshold frequency for this surface?

6. Light of wavelength 5×10^{-7} m falls on a potassium surface whose nature is such that 2 ev is needed to eject an electron. What is the maximum kinetic energy in ev of the photoelectrons that are emitted?

7. Why do you think the wave aspect of light was discovered earlier than its particle aspect?

8. What potential difference must be applied across an x-ray tube for it to emit x-rays with a minimum wavelength of 10^{-11} m?

9. What potential difference must be applied across an x-ray tube for it to emit x-rays with a maximum frequency of 10^{19} cycles/sec?

10. A target is bombarded by 8×10^4 ev electrons. What is the highest frequency present in the emitted x-rays?

11. Electrons are accelerated in television tubes through potential differences of about 10,000 volts. Find the highest frequency of the electromagnetic waves that are emitted when these electrons strike the screen of the tube. Use Fig. 16–5 to determine what type of waves these are.

12. Find the energy and momentum of an x-ray photon whose frequency is 5×10^{18} cycles/sec.

13. Find the energy and momentum of an x-ray photon whose wavelength is 2×10^{-11} m.

14. An x-ray photon of frequency 3×10^{19} cycles/sec is scattered through an angle of 90° after colliding with a stationary electron. (a) What is its new frequency? (b) What is the kinetic energy of the electron after the collision?

15. An x-ray photon of frequency 10^{19} cycles/sec is scattered through an angle of 45° after colliding with a stationary electron. (a) What is its new frequency? (b) What is the kinetic energy of the electron after the collision?

16. An x-ray photon of frequency 1.5×10^{19} cycles/sec undergoes Compton scattering. Its new frequency is 1.2×10^{19} cycles/sec. What is the change in the kinetic energy of the electron it has struck?

17. A photon colliding with an electron at rest gives the electron an energy of 10,000 ev. Find the initial and final frequencies of the photon.

18. Give as many reasons as you can to support (a) the wave theory of light, (b) the particle theory of light.

19. Find the de Broglie wavelength of a 10-ton truck whose speed is 60 mi/hr.

20. Calculate the de Broglie wavelength of (a) an electron whose speed is 1×10^8 m/sec, and (b) an electron whose speed is 2×10^8 m/sec. Use relativistic formulas.

21. Calculate the de Broglie wavelength of a proton whose kinetic energy is 1 Mev ($= 10^6$ ev). This calculation may be made nonrelativistically.

22. (a) How could you experimentally distinguish between a photon whose momentum is 10^{-22} kg·m/sec and an electron whose momentum is 10^{-22} kg·m/sec? (b) What is the energy of each? (c) What is the wavelength of each?

23. How could you experimentally distinguish between an electromagnetic wave whose wavelength is 10^{-11} m and an electron whose de Broglie wavelength is 10^{-11} m?

24. What is the approximate momentum imparted to a proton initially at rest by a measurement which locates its position to 10^{-11} m?

25. (a) An electron is confined in a box 10^{-9} m in length. What is the uncertainty in its velocity? (b) A proton is confined in the same box. What is the uncertainty in its velocity?

26. The electron in a hydrogen atom may be thought of as confined to a region a radius of 5×10^{-11} m from the proton. (a) Use the uncertainty principle to calculate the momentum of the electron, and from this calculate its kinetic energy. (b) What keeps the electron in this region despite its kinetic energy?

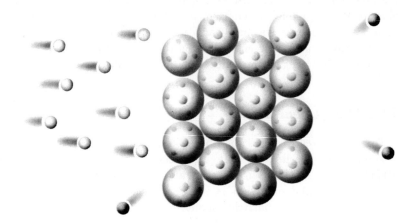

The Atom | 20

By the beginning of this century a substantial body of evidence had been accumulated in support of the idea that the chemical elements consist of atoms. The nature of the atoms themselves, however, was still a mystery, although a significant clue had been discovered. This clue was the fact that electrons are constituents of atoms, which suggests that electrical forces are significant in atomic phenomena. J. J. Thomson, whose work had led to the identification of the electron, proposed in 1898 that atoms are spheres of positively charged matter containing embedded electrons, much as a fruit cake is studded with nuts. Thirteen years later an important experiment was performed whose results forced the abandonment of Thomson's apparently plausible model and its replacement by a model which seemed at the time to raise more questions than it answered. As we shall learn in this chapter, the search for a way of reconciling the experimental data on the atom with the laws of physics has profoundly affected our ideas on the structure of matter.

20–1 The nuclear atom

The most direct method for finding out what is inside a fruit cake is to simply plunge a finger into it. In essence this is the classic experiment performed in 1911 by Geiger and Marsden at the suggestion of Ernest Rutherford to find out what is inside an atom. The probes that they used were fast *alpha particles* spontaneously emitted by certain radioactive elements. For the present all we need know about alpha particles is that their mass is that of a helium atom and that they carry positive charges equal in magnitude to twice the charge on the electron *e;* we shall discuss their properties in more detail in Chapter 23. Geiger and Marsden placed a sample of an alpha-emitting substance behind a lead screen with a small hole in it, as shown in Fig. 20–1, so that a narrow beam of alpha particles is produced. On the other side of a thin metallic foil in the path of the beam they placed a movable zinc sulfide screen which gave off a flash of light when struck by an alpha particle, thus indicating the extent to which the alpha particles are scattered from their original direction of motion. They expected to find that most of the alpha particles go through the foil without being affected by it, with the remainder receiving only slight deflections. This anticipated behavior

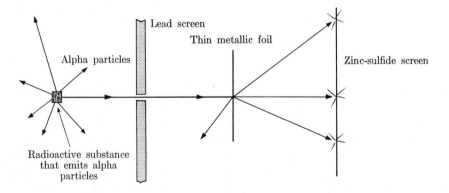

Fɪɢ. 20–1. Diagram of the Rutherford experiment.

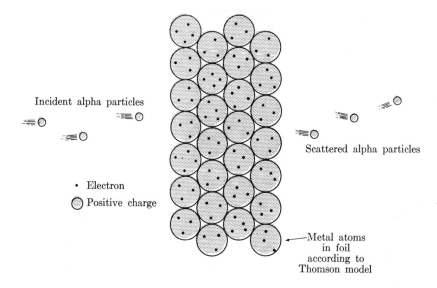

Incident alpha particles

• Electron

◉ Positive charge

Scattered alpha particles

Metal atoms
in foil
according to
Thomson model

FIG. 20-2. According to the Thomson model of the atom, positive charge is spread uniformly throughout its volume with electrons embedded here and there. Only weak electric fields are present within atoms on the basis of this model, and it accordingly predicts very little deflection of alpha particles striking a thin foil. This prediction does not agree with experiment.

follows from the Thomson atomic model, in which the positive and negative electric charges within an atom are assumed to be spread more or less evenly throughout its volume (Fig. 20–2). If the Thomson model is correct, only weak electric forces would be exerted on alpha particles passing through a thin foil, and their momenta would be enough to carry them through with only minor departures from their original paths.

What Geiger and Marsden actually found was that, while most of the alpha particles indeed emerged unaffected from the foil, the others underwent deflections through very large angles, in some cases even being scattered in the backward direction (Fig. 20–1). Since alpha particles are relatively heavy (almost 8000 times more massive than electrons) and have high initial speeds, it was clear that strong forces had to be exerted upon them to cause such marked deflections. To explain the results, Rutherford adopted the hypothesis that an atom is composed of a tiny *nucleus*, in which its positive charge and nearly all its mass are concentrated, with the electrons some distance away (Fig. 20–3). With the atom largely empty space, it is easy to see why most alpha particles proceed right through a thin foil. On the other hand, an alpha particle coming near a nucleus experiences a strong electric force, and is likely to be

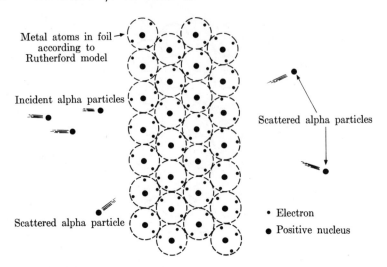

Metal atoms in foil →
according to
Rutherford model

Incident alpha particles

Scattered alpha particles

Scattered alpha particle

• Electron
● Positive nucleus

Fig. 20–3. According to the Rutherford model of the atom, positive charge is concentrated in a tiny nucleus at its center with electrons some distance away. Strong electric fields are present within atoms on the basis of this model, and it accordingly predicts considerable deflection of alpha particles striking a thin foil. This prediction agrees with experiment.

scattered through a large angle. (The atomic electrons, being very light, are readily knocked out of the way by alpha particles, while the situation is reversed in the case of the nuclei, which are heavier than alpha particles.)

Rutherford was able to obtain a formula for the scattering of alpha particles by thin foils on the basis of his hypothesis that agreed with the experimental results. Rutherford is therefore credited with the "discovery" of the nucleus.

20–2 Electron orbits: force balance

The picture of the atom that emerges from Rutherford's work is a tiny, massive nucleus with a positive charge surrounded by distant, light, negatively charged electrons. These electrons cannot be stationary, since the electrostatic force between them and the nucleus would attract them to the latter at once. If the electrons are in motion about the nucleus, however, dynamically stable orbits, like those of the planets about the sun, are possible.

Let us restrict ourselves for the time being to the hydrogen atom, which, having an atomic number of $z = 1$, is the simplest, with but a

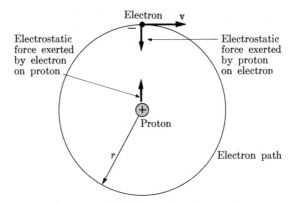

Fɪɢ. 20–4. The hydrogen atom consists of an electron circling a proton. The electrostatic force exerted by the proton on the electron provides the centripetal force required to hold it in a circular path. The proton is nearly 2000 times heavier than the electron, and so its motion under the influence of the electrostatic force the electron exerts is not significant.

single proton as nucleus and one electron around it (Fig. 20–4). Then the centripetal force

$$F_c = \frac{mv^2}{r}$$

holding the electron in an orbit r from the nucleus must be provided by the electrostatic force

$$F_e = \frac{1}{4\pi\epsilon_0}\frac{e^2}{r^2}$$

between them, and

$$\frac{mv^2}{r} = \frac{1}{4\pi\epsilon_0}\frac{e^2}{r^2}. \qquad (20\text{–}1)$$

The electron speed v is therefore related to its orbit radius r by the formula

$$v = \frac{e}{\sqrt{4\pi\epsilon_0 m r}}. \qquad (20\text{–}2)$$

The total energy E of the electron in a hydrogen atom is the sum of its kinetic energy

$$KE = \tfrac{1}{2}mv^2$$

and its electrostatic potential energy

$$PE = -\frac{1}{4\pi\epsilon_0}\frac{e^2}{r}.$$

(The latter formula follows from a calculation of the amount of work

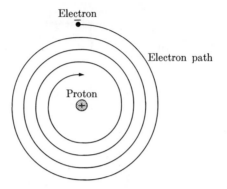

Fig. 20–5. According to electromagnetic theory, an accelerated electron radiates electromagnetic waves. The electron in a hydrogen atom, since it constantly experiences a centripetal acceleration, should therefore radiate energy and spiral into the nucleus.

that would have to be done to bring the electron to an infinite distance from the proton; the potential energy here is a *negative* quantity since, being attracted to the proton, the electron cannot do work on anything outside the atom if it were to be released from its orbital motion.) Hence

$$E = KE + PE = \frac{mv^2}{2} - \frac{1}{4\pi\epsilon_0} \frac{e^2}{r}.$$

Substituting for the speed v from Eq. (20–2),

$$E = \frac{1}{4\pi\epsilon_0} \left(\frac{e^2}{2r} - \frac{e^2}{r} \right) = - \frac{1}{8\pi\epsilon_0} \frac{e^2}{r}. \qquad (20\text{–}3)$$

The total energy of an atomic electron is negative, implying that it is bound to the nucleus. If E were greater than zero, the electron would not be able to remain in a closed orbit about the nucleus.

Experiments indicate that 13.6 ev are required to separate a hydrogen atom into a proton and an electron, which means that its binding energy E is -13.6 ev. Since 13.6 ev $= 2.2 \times 10^{-18}$ joule, we can find the orbital radius of the electron in a hydrogen atom from Eq. (20–3). Since $1/4\pi\epsilon_0 = 9 \times 10^9$ n·m²/coul², this radius is

$$r = - \frac{1}{8\pi\epsilon_0} \frac{e}{E}$$

$$= - \frac{9 \times 10^9 \text{ n·m}^2/\text{coul}^2 \times (1.6 \times 10^{-19} \text{ coul})^2}{2 \times (-2.2 \times 10^{-18} \text{ j})}$$

$$= 5.3 \times 10^{-11} \text{ m.}$$

The above analysis is dynamically flawless, but unfortunately it requires that the electron be continuously accelerated owing to its circular motion. And, according to electromagnetic theory, all accelerated electric charges must radiate electromagnetic waves. It is the acceleration experienced by a fast electron when it strikes a target that causes it to emit x-rays, for instance. An atomic electron circling its nucleus to keep from falling into it by electrostatic attraction therefore cannot (according to electromagnetic theory) help radiating away energy, thereby spiraling inward as r decreases until it is swallowed up by the nucleus (Fig. 20–5). Conventional physics is thus completely unable to account for the observed existence of stable atoms whose electrons remain at great distances from their respective nuclei, and we shall find that the once-revolutionary ideas discussed in the previous chapter must be called upon if we are to understand the atom.

20–3 Atomic spectra

We have already discussed the fact that heated solids emit radiation whose spectra are continuous, with all wavelengths present although in varying intensities. It happens that the observed features of blackbody radiation can be explained on the basis of the quantum theory without our having to consider the details of the radiation process itself; this implies that the nature of the heated solid is irrelevant to the process. Since the atoms of a solid are packed closely together, it is reasonable to suppose that their constituent electrons interact with one another. Hence, when the solid is heated to incandescence, any electron behavior characteristic of a particular atomic species becomes subordinate to the collective behavior of a great many interacting electrons.

At the other extreme, the atoms or molecules in a rarefied gas are so far apart on the average that they can be considered as not interacting with one another at all except during brief, relatively infrequent collisions. Here we would expect any emitted radiation to reflect the properties of the individual atoms or molecules that are present, an expectation that is fulfilled experimentally. When an atomic gas or vapor at somewhat less than atmospheric pressure is "excited" by the passage of an electric current through it, light whose spectrum consists of a limited number of discrete wavelengths is emitted. A typical laboratory arrangement for observing such spectra is sketched in Fig. 20–6. Figure 20–7 shows the atomic spectrum of hydrogen; it is called a *line spectrum* from its appearance. Every element exhibits a unique line spectrum when a sample of it is suitably excited, and its presence in a substance of

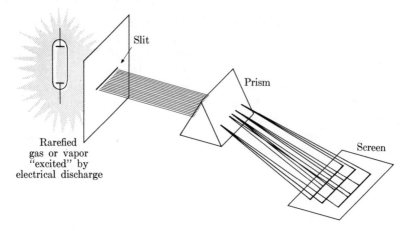

FIG. 20–6. The spectrum of a substance may be obtained by "exciting" a sample of its vapor in a tube by means of an electrical discharge and directing the light it gives off through a prism.

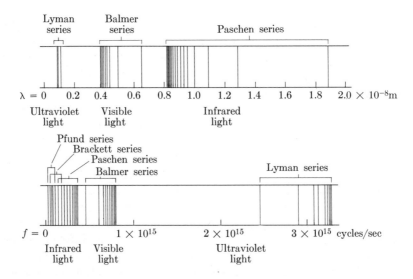

FIG. 20–7. The line spectrum of hydrogen. Note the presence of various series of spectral lines.

FIG. 20–8. A portion of the spectrum of the molecular gas NO showing the presence of bands rather than lines.

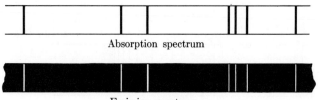

Absorption spectrum

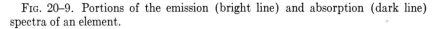

Emission spectrum

FIG. 20–9. Portions of the emission (bright line) and absorption (dark line) spectra of an element.

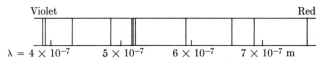

Violet   Red

$\lambda = 4 \times 10^{-7}$ 5×10^{-7} 6×10^{-7} 7×10^{-7} m

FIG. 20–10. The most prominent Fraunhofer lines in the visible part of the solar spectrum.

unknown composition can be ascertained by the appearance of its characteristic wavelengths in the spectrum of the substance.

When a molecular gas or vapor at low pressure is excited, "bands" consisting of many individual lines very close together are found in its spectrum (Fig. 20–8). The presence of bands in molecular spectra arises from interactions among the electrons of the atoms in each molecule; hence molecular spectra, like blackbody spectra, do not provide a great deal of insight into atomic structure.

It is worth noting that while unexcited gases (for instance, atomic hydrogen at room temperature) do not radiate their characteristic spectral lines, they *absorb* light of those particular wavelengths when white light is passed through them. In other words, the *absorption spectrum* of a gas is the same as its *emission spectrum*. Emission spectra consist of bright lines on a dark background, while absorption spectra consist of dark lines on a bright background (Fig. 20–9). The dark *Fraunhofer lines* in the solar spectrum (Fig. 20–10) occur because the luminous sun, which radiates approximately like a blackbody at 5800°K, is surrounded by an envelope of cooler gas that absorbs light of certain characteristic wavelengths.

The early spectroscopists found that the wavelengths present in atomic spectra fall into definite series, all of which can be expressed by similar simple formulas. The first spectral series was discovered by J. J. Balmer in 1885, who studied the wavelengths emitted by hydrogen in the visible part of its spectrum. As shown in Fig. 20–11, the *Balmer series* consists of a number of lines, the one with the longest wavelength being desig-

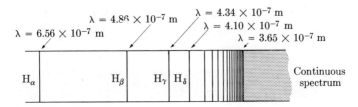

FIG. 20–11. The Balmer series of spectral lines in atomic hydrogen.

nated H_α, the next H_β, and so on. As the wavelength decreases, the lines are closer together and weaker in intensity until the *series limit* at a wavelength of 3.65×10^{-7} meter is reached, beyond which there are no more lines but only a faint continuous spectrum. Balmer's empirical formula for this series is

$$\text{Balmer:} \quad \frac{1}{\lambda} = R\left(\frac{1}{2^2} - \frac{1}{n^2}\right), \qquad n = 3, 4, 5, \ldots, \qquad (20\text{–}4)$$

where R, called the *Rydberg constant*, has the value

$$R = 1.097 \times 10^7 \text{ m}^{-1}.$$

The H_α line corresponds to $n = 3$, the H_β line to $n = 4$, and so on. The series limit corresponds to $n = \infty$, so that its wavelength is $4/R$, in agreement with experiment.

The Balmer series only includes wavelengths in the visible part of the hydrogen spectrum. However, the spectral lines of hydrogen in the ultraviolet and infrared have been found to comprise four other series (Fig. 20–7). In the ultraviolet the *Lyman series* is specified by

$$\text{Lyman:} \quad \frac{1}{\lambda} = R\left(\frac{1}{1^2} - \frac{1}{n^2}\right), \qquad n = 2, 3, 4, \ldots, \qquad (20\text{–}5)$$

while in the infrared there are the three series

$$\text{Paschen:} \quad \frac{1}{\lambda} = R\left(\frac{1}{3^2} - \frac{1}{n^2}\right), \qquad n = 4, 5, 6, \ldots, \qquad (20\text{–}6)$$

$$\text{Brackett:} \quad \frac{1}{\lambda} = R\left(\frac{1}{4^2} - \frac{1}{n^2}\right), \qquad n = 5, 6, 7, \ldots, \qquad (20\text{–}7)$$

$$\text{Pfund:} \quad \frac{1}{\lambda} = R\left(\frac{1}{5^2} - \frac{1}{n^2}\right), \qquad n = 6, 7, 8, \ldots. \qquad (20\text{–}8)$$

In the above equations the value of R is the same as in Eq. (20–4).

The remarkable regularities in the hydrogen spectrum are not able to lead us directly to an elucidation of the structure of the hydrogen atom, but they do present a definite, quantitative test for any theory of this structure.

20–4 The Bohr atom

The first theory of the atom to yield results that agree with our observations was proposed by Niels Bohr in 1913. By applying quantum ideas to atomic structure, Bohr arrived at a detailed model which, though subsequently replaced by an abstract description of greater accuracy and usefulness, remains today as the mental image most scientists have of the atom. Our derivation of the Bohr atom differs slightly from that originally given by Bohr, though all the results are identical.

As we saw, the ordinary laws of physics cannot account for the stability of the hydrogen atom, the simplest of all, whose electron must be whirling around the nucleus to keep from being pulled into it and yet must be radiating electromagnetic energy continuously. However, since other phenomena apparently impossible to understand—the photoelectric effect, for instance—found complete explanation in terms of quantum concepts, it is appropriate for us to inquire whether this might also be true for the atom. Let us begin by looking into the wave properties of the electron in a hydrogen atom. The de Broglie wavelength λ of any moving particle of mass m and speed v is

$$\lambda = \frac{h}{mv},$$

where h is Planck's constant. Now the speed of the electron in a hydrogen atom is, by Eq. (20–2),

$$v = \frac{e}{\sqrt{4\pi\epsilon_0 mr}},$$

and so its wavelength is

$$\lambda = \frac{h}{e}\sqrt{4\pi\epsilon_0 r/m}. \tag{20–9}$$

Substituting 5.3×10^{-11} meter for the radius r of the electron orbit, we find, again setting $1/4\pi\epsilon_0 = 9 \times 10^9$ n·m²/coul²,

$$\lambda = \frac{6.6 \times 10^{-34}\,\text{j·sec}}{1.6 \times 10^{-19}\,\text{coul}} \sqrt{\frac{5.3 \times 10^{-11}\,\text{m}}{9 \times 10^9\,\text{n·m}^2/\text{coul}^2 \times 9.1 \times 10^{-31}\,\text{kg}}}$$

$$= 33 \times 10^{-11}\,\text{m}.$$

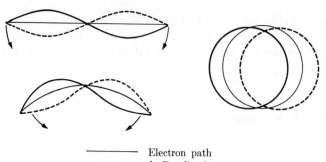

———————— Electron path

—— ——— de Broglie electron wave

FIG. 20–12. The electron orbit in a hydrogen atom is exactly one de Broglie wavelength in circumference.

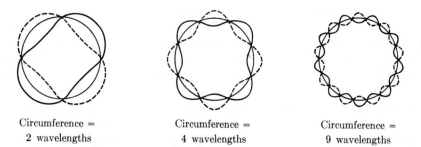

Circumference =
2 wavelengths

Circumference =
4 wavelengths

Circumference =
9 wavelengths

FIG. 20–13. Three possible modes of vibration of a wire loop.

This is a most exciting result, because the electron orbit is exactly

$$2\pi r = 33 \times 10^{-11}\ \text{m}$$

in circumference. Therefore the orbit of the electron in a hydrogen atom corresponds to one complete electron wave joined on itself (Fig. 20–12).

The fact that the electron orbit in a hydrogen atom is one electron wavelength in circumference is just the clue we need to construct a theory of the atom. If we examine the vibrations of a wire loop, as in Fig. 20–13, we see that their wavelengths always fit an integral number of times into the loop's circumference, each wave joining smoothly with the next. In the absence of dissipative effects, such vibrations would persist indefinitely. Why are these the only vibrations possible in a wire loop? A fractional number of wavelengths cannot be fitted into the loop and still allow each wave to join smoothly with the next (Fig. 20–14); the result would be destructive interference as the waves travel around the loop, and the vibrations would die out rapidly. By considering the

Fɪɢ. 20–14. Unless a whole (integral) number of wavelengths fit into the wire loop, destructive interference causes the vibrations to die out rapidly.

behavior of electron waves in the hydrogen atom analogous to the vibrations of a wire loop, then, we may postulate that

An electron can circle an atomic nucleus indefinitely without radiating energy if its orbit is an integral number of electron wavelengths in circumference.

[The electron speed must be that given by Eq. (20–2), of course, in order that the electrostatic attraction of the nucleus not pull the electron into it.]

The above postulate is the decisive one in our understanding of the atom. We note that it combines both the particle and wave characters of the electron into a single statement; while we can never observe these antithetical characters at the same time, they are inseparable in nature.

It is easy to express in a formula the condition that an integral number of electron wavelengths fit into the electron's orbit. The circumference of a circular orbit of radius r is $2\pi r$, and so the condition for orbit stability is

$$n\lambda = 2\pi r_n, \qquad n = 1, 2, 3, \ldots, \qquad (20\text{–}10)$$

where r_n designates the radius of the orbit that contains n wavelengths. The quantity n is called the *quantum number* of the orbit. Substituting for λ the electron wavelength given by Eq. (20–9), we have

$$\frac{nh}{e} \sqrt{4\pi\epsilon_0 r_n/m} = 2\pi r_n,$$

and so the stable electron orbits are those whose radii are given by

$$r_n = \frac{\epsilon_0 n^2 h^2}{\pi m e^2}, \qquad n = 1, 2, 3, \ldots \qquad (20\text{–}11)$$

The innermost orbit has the radius

$$r_1 = 5.3 \times 10^{-11} \text{ m,}$$

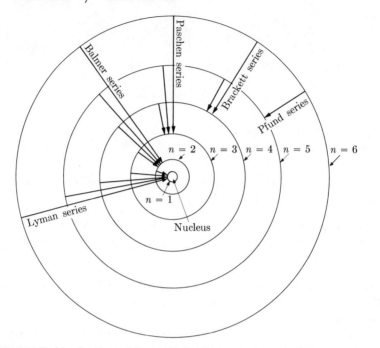

Fig. 20-15. Stable electron orbits in the hydrogen atom according to the Bohr theory. Some of the transitions between orbits that give rise to spectral lines are indicated.

in agreement with our previous calculation. The other radii are given in terms of r_1 by the formula

$$r_n = n^2 r_1,$$

so that the spacing between adjacent orbits increases progressively. (Fig. 20-15).

20-5 Energy levels and spectra

The energy of the electron is not the same in the various permitted orbits. The electron energy E_n is given in terms of the orbit radius r_n by Eq. (20-3) as

$$E_n = -\frac{1}{8\pi\epsilon_0}\frac{e^2}{r_n}\ ;$$

substituting for r_n from Eq. (20-11), we see that

$$E_n = -\frac{me^4}{8\epsilon_0^2 n^2 h^2}, \qquad n = 1, 2, 3, \ldots. \tag{20-12}$$

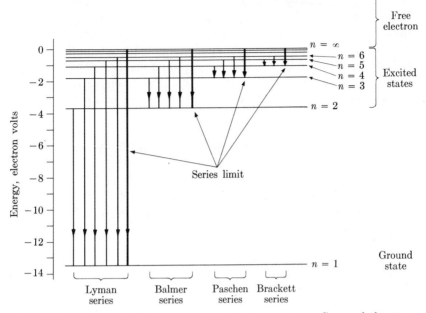

FIG. 20-16. Energy-level diagram of the hydrogen atom. Some of the transitions that give rise to spectral lines are indicated.

The energies specified by Eq. (20-12), called the *energy levels* of the hydrogen atom, are plotted in Fig. 20-16. These levels are all negative, signifying that the electron does not have enough energy to escape from the atom. The lowest energy level E_1, corresponding to quantum number $n = 1$, is called the *ground state* of the atom; the higher levels E_2, E_3, E_4, ... are called excited states. As the quantum number n increases, the corresponding energy E_n approaches closer and closer to zero; in the limit of $n = \infty$, $E_\infty = 0$ and the electron is no longer bound to the nucleus to form an atom. An energy greater than zero signifies an unbound electron, which, since it has no closed orbit that must satisfy quantum conditions, may have any positive energy value whatever.

While everything we have said thus far about the hydrogen atom seems perfectly reasonable, we must nevertheless find a way of directly confronting the equations we have developed with experiment. A particularly striking experimental result is that excited atoms emit line spectra. Can we show that these spectra are a natural consequence of our atomic model?

The presence of a series of definite, discrete energy levels in the hydrogen atom suggests a connection with line spectra. Let us tentatively assert that when an electron in an excited state drops to a lower state, the difference in energy between the states is emitted as a single photon of light. Because electrons cannot, according to our model, exist in an atom except in certain specific energy levels, a rapid "jump" from one level to the other, with the difference in energy being given off all at once in a photon rather than in some more gradual manner, fits in well with this model. If the quantum number of the initial (higher energy) state is n_i and the quantum number of the final (lower energy) state is n_f, what we assert is that

$$\text{Initial energy} - \text{final energy} = \text{quantum energy}$$

$$E_i - E_f = hf, \tag{20-13}$$

where f is the frequency of the emitted photon.

The energies of the initial and final states of the electron in a hydrogen atom are, from Eq. (20–12), given by

$$\text{Initial energy} = E_i = -\frac{me^4}{8\epsilon_0^2 n_i^2 h^2},$$

$$\text{Final energy} = E_f = -\frac{me^4}{8\epsilon_0^2 n_f^2 h^2}.$$

Hence the energy difference between these states is

$$E_i - E_f = \frac{me^4}{8\epsilon_0^2 h^2}\left(\frac{-1}{n_i^2}\right) - \left(\frac{-1}{n_f^2}\right)$$

$$= \frac{me^4}{8\epsilon_0^2 h^2}\left(\frac{1}{n_f^2} - \frac{1}{n_i^2}\right),$$

corresponding to a photon frequency f of

$$f = \frac{E_i - E_f}{h} = \frac{me^4}{8\epsilon_0^2 h^3}\left(\frac{1}{n_f^2} - \frac{1}{n_i^2}\right).$$

In terms of photon wavelength λ, since

$$\lambda = \frac{c}{f},$$

we have

$$\frac{1}{\lambda} = \frac{f}{c} = \frac{me^4}{8\epsilon_0^2 ch^3}\left(\frac{1}{n_f^2} - \frac{1}{n_i^2}\right). \tag{20-14}$$

Equation (20–14) predicts that the radiation emitted by excited hydrogen atoms should consist of certain wavelengths only. These wavelengths, furthermore, should fall into definite series depending upon the quantum number n_f of the final energy level of the electron. Since the initial quantum number n_i must always be greater than the final quantum number n_f in each case, in order that there be an excess of energy to be given off as a photon, the calculated formulas for the first five series are

$$n_f = 1: \quad \frac{1}{\lambda} = \frac{me^4}{8\epsilon_0^2 ch^3}\left(\frac{1}{1^2} - \frac{1}{n^2}\right), \qquad n = 2, 3, 4, \ldots \qquad (20\text{--}15)$$

$$n_f = 2: \quad \frac{1}{\lambda} = \frac{me^4}{8\epsilon_0^2 ch^3}\left(\frac{1}{2^2} - \frac{1}{n^2}\right), \qquad n = 3, 4, 5, \ldots \qquad (20\text{--}16)$$

$$n_f = 3: \quad \frac{1}{\lambda} = \frac{me^4}{8\epsilon_0^2 ch^3}\left(\frac{1}{3^2} - \frac{1}{n^2}\right), \qquad n = 4, 5, 6, \ldots \qquad (20\text{--}17)$$

$$n_f = 4: \quad \frac{1}{\lambda} = \frac{me^4}{8\epsilon_0^2 ch^3}\left(\frac{1}{4^2} - \frac{1}{n^2}\right), \qquad n = 5, 6, 7, \ldots \qquad (20\text{--}18)$$

$$n_f = 5: \quad \frac{1}{\lambda} = \frac{me^4}{8\epsilon_0^2 ch^3}\left(\frac{1}{5^2} - \frac{1}{n^2}\right), \qquad n = 6, 7, 8, \ldots \qquad (20\text{--}19)$$

These series agree in form with the empirical spectral series we discussed earlier, with the Lyman series, Eq. (20–5), corresponding to $n_f = 1$, the Balmer series, Eq. (20–4), corresponding to $n_f = 2$, the Paschen series, Eq. (20–6), corresponding to $n_f = 3$, and so on!

Only one thing remains to be determined before we can consider our assertion that line spectra originate in electron transitions from high to low energy states as verified: does the constant term in Eqs. (20–15) through (20–19) have the same value as the Rydberg constant R of the empirical Eqs. (20–4) through (20–8)? Let us compute the value of this constant term. It is

$$\frac{me^4}{8\epsilon_0^2 ch^3}$$

$$= \frac{9.1 \times 10^{-31}\text{ kg} \times (1.6 \times 10^{-19}\text{ coul})^4}{8 \times (8.85 \times 10^{-12}\text{ coul}^2/\text{n·m}^2)^2 \times 3 \times 10^8\text{ m/sec} \times (6.6 \times 10^{-34}\text{ j·sec})^3}$$

$$= 1.097 \times 10^7\text{ m}^{-1},$$

which is the same as R. Our theory of the hydrogen atom, which is essentially that developed by Bohr in 1913, therefore agrees both qualitatively and quantitatively with experiment. Figure 20–16 shows schematically how spectral lines are related to atomic energy levels; Fig. 20–15 indicates the transitions in terms of Bohr orbits.

20–6 Atomic excitation

There are two principal mechanisms by which an atom may be excited to an energy level above that of its ground state and thereby become capable of radiating. One of these mechanisms is a collision with another atom during which part of their kinetic energy is transformed into electron energy within either or both of the participating atoms. An atom excited in this way will then lose its excitation energy by emitting one or more photons in the course of returning to its ground state (Fig. 20–17). In an electric discharge in a rarefied gas, an electric field

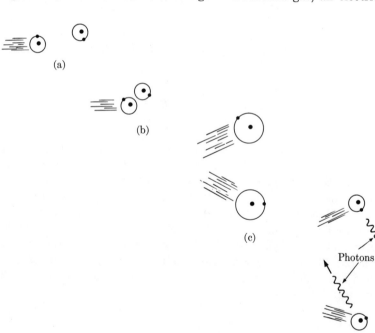

Fig. 20–17. Excitation by collision. In (a) both atoms are in their ground states. During the collision (b) some kinetic energy is transformed into excitation energy, and in (c) the atoms are both in excited states. In (d) they have returned to their ground states by emitting photons.

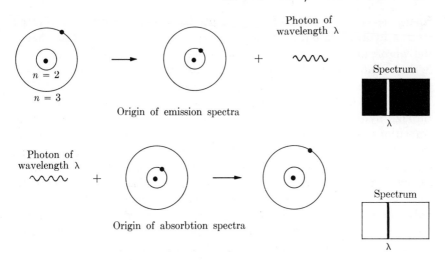

Origin of emission spectra

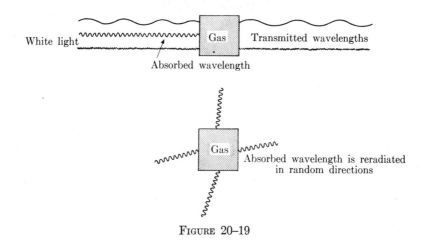

FIG. 20–18. The origins of emission and absorption spectra.

FIGURE 20–19

accelerates electrons and charged atoms and molecules (whose charge arises from either an excess or a deficiency in the electrons required to neutralize the positive charge of their nuclei) until their kinetic energies are sufficient to excite atoms they happen to collide with. A neon sign is a familiar example of how applying a strong electric field between electrodes in a gas-filled tube leads to the emission of the characteristic spectral radiation of that gas, which happens to be orange light in the case of neon.

Another excitation mechanism is the absorption by an atom of a photon of light whose energy is just the right amount to raise it to a higher energy level. A photon of wavelength 6.565×10^{-7} meter is emitted when a hydrogen atom in the $n = 3$ state drops to the $n = 2$ state; hence the absorption of a photon of wavelength 6.565×10^{-7} meter by a hydrogen atom initially in the $n = 2$ state will bring it up to the $n = 3$ state. This process explains the origin of absorption spectra (Fig. 20–18). When white light (in which all wavelengths are present) is passed through hydrogen gas, photons of those wavelengths that correspond to transitions between hydrogen energy levels are absorbed. The resulting excited hydrogen atoms reradiate their excitation energy almost at once, but these photons come off in random directions, not all in the same direction as in the original beam of white light (Fig. 20–19). The dark lines in an absorption spectrum are therefore never totally dark, but only appear so by contrast with the bright background of transmitted light. We would expect the absorption and emission spectra of a particular substance to be identical, which is what we find.

IMPORTANT TERMS

In the *Rutherford model of the atom* a tiny *nucleus*, in which its positive charge and nearly all of its mass are concentrated, is at the center of the atom, with its electrons some distance away. In this model the atom is largely empty space.

An *emission spectrum* consists of the various wavelengths of light emitted by an excited substance; it may be a *continuous spectrum*, in which all wavelengths are present, or a *bright line spectrum*, in which only a few wavelengths characteristic of the individual atoms of the substance are present.

An *absorption spectrum* results when white light is passed through a cool gas; it is a *dark line spectrum* because it appears as a series of dark lines on a bright background, with the lines representing characteristic wavelengths absorbed by the gas.

The lines in the spectrum of an element fall into several *spectral series* in which the wavelengths of the various lines are related by simple formulas. The spectral series of hydrogen are called the Lyman, Balmer, Paschen, Brackett, and Pfund series, which range in that order from the ultraviolet to the infrared ends of the spectrum.

According to the *Bohr theory of the atom,* an electron can circle an atomic nucleus indefinitely without radiating energy if its orbit is an integral number of electron wavelengths in circumference. The number of wavelengths that fit into a particular permitted orbit is called the *quantum number* of that orbit. The electron energies corresponding to the various quantum numbers constitute the *energy levels* of the atom, of which the lowest is the *ground state* and the rest are *excited states*. When an electron in an excited state drops to a lower state, the difference in energy between the states is emitted as a single photon of light; when a photon of the same wavelength is absorbed, the electron goes from the lower to the higher state. The above two processes account for the properties of emission and absorption spectra respectively.

PROBLEMS

1. In what ways do the Thomson and Rutherford atomic models agree? In what ways do they disagree?

2. In the Bohr theory of the hydrogen atom the electron is in constant motion. How is it possible for such an electron to have a *negative* amount of energy?

3. Calculate the speed of the electron in the innermost Bohr orbit of a hydrogen atom.

4. (a) Calculate the de Broglie wavelength of the earth. (b) What is the quantum number that characterizes the earth's orbit about the sun? (The earth's mass is 6.0×10^{24} kg, its orbital radius is 1.5×10^{11} m, and its orbital speed is 3×10^4 m/sec.)

5. Explain why the spectrum of hydrogen has many lines, although a hydrogen atom contains only one electron.

6. A beam of electrons whose energy is 13 ev is used to bombard gaseous hydrogen. What wavelengths of light will be emitted?

7. A beam of electrons is used to bombard gaseous hydrogen. What is the minimum energy in ev the electrons must have if the first line of the Balmer series, corresponding to a transition from the $n=3$ state to the $n=2$ state, is to be emitted?

8. What is the shortest wavelength present in the Paschen series of spectral lines?

9. What is the shortest wavelength present in the Brackett series of spectral lines?

10. A proton and an electron, both at rest initially, combine to form a hydrogen atom in the ground state. A single photon is emitted in this process. What is its wavelength?

11. How much energy (in joules and in electron volts) is required to remove the electron from a hydrogen atom when it is in the $n = 5$ state?

12. To what temperature must a hydrogen gas be heated if the average molecular kinetic energy is to equal the binding energy of the hydrogen atom?

13. When radiation with a continuous spectrum is passed through a volume of hydrogen gas whose atoms are all in their ground state, which spectral series will be present in the resulting absorption spectrum?

14. An electron spends about 10^{-8} sec in an excited state before it drops to a lower state by giving up energy in the form of a photon. How many revolutions does an electron in the $n = 2$ state of a hydrogen atom make before dropping to the $n = 1$ state?

15. (a) Find the wavelength of the photon emitted when a hydrogen atom goes from the state $n=4$ to $n=1$. (b) Find the momentum of this photon. (c) Find the recoil speed of the hydrogen atom after it has emitted this photon.

16. Calculate the average kinetic energy per molecule in a gas at room temperature (20°C), and show that this is much less than the energy required to raise a hydrogen atom from its ground state ($n = 1$) to its first excited state ($n = 2$).

17. What kind of spectrum is observed in solar radiation during an eclipse?

18. How can the narrowness of spectral lines, which indicates that the responsible photons have very precisely defined energies, be reconciled with the version of the uncertainty principle that states that $\Delta E \Delta t = h$?

Complex Atoms 21

21–1 Introduction

The Bohr theory of the atom that we have sketched is indeed impressive in its agreement with experiment, but it has certain serious limitations. For one thing, the careful study of spectral lines shows that many of them actually consist of two or more separate lines that are close together, as in Fig. 21–1, something that the Bohr theory cannot account for. While correctly predicting the wavelengths of the spectral lines in hydrogen, which has but a single atomic electron, the Bohr theory fails when attempts are made to apply it to more complex atoms.

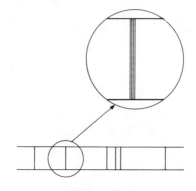

FIGURE 21–1

Even in hydrogen it is not possible to calculate from the Bohr theory the relative probabilities of various transitions between energy levels, for instance whether it is more likely that an atom in the $n = 3$ state will go directly to the $n = 1$ state or instead first drop to the $n = 2$ state. (In other words, we cannot find from the Bohr theory which of the spectral lines of hydrogen will show up brightest in an emission spectrum and which will be faint.)

In cataloging these objections to the Bohr theory, the intent is not to detract from its eminence in the history of science, which is certainly secure, but instead to emphasize that a more general approach capable of wider application is necessary. Such an approach was developed in the 1920's by Schrödinger, Heisenberg, and others under the name of the *quantum theory of the atom*. Instead of trying to visualize an atomic electron as a kind of cross between a particle and a wave and thinking of it as occupying one of various possible orbits, the quantum theory of the atom avoids all reference to anything not capable of direct measurement and restricts itself only to such observable quantities as photon energies. In the Bohr theory we compute the radius of an electron orbit from a knowledge of its de Broglie wavelength, which depends upon the electron momentum; however, as we learned in Chapter 19, the position (and hence orbital radius) and momentum of an electron can never be simultaneously determined accurately, and so *even in principle* the Bohr theory cannot be subjected to an experimental test. The quantum theory sacrifices such intuitively accessible notions as that of electrons circling a nucleus like planets around the sun in favor of a wholly abstract mathematical formulation each of whose statements can be verified. Freed from the necessity of working in terms of a "model" of any kind, the quantum theory is able to tackle successfully a broad range of atomic problems.

21-2 Quantum numbers

In the Bohr model of the hydrogen atom, the motion of the orbital electron is essentially one-dimensional. There the electron is regarded as being confined to a definite circular orbit, and the only quantity that changes as it revolves around its nucleus is its position on this fixed circle. A single quantum number is all that is required to specify the physical state of such an electron. In the more general quantum theory of the atom, the electrons have no such spatial restriction, and three different coordinates are needed to describe their motion. When the theory is worked out, three quantum numbers (one for each coordinate) turn out to be necessary instead of the single one of the Bohr theory. These are the *total quantum number n*, the *orbital quantum number l*, and the *magnetic quantum number m_l*.

The value of n is the chief factor that governs the total energy of an electron bound to a nucleus. The energy levels of a hydrogen atom are given by the formula

$$E_n = -\frac{me^4}{8\epsilon_0^2 h^2 n^2},\tag{21-1}$$

the same result found by the Bohr theory.

A more novel conclusion of the quantum theory of the atom is the manner in which the angular momentum **L** of an electron is quantized. Angular momentum is a vector quantity, not a scalar quantity like energy, and has both magnitude and direction. The orbital quantum number l governs the magnitude L of an electron's angular momentum, which is restricted to

$$L = \sqrt{l(l+1)}\,\frac{h}{2\pi}.\tag{21-2}$$

The orbital quantum number l can be 0 or any integer up to $n-1$; that is,

$$l = 0, 1, 2, \ldots, (n-1).$$

When $l = 0$, the angular momentum **L** $= 0$ also. Here is a significant difference from the Bohr theory, since an electron revolving in a circle must possess angular momentum, a requirement absent from the quantum theory.

The term quantization means that a certain quantity can have only certain discrete values. When we say that the direction of the angular momentum vector **L** is quantized, we mean that it can have only certain orientations in space. The space quantization of an atomic electron follows from the fact that, if such an electron has angular momentum, it

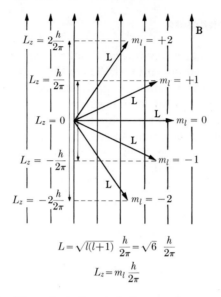

$$L = \sqrt{l(l+1)}\ \frac{h}{2\pi} = \sqrt{6}\ \frac{h}{2\pi}$$

$$L_z = m_l\frac{h}{2\pi}$$

FIG. 21–2. Possible orientations of the angular momentum vector **L** in a magnetic field **B** when $l = 2$.

behaves like an electric current loop and interacts with an external magnetic field in a manner similar to that of a bar magnet in such a field. The larger the angular momentum **L**, the stronger the equivalent bar magnet. The potential energy of a bar magnet in a magnetic field **B** varies with the strength of the magnet and with its direction relative to **B**; the energy is least when the magnet is parallel to the field, and is most when it is antiparallel. The magnetic quantum number m_l determines the angle between **L** and **B** and hence governs the extent of the magnetic contribution to the total energy of the atom when it is in a magnetic field. Thus an atomic electron characterized by a certain value of m_l will assume a certain corresponding orientation of its angular momentum **L** relative to a magnetic field when placed in such a field. The orientation is specified in terms of the component of **L** parallel to the magnetic field. If we denote this component by L_z, as in Fig. 21–2, its possible values are

$$L_z = m_l\ \frac{h}{2\pi}, \tag{21–3}$$

where m_l can be any integer from $-l$ through 0 to $+l$. That is,

$$m_l = -\,l,\, -\,(l-1),\, \ldots,\, 0,\, \ldots,\, +\,(l-1),\, +\,l.$$

An electron for which $l = 2$, for instance, could have a magnetic quantum

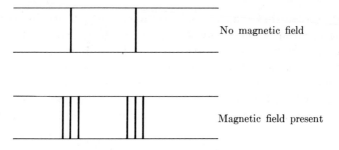

No magnetic field

Magnetic field present

FIG. 21-3. The Zeeman effect.

number of -2, -1, 0, $+1$, or $+2$. When $m_l = 0$, $L_z = 0$ also, and $\mathbf{L}$ will be perpendicular to $\mathbf{B}$ whenever the atom is in a magnetic field. The angular momentum vector $\mathbf{L}$ can never be exactly parallel to $\mathbf{B}$, even when $m_l = l$, since comparing Eqs. (21-2) and (21-3) shows that L_z is always less than L.

Because the magnetic quantum number m_l has several possible values for values of n other than $n = 1$, the presence of an external magnetic field splits the energy levels of a particular atom into two or more sublevels. The emission spectrum of an element in a magnetic field accordingly differs from its ordinary spectrum in that the spectral lines of the latter now have several components whose spacing varies with the magnetic flux density (Fig. 21-3). This phenomenon is known as the *Zeeman effect*. For example, a field of 1 weber/m² leads to an energy difference of 5.8×10^{-4} ev between adjacent sublevels of different m_l, which is observed as a wavelength difference between the split components of a given spectral line of order of magnitude 0.1%.

21-3 Electron spin

We mentioned at the start of this chapter that one of the problems facing the atomic theorist is the observed splitting of many spectral lines into several components close together, as in Fig. 21-1. For example, the first line of the Balmer series of hydrogen (Fig. 20-11), which both theory and experiment place at a wavelength of 6.56280×10^{-7} m, actually consists of a pair of lines 0.00014×10^{-7} m apart. Furthermore, many cases of Zeeman splitting in magnetic fields cannot be explained solely in terms of the space quantization of angular momentum. In 1925 it was pointed out that the fine structure of spectral lines would occur if the electron behaves like a charged sphere spinning on its axis rather than like a single point charge; several years later the British physicist P.A.M. Dirac was able to show on the basis of a relativistic version of quantum theory that the electron *must* have the spin attributed to it. A spinning electron is effectively a tiny bar magnet, and it interacts with the magnetic field produced by its own orbital motion in an atom as well as with any magnetic fields originating outside the atom. Electron spin is described by the spin quantum number s, whose sole value is $s = \frac{1}{2}$. The magnitude of the intrinsic spin angular momentum of the electron is given by a formula similar to that describing the orbital angular momentum, namely

$$L_s = \sqrt{s(s+1)}\,\frac{h}{2\pi} = \frac{\sqrt{3}}{2}\,\frac{h}{2\pi}. \tag{21-4}$$

The space quantization of electron spin is described by the *spin magnetic quantum number*, m_s, whose values are $+\frac{1}{2}$ and $-\frac{1}{2}$. The component of spin angular momentum along a magnetic field is $m_s\, h/2\pi$, so that it can be either $+\frac{1}{2}\, h/2\pi$ *or* $-\frac{1}{2}\, h/2\pi$. Because the energy of an electron is different in each orientation of its spin owing to the presence of the magnetic field of its orbital motion, the various energy levels of an atom are split into sublevels, and the presence of these sublevels is responsible for the observed fine structure of spectral lines. (In its own frame of reference the electron "sees" the positively charged nucleus revolving around *it*, and it accordingly experiences a magnetic field like that present at the center of a loop of electric current.) The notion of electron spin is also able to explain those aspects of the Zeeman effect that cannot be understood on the basis of the space quantization of angular momentum alone.

21–4 The exclusion principle

The introduction of electron spin into the theory of the atom means that a total of four quantum numbers, n, l, m_l, and m_s, is required to describe each possible state of an atomic electron. All atoms other than hydrogen have more than one electron, and it is appropriate to inquire into their ground state configurations. Do all the electrons in a complex atom have the same sets of quantum numbers, or are the sets all different, or what? This question was answered by Wolfgang Pauli, whose *exclusion principle* states that

No two electrons in an atom can exist in the same quantum state.

That is, each electron in a complex atom must have a different set of the quantum numbers n, l, m_s, and m_l. The exclusion principle can be generalized to refer to the electrons in *any* small region of space, regardless of whether or not they constitute an atom. We shall make use of this generalization in the next chapter.

Pauli was led to the exclusion principle by a study of atomic spectra. It is possible to determine the quantum states of an atom empirically by analyzing its spectrum, since states with different quantum numbers differ in energy (even if only slightly) and the various wavelengths present correspond to transitions between these states. Pauli found several lines missing from the spectra of every element except hydrogen; the missing lines correspond to transitions to and from atomic states having certain sets of quantum numbers. Every one of these absent states has two or more electrons with identical sets of quantum numbers, a result that is expressed in the exclusion principle. Hydrogen, with a single electron, naturally has no absent states.

The Pauli exclusion principle is the final piece of information we require in order to understand atomic structure. Before we actually go into the electron configurations of the various atoms, however, it will be useful to review the periodic law of chemistry.

21–5 The periodic law

Certain elements resemble one another so closely in their chemical and physical properties that it is natural to think of them as forming a "family." Three particularly striking examples of such families are the *halogens,* the *inert gases,* and the *alkali metals.* The members of these groups of elements are listed in Table 21–1 with their atomic numbers.

The halogens are nonmetallic elements with a high degree of chemical activity. At room temperature fluorine and chlorine are gases, bromine a liquid, and iodine and astatine solids. The halogens have valences of −1, and form diatomic molecules in the vapor state. The inert gases, as their name suggests, are inactive chemically: they form no compounds with other elements, and their atoms do not join together into molecules.

TABLE 21–1

Three families of elements*

Halogens	Inert gases	Alkali metals
	(2) Helium	(3) Lithium
(9) Fluorine	(10) Neon	(11) Sodium
(17) Chlorine	(18) Argon	(19) Potassium
(35) Bromine	(36) Krypton	(37) Rubidium
(53) Iodine	(54) Xenon	(55) Cesium
(85) Astatine	(86) Radon	(87) Francium

* The atomic numbers of the elements are in parentheses.

TABLE 21-2

Periodic classification of the elements

Group→ Period↓	I	II											III	IV	V	VI	VII	0
1	1.0080 H 1																	4.003 He 2
2	6.940 Li 3	9.013 Be 4											10.82 B 5	12.011 C 6	14.008 N 7	16.0000 O 8	19.000 F 9	20.183 Ne 10
3	22.991 Na 11	24.32 Mg 12											26.98 Al 13	28.09 Si 14	30.975 P 15	32.066 S 16	35.457 Cl 17	39.944 Ar 18
4	39.100 K 19	40.08 Ca 20	44.96 Sc 21	47.90 Ti 22	50.95 V 23	52.01 Cr 24	54.94 Mn 25	55.85 Fe 26	58.94 Co 27	58.71 Ni 28	63.54 Cu 29	65.38 Zn 30	69.72 Ga 31	72.60 Ge 32	74.91 As 33	78.96 Se 34	79.916 Br 35	83.80 Kr 36
5	85.48 Rb 37	87.63 Sr 38	88.92 Y 39	91.22 Zr 40	92.91 Nb 41	95.95 Mo 42	(99) Tc 43	101.1 Ru 44	102.91 Rh 45	106.4 Pd 46	107.88 Ag 47	112.41 Cd 48	114.82 In 49	118.70 Sn 50	121.76 Sb 51	127.61 Te 52	126.91 I 53	131.3 Xe 54
6	132.91 Cs 55	137.36 Ba 56	* 57–71	178.5 Hf 72	180.95 Ta 73	183.86 W 74	186.22 Re 75	190.2 Os 76	192.2 Ir 77	195.09 Pt 78	197.0 Au 79	200.61 Hg 80	204.39 Tl 81	207.21 Pb 82	209.00 Bi 83	210 Po 84	(210) At 85	222 Rn 86
7	(223) Fr 87	226.05 Ra 88	† 89															

*Rare-earth metals														
138.92 La 57	140.13 Ce 58	140.92 Pr 59	144.27 Nd 60	(145) Pm 61	150.35 Sm 62	152.0 Eu 63	157.26 Gd 64	158.93 Tb 65	162.51 Dy 66	164.94 Ho 67	167.27 Er 68	168.94 Tm 69	173.04 Yb 70	174.99 Lu 71

†Actinide metals														
227 Ac 89	232.05 Th 90	231 Pa 91	238.07 U 92	(237) Np 93	(244) Pu 94	(243) Am 95	(245) Cm 96	(249) Bk 97	(249) Cf 98	(254) E 99	(252) Fm 100	(256) Mv 101	(253?) No 102	

The inert gases have valences of 0. The alkali metals, like the halogens, are chemically very active, but they are active as reducing agents rather than as oxidizing agents. They are soft, not very dense, and have low melting points (all but lithium are liquid below 100°C). The alkali metals have valences of +1.

A curious feature of the three groups listed above is that, while the atomic numbers of the member elements of each group bear no obvious relation to one another, each inert gas is preceded in atomic number by a halogen (except in the case of helium) and followed by an alkali metal. Thus fluorine, neon, and sodium have the atomic numbers 9, 10, and 11, respectively, a sequence that persists through astatine (85), radon (86), and francium (87). If we list *all* of the elements in order of their atomic number, *elements with similar properties recur at regular intervals.* This observation, first formulated in detail by Dmitri Mendeleev about 1869, is known as the *periodic law.*

A periodic table is a listing of the elements according to atomic number in a series of rows such that elements with similar properties form vertical columns. Table 21–2 is perhaps the most common form of periodic table; the number above the symbol of each element is its atomic weight, and the number below the symbol is its atomic number. The elements whose atomic weights appear in parentheses are radioactive and are not found in nature, but have been prepared in nuclear reactions. The atomic weight in each such case is the mass number of the longest-lived isotope of the element.

The columns in the periodic table are called *groups*. We recognize group I as the alkali metals plus hydrogen, group VII as the halogens, and group 0 as the inert gases. In addition to the elements forming the eight principal groups there are a number of *transition elements* falling between groups II and III. The transition elements are metals which share certain general properties: most are hard and brittle, have high melting points, exhibit several different valences, and form compounds that are paramagnetic. The rows in the periodic table are called *periods*. Each period starts with an active alkali metal and proceeds through less active metals to weakly active nonmetals to an active halogen and an inactive inert gas. The transition elements in each period may be very much alike; the rare earths and actinides are so much alike that they are usually considered as separate categories.

For nearly a century the periodic law has been a mainstay of the chemist by permitting him to predict the behavior of undiscovered elements and by providing a framework for organizing his knowledge. It is one of the triumphs of the quantum theory of the atom that it enables us to account for the periodic law in complete detail without invoking any new assumptions or postulates.

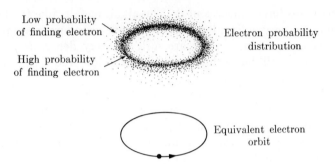

Low probability
of finding electron

Electron probability
distribution

High probability
of finding electron

Equivalent electron
orbit

FIGURE 21–4

21–6 Atomic structure

A physical system is stable when its energy is a minimum, which means that in the normal configuration of an atom its constituent electrons are present in the lowest energy states available to them. The electrons in an atom that share the same total quantum number n are said to occupy the same *shell*. This is a useful description, since such electrons average about the same distance from the nucleus and have comparable, though not identical, energies. Electrons in a given shell that also have the same orbital quantum number l are said to occupy the same *subshell*. In complex atoms the various subshells of the same shell vary in energy because electrons with different angular momenta pursue different orbits around the nucleus. (Strictly speaking, electrons do not move in specific orbits. However, the likelihood of their presence varies with position in such a way that a line drawn through the points of greatest probability may be thought of as an orbit. See Fig. 21–4.) In a complex atom the outer electrons are partially shielded from the full nuclear charge of $+Ze$ by the inner electrons, and are therefore less tightly bound than the latter. The extent of this shielding on a particular electron depends upon the shape of its orbit, which in turn depends upon its orbital quantum number l. When l is large, the corresponding orbit is more or less circular, while when it is small the orbit is pronouncedly elliptical. In an elliptical orbit an electron spends part of its time near the nucleus without being shielded by other electrons, as in Fig. 21–5, and its energy is correspondingly lower than that of an electron in the same shell whose orbit is circular. The general result, then, is that the energies of the various sublevels increase as l increases.

We shall now construct the periodic table of the elements with the help of the above considerations and the exclusion principle. Our procedure will be to investigate the status of a new electron added to an existing

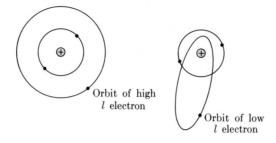

Orbit of high
l electron

Orbit of low
l electron

FIGURE 21-5

electronic structure. (Of course, the nuclear charge must also increase by $+e$ each time this is done.) In the simplest case we add an electron to a hydrogen atom $(Z = 1)$ to give a helium atom $(Z = 2)$. Both electrons in helium fall into the same $n = 1$ shell. Since $l = 0$ is the only value l can have when $n = 1$, both electrons have $l = m_l = 0$. The exclusion principle is not violated here since one electron can have the spin magnetic quantum number $m_s = +\frac{1}{2}$ while the other has $m_s = -\frac{1}{2}$. It is customary to describe this situation by saying that the electrons have *opposite spins*.

Because no more than two electrons can occupy the $n = 1$ shell, helium atoms have *closed shells*. The characteristic properties of closed shells and subshells are that the orbital and spin angular momenta of their constituent electrons cancel out independently and that their effective electric charge distributions are perfectly symmetrical. As a result atoms with closed shells do not tend to interact with other atoms, which we know to be true of helium.

Lithium, with $Z = 3$, has one more electron than helium. There is no room left in the $n = 1$ shell, and so the additional electron goes into the $l = 0$ subshell of the $n = 2$ shell. The outer electron is relatively far from the nucleus in this shell, and is much less tightly bound. The chemical activity of lithium is a consequence of the low binding energy of this electron, which is readily lost to other atoms.

The next element, beryllium, has two electrons of opposite spin in the $l = 0$ subshell of the $n = 2$ shell. Boron, with $Z = 5$, has an electron in the $l = 1$ subshell as well as two in the $l = 0$ one. The $l = 1$ subshell can contain a total of six electrons, corresponding to two electrons of opposite spin in the $m_l = +1$, $m_l = 0$, and $m_l = -1$ states. This subshell is closed (and the $n = 2$ shell also closed) in neon, whose atomic number is 10. We therefore expect neon to be chemically inert, as indeed it is. Fluorine, the element just before neon in the periodic table, lacks but one electron of having a closed outer shell. Just as lithium tends to lose its single outermost electron in interacting with other elements, thereby leaving it with a

TABLE 21-3

Electron configurations
of the thirty-six lightest elements

Atomic number	Symbol	Element	Electron configuration of atom						
			$n=1$ $l=0$	$n=2$ $l=0$	$n=2$ $l=1$	$n=3$ $l=0$	$n=3$ $l=1$	$n=3$ $l=2$	$n=4$ $l=0, l=1, \ldots$
1	H	Hydrogen	1						
2	He	Helium	2						
3	Li	Lithium	2	1					
4	Be	Beryllium	2	2					
5	B	Boron	2	2	1				
6	C	Carbon	2	2	2				
7	N	Nitrogen	2	2	3				
8	O	Oxygen	2	2	4				
9	F	Fluorine	2	2	5				
10	Ne	Neon	2	2	6				
11	Na	Sodium	2	2	6	1			
12	Mg	Magnesium	2	2	6	2			
13	Al	Aluminum	2	2	6	2	1		
14	Si	Silicon	2	2	6	2	2		
15	P	Phosphorus	2	2	6	2	3		
16	S	Sulfur	2	2	6	2	4		

17	Cl	Chlorine	2	2	6	2	5			
18	Ar	Argon	2	2	6	2	6			
19	K	Potassium	2	2	6	2	6		1	
20	Ca	Calcium	2	2	6	2	6		2	
21	Sc	Scandium	2	2	6	2	6	1	2	
22	Ti	Titanium	2	2	6	2	6	2	2	
23	V	Vanadium	2	2	6	2	6	3	2	
24	Cr	Chromium	2	2	6	2	6	5	1	
25	Mn	Manganese	2	2	6	2	6	5	2	
26	Fe	Iron	2	2	6	2	6	6	2	
27	Co	Cobalt	2	2	6	2	6	7	2	
28	Ni	Nickel	2	2	6	2	6	8	2	
29	Cu	Copper	2	2	6	2	6	10	1	
30	Zn	Zinc	2	2	6	2	6	10	2	
31	Ga	Gallium	2	2	6	2	6	10	2	1
32	Ge	Germanium	2	2	6	2	6	10	2	2
33	As	Arsenic	2	2	6	2	6	10	2	3
34	Se	Selenium	2	2	6	2	6	10	2	4
35	Br	Bromine	2	2	6	2	6	10	2	5
36	Kr	Krypton	2	2	6	2	6	10	2	6

closed shell configuration, fluorine tends to gain a single electron to close its outer shell. The very different behavior of the alkali metals, the inert gases, and the halogens which we discussed in the previous section thus find explanation in terms of their respective atomic structures.

Table 21–3 shows the electron configurations of the 36 lightest elements. We note that the sequence of electron addition becomes irregular with potassium, which has an electron in its $n = 4$ shell even though its $n = 3$ shell is incomplete. The origin of this apparent anomaly is that electrons in the $n = 4$, $l = 0$ subshell have less energy, and therefore are more tightly bound, than those in the $n = 3$, $l = 2$ subshell, since the orbits of the former electrons pass closer to the nucleus than the more circular orbits of the latter electrons. The energy difference between the $n = 3$, $l = 2$ and $n = 4$, $l = 0$ subshells is actually quite small, as we can see from the configurations of chromium ($Z = 24$) and copper ($Z = 29$). In both of these elements an additional electron is present in the $n = 3$, $l = 2$ subshell at the expense of the $n = 4$, $l = 0$ subshell, leaving a vacancy in the latter that is filled in the succeeding element.

The ferromagnetic properties of iron, cobalt, and nickel are a consequence of their unfilled $n = 3$, $l = 2$ subshells; without violating the exclusion principle, the electrons in these subshells do *not* pair off, and their spins do not cancel out. In the $n = 3$, $l = 2$ subshell of iron, for instance, five of the six electrons have parallel spins, leaving each iron atom with a strongly magnetic character.

As we can see, the quantum theory of the atom is one of the most powerful and fruitful approaches yet devised for understanding the properties of matter.

IMPORTANT TERMS

In the *quantum theory of the atom* only experimentally measurable quantities are considered, and no use is made of mechanical models that contradict the uncertainty principle. This theory is able to successfully account for a wide range of atomic phenomena which the Bohr theory cannot explain.

Four quantum numbers are needed to specify each quantum state in an atom. These are the *total quantum number n*, the *orbital quantum number l*, the *magnetic quantum number m_l*, and the *spin magnetic quantum number m_s*.

Space quantization refers to the fact that the angular momentum vector of an electron can assume only certain orientations in a magnetic field.

The intrinsic angular momentum of an electron is called its *spin*, since an electron behaves much like a charged sphere spinning on its axis.

According to the Pauli *exclusion principle,* no two electrons in an atom can exist in the same quantum state.

The *periodic law* of chemistry states that if we list the elements in order of atomic number, elements with similar properties recur at regular intervals. The quantum theory of the atom together with the exclusion principle is able to explain the origin of the periodic law.

The electrons in an atom that have the same total quantum number n are said to occupy the same *shell.* Electrons in a given shell which have the same orbital quantum number l are said to occupy the same *subshell.* Shells and subshells containing the maximum number of electrons permitted by the exclusion principle are *closed.* Atoms whose subshells are all closed possess unusual stability.

PROBLEMS

1. Show that the electron structure of the inert gas xenon ($Z = 54$) contains only closed subshells.

2. Show that the electron structure of the alkali metal francium ($Z = 87$) has a single electron in its outermost shell.

3. How many electrons do the elements in group II of the periodic table have in their outermost shells?

4. Find the maximum number of electrons that can be present in each of the subshells of the $n = 4$ shell.

5. Show that the maximum number of electrons in a shell of given n is $2n^2$. (*Hint:* Multiply the average number of electrons in the subshells of the nth shell by the number of subshells in that shell.)

6. How many elements would there be if atoms with occupied electron shells up through $n = 6$ could exist?

7. The energy required to remove an electron from an atom is called its ionization energy. Account for the fact that the ionization energies of the elements of atomic numbers 20 through 29 are very nearly the same, although wide variations are found in the ionization energies of other sequences of elements.

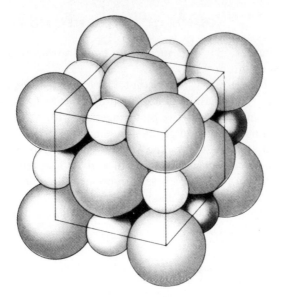

Molecules and Solids | 22

Molecules and solids both exist by virtue of the strong interactions that occur between atoms under certain circumstances. The same physical processes that can bind several atoms together into a molecule can also bind an unlimited number of them together into a solid. In this chapter we shall see how the modern theory of the atom is able to account for the properties of molecules and solids.

22–1 Ionic binding

Atoms are most stable when their electron shells are closed, and tend to gain or lose electrons in order to secure closed shells by joining with other atoms. Two molecular binding mechanisms are known, *ionic* and *covalent*. In ionic binding, electrons are effectively transferred from one atom to another, while in covalent binding, two atoms may share one or more pairs of electrons in order to provide each with a closed outer electron shell. In some molecules one or the other mechanism predominates, while in others both participate to comparable extents in holding molecules together.

Let us see how ionic binding functions in the case of NaCl, the sodium chloride molecule of ordinary table salt. The work that must be done to remove an electron from an atom is called the *ionization energy* of that atom. An atom with just one electron in its outer shell has a relatively low ionization energy which reflects the shielding effect of the inner electrons that cancel out most of the charge of the nucleus (Fig. 22–1). Such an atom tends to lose the outer electron and thereby assume an electron configuration composed solely of closed shells. At the other extreme, an atom lacking just one electron for the completion of its outer shell has a relatively high ionization energy which reflects the greater attractive force exerted by the nucleus when the nuclear charge is less effectively shielded by the intervening electrons. Thus sodium, with filled $n = 1$ and $n = 2$ shells and 1 electron in its $n = 3$ shell, has an ionization energy of 5.1 ev, while chlorine, also with filled $n = 1$ and $n = 2$ shells but with 7 electrons in its $n = 3$ shell, has an ionization energy of 13.0 ev.

Another atomic property relevant to ionic binding is *electron affinity*. The electron affinity of an atom is the amount of energy released when it acquires an additional electron. Since this amount of work must be done to remove the additional electrons after it is in place, electron affinity is a measure of the tendency of an atom to attract electrons in excess of its normal complement. As we would expect, atoms with high ionization energies have high electron affinities, while those with low ionization energies have low electron affinities. Atoms of the former kind, such as chlorine, tend to form negative ions by picking up electrons, and atoms of the latter kind, such as sodium, tend to form positive ions by losing electrons. An ionic bond between an atom of each kind occurs when one or more electrons pass from the atom with the low ionization energy to the one with the high electron affinity, since the resulting ions, both with

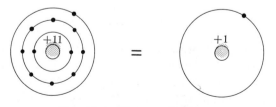

FIG. 22–1. Electron shielding.

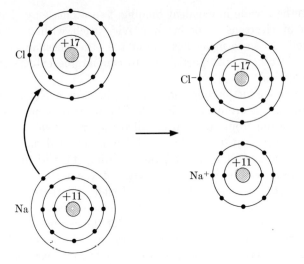

Fig. 22–2. Electron transfer in NaCl. The resulting ions attract each other electrostatically.

complete outer shells, then attract each other electrostatically (Fig. 22–2). The bond is stable if more energy is needed to separate the ions than can be supplied by the return of the transferred electron or electrons.

As we said, the ionization energy of sodium is 5.1 ev, so that 5.1 ev of work is required to remove its single outer electron. The electron affinity of chlorine is 3.8 ev, so that 3.8 ev of energy is liberated when an electron is added to fill the vacancy in its outer shell. Transferring an electron from a sodium atom to a chlorine atom to form Na^+ and Cl^- ions therefore involves a net energy expenditure of 5.1 ev − 3.8 ev or 1.3 ev:

$$Na + Cl + 1.3 \, ev \rightarrow Na^+ + Cl^-. \qquad (22\text{–}1)$$

Now we consider the electrostatic potential energy of a charge of $+e$, corresponding to the Na^+ ion, that is a distance r from a charge of $-e$, corresponding to the Cl^- ion. This energy is

$$PE = -\frac{e^2}{4\pi\epsilon_0 r}; \qquad (22\text{–}2)$$

the closer together the charges, the more negative their potential energy PE becomes. We recall that a system whose total energy is negative is stable, since work must be done on it from the outside to break it up by making its total energy positive. The PE of a system of a $+e$ and a $-e$ charge is -1.3 ev when the charges are 11×10^{-10} meter apart; hence

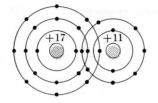

FIGURE 22-3

the total energy of a Na^+ ion and a Cl^- ion is zero when the ions are separated by this distance. When the ions are still closer together, their total energy is negative, and external energy is needed to pull them apart. The actual separation of Na^+ and Cl^- in a NaCl molecule is 2.4×10^{-10} meter, so that this molecule is stable. The work that must be done to break a NaCl molecule into Na and Cl atoms is 4.2 ev; that is,

$$NaCl + 4.2\ ev \rightarrow Na + Cl. \qquad (22-3)$$

There are two different repulsive effects that keep the Na^+ and Cl^- ions from being closer than 2.4×10^{-10} meter. Both effects arise because the electron structures of the ions mesh together when they are too near one another (Fig. 22-3). First, when such meshing occurs, the Na and Cl nuclei cease being shielded by their electrons and repel electrostatically. Second but more important, such meshing means that the Na^+ and Cl^- electrons no longer constitute separate atomic systems but instead constitute a single one. According to the Pauli exclusion principle, no two electrons in the same system can have identical sets of quantum numbers, and so some of the electrons must go into quantum states of higher energy than they otherwise would occupy. The result of both effects is that work must be done to push the ions together to make their electron structures overlap. The various attractive and repulsive forces cancel out at a separation of 2.4×10^{-10} meter, which is why the Na^+ and Cl^- ions in a NaCl molecule are this far apart.

22-2 Covalent binding

In covalent binding the atoms composing a molecule share one or more pairs of electrons. Let us examine the origin of the covalent bond between the hydrogen atoms in a H_2 molecule, where ionic binding can play no part. If the shared electrons circulate around the H nuclei as shown in Fig. 22-4, they spend more time on the average between the nuclei than they do on the outside. The result is an effective negative charge between the positive H nuclei, and the attractive force this charge exerts on the nuclei is more than enough to counterbalance the direct repulsion between them. If the nuclei are too close together, however, their repulsion becomes dominant and the molecule is not stable. The balance between attractive and repulsive forces occurs at a separation of 0.74×10^{-10} meters where the total energy of the H_2 molecule is -4.5 ev. Hence 4.5 ev of work must be done to break a H_2 molecule into two H atoms:

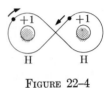

FIGURE 22-4

$$H_2 + 4.5 \text{ ev} \rightarrow H + H. \qquad (22\text{--}4)$$

According to the Pauli exclusion principle, two electrons can jointly occupy the same region while in the lowest possible energy state when their spin quantum numbers are different, one having $s = +\frac{1}{2}$ and the other $s = -\frac{1}{2}$. In other words, the two electrons must have opposite spins. If the spins were the same, the electrons would be restricted to spending most of their time near the far ends of the H atoms where they interact least. The result would be both a direct repulsion between the nuclei and a repulsion induced by the outward attraction of the electrons, making a stable molecule impossible.

Carbon atoms tend to form four covalent bonds at the same time, since they have four electrons in their outer shells and these shells lack four electrons for completion. Various distributions of these bonds are possible, including bonds between adjacent carbon atoms. Chemists usually represent covalent bonds by using a dash for each shared pair of electrons. The structural diagrams of three common covalent molecules (methane, carbon dioxide, and acetylene) illustrate the different bonds in which carbon can participate to form a covalent molecule:

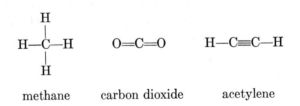

Carbon atoms are so versatile in forming covalent bonds with each other as well as with other atoms that literally millions of carbon compounds are known, some whose molecules contain tens of thousands of atoms. Such compounds were once thought to originate only in living things, and their study is accordingly known even today as organic chemistry.

22-3 Polar molecules

When atoms join together to form a molecule, their inner, complete electron shells undergo little change. In the limit of pure covalent binding, the atoms in a covalent molecule share their outer electrons evenly, and on the average remain neutral. In the opposite limit of pure ionic binding, however, electrons are transferred from certain atoms to other atoms in the molecule so that all have complete outer electron shells. The resulting molecule contains ions of opposite charge in different locations (Fig. 22–5). A molecule of this kind, as we learned in Chapter 13, is called a *polar molecule,* while a molecule whose charge distribution is symmetric about its center is called a *nonpolar molecule.* Ionic molecules are polar, as we expect, but so are many covalent molecules: while their component atoms share electrons, the electrons may spend more time in the vicinity of some atoms than in the vicinity of others. Thus the covalent molecules HCl is polar because a chlorine atom has greater attraction for an electron than a hydrogen atom, while the covalent molecule H_2 is nonpolar because both hydrogen atoms have equal attractions for electrons.

The fact that polar molecules exist helps to explain a number of familiar phenomena. The behavior of compounds in solution is a good example. Water readily dissolves such compounds as salt and sugar, but cannot dissolve fats or oils. Gasoline readily dissolves fats and oils, but cannot dissolve salt or sugar. The key to these differences lies in the strongly polar nature of water molecules and the nonpolar nature of gasoline molecules. Water molecules tend to form aggregates under the influence of the electric forces between the ends of adjacent molecules, as shown in Fig. 22–6. Polar molecules of other substances, such as salt and sugar, can join in these aggregates, and are therefore easily dissolved by water. The nonpolar molecules of fats and oils, however, do not interact with water molecules. If samples of oil and water are mixed together, the attraction of water molecules for one another acts to squeeze out the oil molecules, and the mixture soon separates into layers of each substance. Fat and oil molecules dissolve only in liquids whose molecules are similar to theirs, which is why gasoline is a solvent for these compounds.

NaCl

FIG. 22–5. Polar molecule.

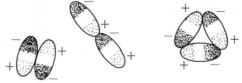

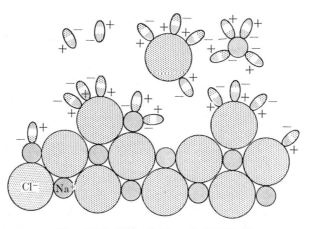

FIG. 22–6. Water molecules.

FIG. 22–7. The solution of solid NaCl.

When an ionic compound dissolves in a strongly polar liquid like water, the forces attracting the ions to water molecules usually are greater than the forces holding the ions together. If a crystal of such a compound is placed in water, water molecules cluster around the crystal ions with their positive ends toward negative ions and their negative ends toward positive ions. The attraction of several water molecules is sufficiently great to extract the ion from the crystal lattice, and it moves into solution surrounded by water molecules (Fig. 22–7). The resulting solution contains ions rather than molecules of the dissolved compound. Substances that separate into free ions when dissolved in water are called *electrolytes* since they are able to conduct electric current by the migration of positive and negative ions. Figure 22–8 shows how a NaCl solution conducts electricity. All ionic compounds soluble in water and some polar covalent compounds, such as HCl, are electrolytes. Other covalent compounds, such as sugar, are nonelectrolytes even though they are soluble in water.

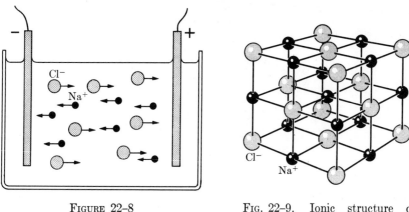

FIGURE 22-8

FIG. 22-9. Ionic structure of NaCl crystal.

Since the outer electron structure of an ion may be very different from that of a neutral atom of the same species, it is not surprising that the ions of an element may have very different behavior from that of its atoms or molecules. Thus gaseous chlorine is greenish in color, has a strong, irritating taste, and is very active chemically, while a solution of chlorine ions is colorless, has a mild, pleasant taste, and is only feebly active. The contrast in chemical activity, of course, follows from the incomplete outer electron shell in a chlorine atom while this shell is complete in a chlorine ion.

22–4 The structure of solids

Most solids are crystalline in nature, with their constituent atoms arranged in regular, repeated patterns. Sometimes the macroscopic shape of a solid mirrors its crystalline structure, as in salt crystals or snowflakes, but this is far from being generally true. The ionic and covalent bonds between atoms that are responsible for the formation of molecules also act to hold many crystalline solids together. Figure 22–9 shows the structure of a NaCl crystal; the small spheres represent Na^+ ions and the large spheres Cl^- ions. Each ion behaves essentially like a point charge, and thus tends to attract to itself as many ions of opposite sign as can fit around it. The latter ions, of course, repel one another, and so the resulting crystal is an equilibrium configuration in which the various attractions and repulsions balance out. In a NaCl crystal each Na^+ ion is surrounded by six Cl^- ions and vice-versa. In crystals having different structures the number of "nearest neighbors" around each ion may be 3,

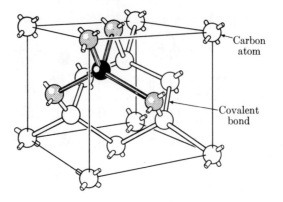

Fig. 22–10. Covalent structure of diamond. Each carbon atom shares electron pairs with four other carbon atoms.

4, 6, 8, or 12. Ionic bonds are usually fairly strong, and consequently ionic crystals are strong, hard, and have high melting points.

In covalent crystals the attractive interatomic forces arise from the sharing of electrons between atoms. Figure 22–10 shows the array of carbon atoms in a diamond crystal, with each carbon atom sharing electron pairs with the four other carbon atoms adjacent to it. All of the electrons in the outer shells of the carbon atoms participate in the binding, and it is therefore not surprising that diamond is extremely hard and must be heated to over 6000°F before its crystal structure is disrupted. Other covalent crystals are those of silicon, silicon carbide, and germanium. As in the case of molecular binding, many crystalline bonds are partly ionic and partly covalent in origin. An example of such mixed binding is quartz (SiO_2).

A third important type of cohesive force in crystalline solids is the *metallic bond*, which has no molecular counterpart. A characteristic property of all metal atoms is the presence of only a few electrons in their outer shells, and these electrons can be detached relatively easily to leave behind positive ions. According to the theory of the metallic bond, a metal in the solid state consists of an assembly of atoms that have given up their outermost electrons to a common "gas" of freely moving electrons that pervades the entire metal. The electrostatic interaction between the positive ions and the negative electron gas holds the metal together. This theory has many attractive features. The high electrical and thermal conductivity of metals follows from the ability of the free electrons to migrate through their crystal structures, while all of the electrons in ionic and covalent crystals are bound to particular atoms or pairs of atoms. Unlike other crystals, metals may be deformed without fracture, because the electron gas permits the atoms to slide past one another by acting as a kind of lubricant. While certain solids, such as rubber or asphalt, that are amorphous rather than crystalline in structure can also be deformed readily, they lack the strength and hardness conferred by the metallic bond. Furthermore, since the atoms in a metal interact through the medium of a common electron gas, the properties of mixtures of different metal atoms should not depend critically on the relative proportions of each kind of atom provided their sizes are similar. This prediction is fulfilled in the observed behavior of alloys, in contrast to the specific atomic proportions characteristic of ionic and covalent solids.

The opacity and metallic luster exhibited by metals may also be traced to the gas of free electrons that pervades them. When light of any frequency shines on a metal, the free electrons are set in vibration by the oscillating electromagnetic fields and thereby absorb the light. The oscillating electrons themselves then act as sources of light, sending out electromagnetic waves of the original frequency in all directions. Those waves that happen to be directed back toward the metal surface are able to escape, and their emergence gives the metal its lustrous appearance. If the metal surface is smooth, the reradiated waves appear to us as a reflection of the original incident light.

22–5 Energy bands

When atoms are brought as close together as those in a crystal, they interact with one another to such an extent that their outer electron shells constitute a single system of electrons common to the entire array of atoms. The Pauli exclusion principle prohibits more than two electrons (one with each spin) in any energy level of a system. This principle is obeyed in a crystal because the energy levels of the outer electron shells of the various atoms are all slightly altered by their mutual interaction. (The inner shells do not interact and therefore do not undergo a change.) As a result of the shifts in the energy levels, an *energy band* exists in a crystal in place of each sharply defined energy level of its component atoms (Fig. 22–11). While these bands are actually composed of a multitude of individual energy levels, as many as there are atoms in the crystal, the levels are so near one another as to form a continuous distribution.

The energy bands in a crystal correspond to energy levels in an atom, and an electron in a crystal can only have an energy that falls within one of these bands. The various energy bands in a crystal may or may not overlap, depending upon the composition of the crystal (Fig. 22–12). If they do not overlap, the gaps between them represent energy values

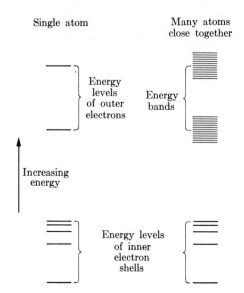

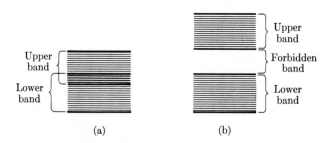

FIG. 22–11. Energy bands replace energy levels of outer electrons in assembly of atoms close together.

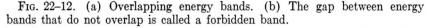

FIG. 22–12. (a) Overlapping energy bands. (b) The gap between energy bands that do not overlap is called a forbidden band.

which electrons in the crystal cannot have. The gaps are accordingly known as *forbidden bands*.

The energy bands we have been speaking of contain all of the possible energies that can be possessed by electrons. The electrical properties of a crystalline solid depend upon both its energy-band structure and the way in which the bands are normally occupied by electrons. We shall examine a few specific cases to see how the energy band approach accounts for the observed electrical behavior of such solids.

Figure 22–13 shows the energy bands of solid sodium. Sodium atoms have but one electron in their outer shells. This means that the upper

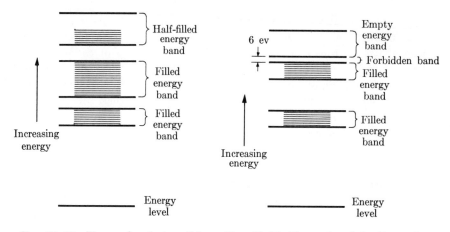

FIG. 22–13. Energy bands in solid sodium.   FIG. 22–14. Energy bands in diamond.

energy band in a sodium crystal is only half filled with electrons, since each level within the band, like each level in the atom, is capable of containing *two* electrons. When an electric field is established in a sodium crystal, electrons readily acquire the small additional energy they need to move up in their energy band. The additional energy is in the form of kinetic energy, and the moving electrons constitute an electric current. Sodium is therefore a good conductor, as are other crystalline solids with energy bands that are only partially filled.

Figure 22–14 shows the energy-band structure of diamond. The two lower energy bands are completely filled, and there is a gap of 6 ev between the top of the higher of these bands and the empty band above it. Hence a minimum of 6 ev of additional energy must be given to an electron in a diamond crystal if it is to be capable of free motion, since it cannot have an energy lying in the forbidden band. Such an energy increment cannot readily be imparted to an electron in a crystal by an electric field. A moving electron in a crystal that is not a member of a free electron gas collides with an atom or another electron after an average of only about 10^{-8} meter of travel, whereupon it loses much of the energy it gained from any electric field during its motion. The enormous electric field intensity of 6×10^8 volts/m is required if an electron is to gain 6 ev in a path length of 10^{-8} meter, well over 10^{10} times greater than the electric intensity needed to cause a current to flow in sodium. Diamond is therefore considered an electrical insulator.

Silicon has a crystal structure similar to that of diamond, and, as in diamond, a gap separates the top of a filled energy band from an empty higher band. However, while the gap is 6 ev wide in diamond, it is only

1.1 ev wide in silicon. At very low temperatures silicon is hardly better than diamond as a conductor, but at room temperature a small proportion of its electrons can possess enough kinetic energy of thermal origin to exist in the higher band. These few electrons are sufficient to permit a limited amount of current to flow when an electric field is applied. Thus silicon has a resistivity intermediate between those of conductors (such as sodium) and those of insulators (such as diamond), and is classified as a *semiconductor*.

22–6 Impurity semiconductors

The conductivity of semiconductors is markedly affected by slight amounts of impurity. Suppose we add several arsenic atoms to a silicon crystal. Arsenic atoms have five electrons in their outermost shells, while silicon atoms have four. When an arsenic atom replaces a silicon atom in a silicon crystal, four of its electrons participate in covalent bonds with its nearest neighbors. The remaining electron needs very little energy to be detached and move about freely in the crystal. In an energy-band diagram, as in Fig. 22–15(a), the effect of arsenic as an impurity is to supply energy levels just below the band in which electrons must be present for conduction to take place. These levels are called *donor levels*, and the material is called an *n-type* semiconductor because electric current in it is carried by the motion of negative charges.

If we add gallium atoms to a silicon crystal, a different effect occurs. Gallium atoms have only three electrons in their outer shells, and their presence leaves vacancies called *holes* in the electron structure of the crystal. An electron requires little energy to move into a hole, but as it does so it leaves a new hole in its previous location. When an electric field is applied to a silicon crystal in which gallium is present as an im-

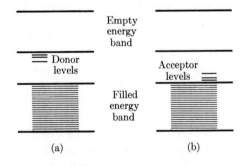

FIG. 22–15. (a) Donor levels due to arsenic atoms in a silicon crystal. (b) Acceptor levels due to gallium atoms in a silicon crystal.

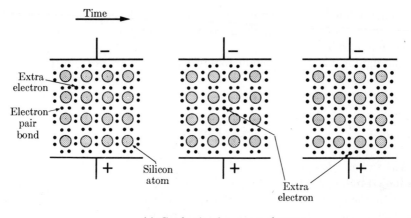

(a) Conduction by excess electrons

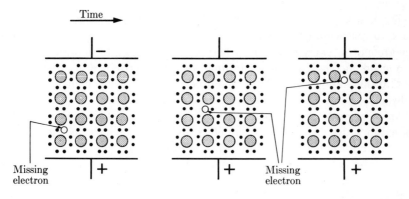

(b) Conduction by "holes"

FIG. 22–16. (a) Current in an *n*-type semiconductor is carried by excess electrons that do not fit into the electron bond structure of the crystal. (b) Current in a *p*-type semiconductor is carried by the motion of "holes," which are sites of missing electrons. Holes move toward the negative electrode as a succession of electrons move into them.

purity, electrons move toward the positive electrode by successively filling holes. The flow of current here is best described in terms of the motion of the holes, which behave as though they are positive charges since they move toward the negative electrode (Fig. 22–16). A material of this kind is therefore called a *p-type* semiconductor. In the energy band diagram of Fig. 22–15(b), we see that the effect of gallium as an impurity is to provide energy levels, called *acceptor levels*, just above the highest filled band; electrons that enter these levels leave behind unoccupied levels in the formerly filled band which make possible the conduction of current.

The significance of semiconductors in technology arises from the degree of control of electric current that can be accomplished by combining *n*- and *p*-type semiconductors in various ways.

IMPORTANT TERMS

The two chief types of molecular bond are the *ionic* and the *covalent*. In ionic binding electrons are transferred from one atom to another, and the two then attract each other electrostatically. In covalent binding two atoms share one or more pairs of electrons, and the presence of these electrons between the atoms also acts to hold them together electrostatically. Often both mechanisms participate in holding molecules together.

The *ionization energy* of an atom is the amount of work that must be done to remove an electron from it. The *electron affinity* of an atom is the amount of energy released when it acquires an additional electron.

A *polar molecule* is one containing an asymmetric charge distribution, so that one end is positive and the other negative even though the molecule as a whole is electrically neutral.

An *electrolyte* is a substance that separates into free ions when dissolved in water. A solution of an electrolyte can conduct electric current.

The *metallic bond* which holds metal atoms together in the solid state arises from a "gas" of freely moving electrons that pervades the entire metal.

Because the atoms in a crystal are so close together, the energy levels of their outer electron shells are altered slightly to produce *energy bands* characteristic of the entire crystal in place of the individual sharply defined energy levels of the separate atoms. Gaps between energy bands represent energies forbidden to electrons in the crystal and are called *forbidden bands*.

Semiconductors are intermediate in their ability to carry electric current between conductors and insulators. An *n-type* semiconductor is one in which electric current is carried by the motion of electrons. A *p-type* semiconductor is one in which electric current is carried by the motion of *holes*, which are vacancies in electron structure that behave like positive charges.

PROBLEMS

1. How can you account for the fact that lithium and sodium exhibit similar chemical behavior?

2. How can you account for the fact that fluorine and chlorine exhibit similar chemical behavior?

3. Why do lithium atoms not form molecules with each other in the way that fluorine atoms do?

4. The ionization energies of Li, Na, K, Rb, and Cs are, respectively, 5.4, 5.1, 4.3, 4.2, and 3.9 ev. All are in group I of the periodic table. Account for the decrease in ionization energy with increasing atomic number.

5. The ionization energy of potassium is 4.3 ev and the electron affinity of chlorine is 3.8 ev. (a) What is the net amount of energy required to form a K^+ and Cl^- ion pair from a pair of the same atoms? (b) Considering them as point charges, how close together must a K^+ and a Cl^- ion be if the total energy of the pair is to be zero?

6. The ionization energy of lithium is 5.4 ev and the electron affinity of bromine is 3.5 ev. (a) What is the net amount of energy required to form a Li^+ and Br^- ion pair from a pair of the same atoms? (b) Considering them as point charges, how close together must a Li^+ and a Br^- ion be if the total energy of the pair is to be zero?

7. Why are electrons much more readily liberated from lithium when it is irradiated with ultraviolet light than from fluorine?

8. Why do molecules increase in size with increasing temperature?

9. Why are Na atoms more active chemically than Na^+ ions?

10. Why are Cl atoms more active chemically than Cl^- ions?

11. Give an argument based on the Pauli principle to show that rare gas atoms (He, Ne, Ar, etc.) cannot participate in covalent bonds.

12. Give an argument based upon the Pauli principle to show that the molecule H_3 cannot exist.

13. The forbidden energy band in germanium that lies between the highest filled band and the empty band above it has a width of 0.7 ev. Compare the conductivity of germanium with that of silicon at (a) very low temperatures and (b) room temperature.

14. Does the addition of a small amount of indium to germanium result in the formation of an n- or a p-type semiconductor?

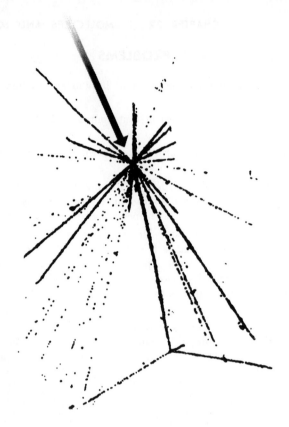

The Nucleus | 23

Until now we have not had to regard the nucleus of an atom as anything but a tiny positively charged lump whose sole function is to provide the atom with most of its mass and to hold its several electrons in place. Since the behavior of atomic electrons is responsible for the behavior of matter in bulk, the properties of matter we have been exploring, save for mass, have nothing directly to do with atomic nuclei. However, for all its seeming passivity, the nucleus turns out to be of supreme importance in the universe: the chemical elements exist by virtue of the ability of nuclei to possess multiple electric charges, and the energy that is involved in nearly all natural processes has its ultimate origin in nuclear reactions and transformations. Only in recent years have tools been available for investigating nuclei in any detail, and their many mysteries are still being plumbed.

23–1 The mass spectrometer

The mass of an atom is one of its most characteristic properties and, as we shall see, an accurate knowledge of atomic masses provides considerable insight into nuclear phenomena. A variety of instruments with the generic name of *mass spectrometers* have been devised to measure atomic masses, and we shall consider the operating principles of a particularly simple one. Figure 23–1 is a schematic diagram of this mass spectrometer. The first step in its operation is to produce ions of the substance under study. If the substance is a gas, ions can be readily formed by electron bombardment, while if it is a solid it is often convenient to incorporate it into an electrode that is used as one terminal of an electric arc discharge. The ions emerge from their source through a slit with the charge $+e$ and are then accelerated by an electric field. (Ions with other charges are sometimes present but are easily taken into account.) When they enter the spectrometer itself the ions as a rule are traveling in slightly different directions with slightly different speeds. A pair of slits serves to collimate the ion beam, which then passes through

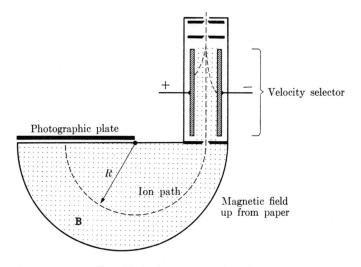

Fig. 23–1. A mass spectrometer.

a *velocity selector*. The velocity selector consists of uniform electric and magnetic fields that are perpendicular to each other and to the beam. The electric field exerts the force

$$F_e = eE$$

on the ions, while the magnetic field exerts the force

$$F_m = Bev$$

on them in the opposite direction. For an ion to reach the slit at the far end of the velocity selector it must suffer no deflection within the selector, which means that only those ions escape for which

$$F_e = F_m \,,$$

$$v = \frac{E}{B}. \tag{23--1}$$

The ions in the beam are now all moving in the same direction with the same velocity. Once past the velocity selector they enter a uniform magnetic field, and follow circular paths whose radii R may be found by equating the magnetic force Bev on them with the centripetal force mv^2/R:

$$Bev = \frac{mv^2}{R} \,,$$

$$R = \frac{mv}{Be}. \tag{23--2}$$

Since v, B, and e are known, a measurement of R yields a value for m, the ion mass. In some spectrometer designs the ions fall upon a photographic plate, permitting R to be determined from the position of their image, while in others B is varied to bring the ion beam to a fixed detector for which R is known.

23–2 Nucleons

What are nuclei composed of? Let us examine a table of the properties of the elements, such as the one in Appendix IV. In this table the elements are listed in the order of their *atomic numbers*, with the atomic number of an element defined as the number of electrons in each of its atoms. Hydrogen, whose atoms have a single electron each, has the atomic number 1; helium, whose atoms have two electrons each, has the atomic number 2; and so on up to nobelium, whose atoms have 102 electrons each and which therefore has the atomic number 102. We notice immediately that atomic mass increases with atomic number, which suggests the hypothesis that all atoms are simply combinations of hydrogen atoms. According to this hypothesis (whose basic idea was put forward about 150 years ago) the helium atom, atomic number 2, should have a nucleus consisting of two protons; the lithium atom, atomic number 3, should have a nucleus consisting of three protons; and so on through the nobelium atom, atomic number 102, which should have a nucleus consisting of 102 protons. However, atomic masses do *not* increase in steps of one hydrogen atom mass. Helium atoms weigh about four times as much as hydrogen atoms, lithium atoms about seven times as much, and nobelium atoms about 253 times as much. So this hypothesis cannot be correct. But it must have some kind of truth in it, because, with few exceptions, atomic masses *are* very close to being exact multiples of the mass of the hydrogen atom.

What we might then suppose is that there are enough protons in each nucleus to provide for the observed atomic mass, with several electrons also present whose negative charges cancel out the positive charges of the "excess" protons. This notion, which had favor for a while, is untenable on several grounds. We discussed one of the objections in Chapter 19: far too much energy is required to localize an electron within a nucleus, according to the uncertainty principle, for it to be a possible nuclear constituent.

In 1932 the puzzle of the missing nuclear ingredient was solved by Chadwick in England and Joliot and Curie in France. They discovered a very penetrating radiation, which was definitely not electromagnetic in character as are x-rays and gamma rays, and which readily ejected protons from hydrogen-rich materials like paraffin. The great penetrability suggested a neutral particle, which would not be affected by the strong electric fields in matter. The mass of the particle was revealed by the ease with which it could transfer energy to protons in collisions; efficient energy transfer in collisions can take place only when the participants have identical masses or very nearly so. This new particle, called the *neutron*, not only could account for the observations of Chadwick, Joliot, and Curie, but also fits perfectly into the nucleus as the uncharged but massive component required to supplement the proton. Present values for the proton and neutron masses are

$$m_{\text{proton}} = 1.6724 \times 10^{-27} \text{ kg},$$

$$m_{\text{neutron}} = 1.6747 \times 10^{-27} \text{ kg}.$$

Neutrons and protons are jointly called *nucleons*.

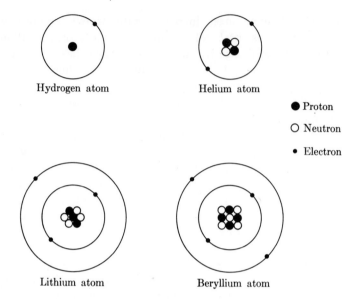

FIG. 23-2. The electronic and nuclear compositions of hydrogen, helium, lithium, and beryllium atoms.

With the help of the neutron, then, the constitution of nuclei more complex than hydrogen can be figured out readily: helium nuclei contain two protons and two neutrons each, lithium nuclei contain three protons and four neutrons each, and so on (Fig. 23-2).

23-3 Isotopes

We are still faced with the question of how to account for atomic masses, which cannot be exactly expressed in terms of combinations of proton and neutron masses. As it happens, the answer was known even before the composition of nuclei was clearly understood. Experimental measurements showed that what we call chemical elements actually each consist of several different components, called *isotopes,* each having the same nuclear charge (and hence the same atomic number) but with different masses. Chlorine, for example, consists of two isotopes, one whose nucleus contains 17 protons and 18 neutrons and another whose nucleus contains 17 protons and 20 neutrons. In nature, chlorine is composed of about 75% of the former isotope and 25% of the latter, yielding a composite atomic mass of 35.46, the value given in Appendix IV.

Because the chemical properties of an element depend upon the distribution of the electrons in its atoms, which in turn depends upon the nuclear charge, nuclear structure beyond the number of protons present has little significance for the chemist. The physical properties of an element, however, depend strongly on the nuclear structures of its isotopes, whose behavior may be very different from one another although chemically they are indistinguishable.

The conventional symbols for isotopes all follow the pattern

$$_nX^m,$$

where X is the chemical symbol of the element, n is its atomic number (equal to the number of electrons per atom or, what is equivalent, the number of protons per nucleus), and m is the *mass number* of the isotope in question, which is defined as the total number of nucleons (neutrons plus protons) in each nucleus. Hence ordinary hydrogen is designated

$$_1H^1,$$

since its atomic number and mass number are both 1, while ordinary helium is designated

$$_2He^4,$$

since its atomic number is 2 and its mass number is 4. The two isotopes of chlorine mentioned above are designated $_{17}Cl^{35}$ and $_{17}Cl^{37}$ respectively.

23–4 Nuclear binding energy

While the discovery of isotopes permits us to understand why certain elements do not have atomic masses that are multiples of the proton and neutron masses, even those elements that seem to fit the picture well have slight mass variations from simple assemblies of these particles. In Table 23–1 is a list of the experimentally determined masses of the neutron and some light isotopes expressed in terms of *atomic mass units*

TABLE 23–1

The masses of the five lightest nuclei and the neutron

Name	*Symbol*	*Mass, amu*
Proton	$_1H^1$	1.007593
Neutron	$_0n^1$	1.008982
Deuteron	$_1H^2$	2.014186
Triton	$_1H^3$	3.016448
Helium-3	$_2He^3$	3.015779
Alpha particle	$_2He^4$	4.003873

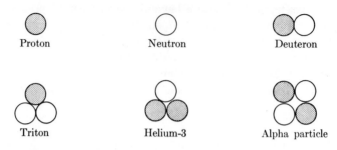

FIG. 23–3. The composition of the six lightest nuclei.

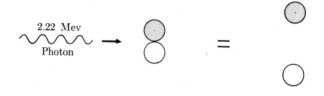

FIG. 23–4. The binding energy of the deuteron is 2.22 Mev, which means that this much energy is required to make up the difference between the deuteron's mass and the combined mass of its constituent neutron and proton. Absorbing 2.22 Mev, for instance by being struck by a 2.22-Mev gamma-ray photon, furnishes a deuteron with enough additional mass to split into a neutron and a proton.

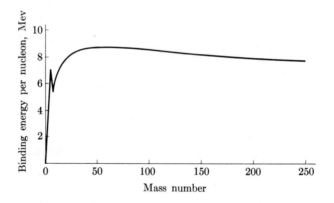

FIG. 23–5. The binding energy per nucleon versus mass number. The higher the binding energy per nucleon, the more stable the nucleus. When a heavy nucleus is split into two lighter ones, a process called *fission*, the greater binding energy of the latter causes the liberation of energy. When two very light nuclei join to form a heavier one, a process called *fusion*, the greater binding energy of the latter again causes the liberation of energy.

(abbreviated amu), equal to one-sixteenth the mass of the $_8O^{16}$ isotope. The value of an amu is

$$1 \text{ amu} = 1.660 \times 10^{-27} \text{ kg.}$$

The deuteron and triton are nuclei of isotopes of hydrogen with mass numbers 2 and 3 respectively, which implies that they consist of one proton plus, for the deuteron, one neutron, and for the triton, two neutrons. Nuclei of $_2He^3$, an isotope of helium, consist of two protons and one neutron each, while ordinary helium nuclei consist of two protons and two neutrons each (Fig. 23–3).

If it is true that a deuteron consists of a neutron plus a proton, having a combined mass of 2.016575 amu, why is this combined mass *more* than the 2.014186 amu mass of the deuteron? The answer is that when nucleons combine to form a nucleus, the total energy of the system (including mass energy) remains constant, although the mass may not. Thus the difference in mass between the constituents of the deuteron and the deuteron itself may appear as additional kinetic energy of the deuteron, beyond that which the neutron and proton brought along with them. Correspondingly, this same mass deficiency must be made up by the addition of energy if it is desired to break up a deuteron.

To put this concept on a quantitative basis, we recall that energy E and mass m are related by the formula

$$E = mc^2.$$

One amu is the energy equivalent of

$$1 \text{ amu} = 1.49 \times 10^{-10} \text{ j} = 9.31 \times 10^8 \text{ ev} = 931 \text{ Mev.}$$

The *binding energy* of the deuteron, which is the energy equivalent of the missing mass, is therefore

$$(2.016575 - 2.014186) \text{ amu} \times 931 \, \frac{\text{Mev}}{\text{amu}} = 2.22 \text{ Mev.}$$

This figure is confirmed by experiments that show that the minimum energy a gamma ray must have in order to disrupt a deuteron is 2.22 Mev (Fig. 23–4).

A very interesting curve results when we plot the binding energy per nucleon (that is, the total binding energy of an isotope divided by the number of neutrons and protons that compose it) versus the mass number of the nucleus (Fig. 23–5). Except for an anomalously high peak for $_2He^4$, this curve is a surprisingly regular one. The middle range of nuclei have the highest binding energy per nucleon values, about 8.75 Mev/nucleon, which means that they require the greatest amount of energy (per nucleon) to disrupt. Such nuclei are therefore the most stable.

A significant feature of nuclear structure is illustrated by this curve. Suppose that we split the nucleus $_{92}U^{235}$, whose binding energy is 7.6 Mev/nucleon, into two fragments. Each fragment will be the nucleus of a much lighter element, and therefore will have a higher binding energy per nucleon than the uranium nucleus. The difference is about 0.8 Mev/nucleon, and so, if such *nuclear fission* were to take place, an energy of

$$0.8 \frac{\text{Mev}}{\text{nucleon}} \times 235 \text{ nucleons} = 190 \text{ Mev}$$

would be given off per splitting. This is a truly immense amount of energy to be produced in a single event. As a comparison, chemical processes involve energies of the order of magnitude of one electron volt per reacting atom, 10^{-8} the energy involved in fission.

Figure 23–5 also shows that if two of the extremely light nuclei are combined to form a heavier one, the higher binding energy of the latter will also result in the evolution of energy. For instance, if two deuterons were to join to make up a $_2He^4$ nucleus, over 23 Mev would be released. This process is known as *fusion,* and, with fission, promises to be the source of more and more of the world's power.

23–5 Nuclear forces

The really surprising thing about nuclei is not that some of them are unstable, but rather that *any* of them are stable in the face of the mutually repulsive electrostatic forces exerted by the nuclear protons upon one another. There must be a nuclear force, imperceptible on a macroscopic scale, which is able to hold nuclei together. Because nuclei larger than about 10^{-15} m in radius do not exist, we may assume that the range of nuclear forces does not exceed this figure. Various experiments, in particular those involving the scattering of one nucleon by another in a collision, indicate that nuclear forces are very strong: they give rise to a binding energy between adjacent nucleons on the average over a million times greater than the 13.6 ev binding energy between the proton and electron in a hydrogen atom. Nuclear forces are evidently enormous. It is interesting that there is evidence to indicate that the proton-proton nuclear force is identical in strength and range to the neutron-proton nuclear force.

If we inquire into the "mechanism" by which two electric charges exert forces upon one another, we find that, from a formal mathematical

FIG. 23–6. Particle exchange can lead to attractive or repulsive forces.

point of view, their interaction can be described in terms of the constant circulation of electromagnetic photons between them. In 1935 the Japanese physicist Yukawa suggested that nuclear forces, too, could be regarded as the result of an interchange of certain particles between nucleons. Today these particles are called π-*mesons*, and may have charges of 0, $+e$, or $-e$. Charged π-mesons have masses of 273 times the electron mass, while neutral π-mesons have masses of 264 times the electron mass.

The crude analogy illustrated in Fig. 23–6 may help in understanding how meson exchange can lead to both attractive and repulsive forces between nuclei. Each child in the figure has a pillow. When the children exchange pillows by snatching them from each other's grasp, the effect is like that of a mutually attractive force. On the other hand, the children may also exchange pillows by throwing them at each other. Here conservation of momentum requires that the children move apart, just as if a repulsive force were present between them.

According to Yukawa's theory, nearby nucleons constantly exchange mesons without themselves being altered. We note that the emission of a meson by a nucleon at rest which does not lose a corresponding amount of mass violates the law of conservation of energy. However, the law of conservation of energy, like all physical laws, deals only with measurable quantities. Because the uncertainty principle restricts the accuracy with which we can perform certain measurements, it limits the range of application of physical laws such as that of energy conservation. We see from the uncertainty principle in the form

$$\Delta E \Delta t \approx h/2\pi$$

that a process can take place in which energy is not conserved by an amount ΔE provided that the time interval Δt in which the process takes place is not more than $h/2\pi\Delta E$. Thus the creation, transfer, and disappearance of a meson do not conflict with the conservation of energy provided that the sequence occurs fast enough.

The most attractive aspect of the meson theory of nuclear forces is that the π-mesons predicted by Yukawa actually exist and have the properties expected of them. For a meson to be produced which can emerge from a nucleus and be detected experimentally, enough energy must be supplied to the nucleus, usually by a collision with another particle, for conversion into meson mass without affecting the invariant masses of its nucleons. When this is done with the help of the energetic particle beams found naturally in cosmic rays or created in laboratory accelerators, mesons can be studied in detail; the results are outlined in Chapter 25. However, despite this success of the meson theory, it has not yet proved useful in understanding nuclear structure. Instead, several very different hypotheses of how nucleons behave within a nucleus seem to be required, depending upon which aspect of nuclear phenomena is involved. For instance, it is known that nuclei have certain energy levels analogous to the energy levels of atomic electrons, and that transitions between these levels can take place upon the absorption or emission by the nucleus of an appropriate amount of energy. In the case of the atom we found that an artificial "model" based on classical concepts, namely the Bohr model, helped clarify what was going on. The nuclear analogy is the *independent particle model*, which supposes that nuclear particles interact with a single force field that pervades the entire nucleus. At the opposite extreme is the *liquid drop model*, whose ability to account for the variation in nuclear binding energy with mass number and for the process of nuclear fission depends upon the assumption that interactions occur only between adjacent nucleons. In this view nucleons resemble the molecules of a liquid, which also interact only with their nearest neighbors, so that a nucleus corresponds to a drop of liquid. Clearly, despite the vast amount of information we have regarding nuclei, the key insight into their nature remains elusive.

IMPORTANT TERMS

The *atomic number* of an element is the number of electrons in each of its atoms or, equivalently, the number of protons in each of its atomic nuclei.

The *neutron* is an electrically neutral particle, slightly heavier than the proton, which is present in nuclei together with protons. Neutrons and protons are jointly called *nucleons*. The *mass number* of a nucleus is the number of nucleons it contains.

Isotopes of an element have the same atomic number but different mass numbers. Isotopic symbols follow the pattern

$$_nX^m,$$

where X is the chemical symbol of the element, n its atomic number, and m the mass number of the particular isotope.

Nuclear masses are expressed in *atomic mass units* (amu), equal to one-sixteenth of the mass of the $_8O^{16}$ isotope. An amu is equal to 1.660×10^{-27} kg.

The *binding energy* of a nucleus is the energy equivalent of the difference between its mass and the sum of the masses of its individual constituent nucleons. This amount of energy must be supplied to the nucleus if it is to be completely disintegrated. The binding energy per nucleon is least for very light and very heavy nuclei; hence the *fusion* of very light nuclei to form heavier ones and the *fission* of very heavy nuclei to form lighter ones are both processes that liberate energy.

According to the *meson theory of nuclear forces*, adjacent nucleons attract each other as the result of a constant exchange of particles called π-*mesons* between them.

Two hypothetical pictures of nuclear structure are the *independent particle model*, according to which each nucleon interacts with a general force field throughout the entire nucleus, and the *liquid drop model*, according to which each nucleon in a nucleus interacts separately with each of its neighbors. Both models are partly successful, but neither can account for the entire range of nuclear properties.

PROBLEMS

1. A velocity selector uses a magnet to produce a flux density of 0.05 weber/m² and a pair of parallel metal plates 1 cm apart to produce a perpendicular electric field. What potential difference should be applied to the plates to permit singly charged ions of speed 5×10^6 m/sec to pass through the selector?

2. A mass spectrometer employs a velocity selector consisting of a magnetic field of flux density 0.04000 weber/m² perpendicular to an electric field of 50,000 volts/m. The same magnetic field is then used to deflect the ions. Find the radius of curvature of singly charged $_3Li^7$ ions of mass 7.0182 amu in this spectrometer.

3. A beam of singly charged boron atoms is accelerated by a potential difference of 3000 volts and enters a magnetic field of flux density 0.2000 weber/m² perpendicular to the field. The ions are deflected through 180° and fall upon a photographic plate. (a) How far apart will the $_5B^{10}$ (10.016 amu) and $_5B^{11}$ (11.012) isotopes strike the photographic plate? (b) The boron found in nature is a mixture of the above isotopes, and has a composite atomic weight of 10.82 amu. Find the relative abundance of each isotope in natural boron.

4. State the number of neutrons and protons in each of the following nuclei: $_3Li^6$; $_6C^{13}$; $_{15}P^{31}$; $_{40}Zr^{94}$; $_{56}Ba^{137}$.

5. State the number of neutrons and protons in each of the following nuclei: $_5Be^{10}$; $_{10}Ne^{22}$; $_{16}S^{36}$; $_{38}Sr^{88}$; $_{72}Hf^{180}$.

6. The mass of $_{10}Ne^{20}$ is 19.9988 amu. What is its binding energy?

7. The mass of $_{17}Cl^{35}$ is 34.9800 amu. What is its binding energy?

8. Find the binding energy of the triton.

9. What is the minimum energy a gamma-ray photon must have if it is to split an alpha particle into (a) a triton and a proton, and (b) a helium-3 nucleus and a neutron?

10. Compare the electrostatic potential energy of two protons 6×10^{-15} m apart with the binding energy per nucleon of nuclei with $A \approx 60$. (Such nuclei are about 6×10^{-15} m in radius.)

11. The distance between the two protons in a $_2He^3$ nucleus is roughly 1.7×10^{-15} m. (a) Calculate the electrostatic potential energy of these protons. (b) Show that this energy is of the right order of magnitude to account for the difference in binding energy between $_1H^3$ and $_2He^3$. What conclusion can be drawn from this result about the dependence of nuclear forces upon electric charge?

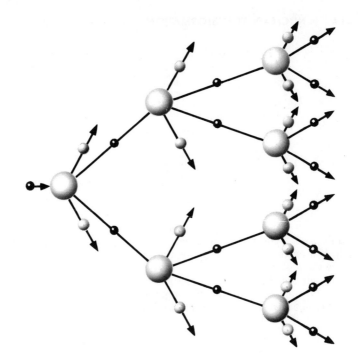

Nuclear Transformations | 24

While atomic nuclei are the most stable composite bodies known, they may nevertheless undergo transformations in which the identity of a particular nucleus changes. These changes may be spontaneous, as in radioactive decay, or induced, as in reactions between colliding nuclei. Aside from the data they provide on nuclear structure, nuclear transformations are noteworthy because of their role in the evolution of the universe: through them the primeval hydrogen of the universe coalesced into more complex elements, and they are responsible for virtually all of the energy that powers the universe. Today the laboratory of the nuclear physicist is as essential to astronomy as is the telescope.

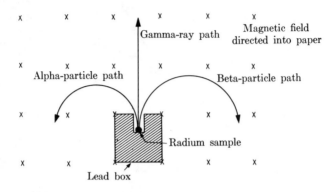

Fɪɢ. 24–1. The radiations from a radium sample may be analyzed with the help of a magnetic field. In the figure the direction of the field is into the paper: hence the positively charged alpha particles (actually helium nuclei) are deflected to the left and the negatively charged beta particles (actually electrons) to the right. Gamma rays, which are composed of energetic photons, carry no charge and are not affected by the magnetic field.

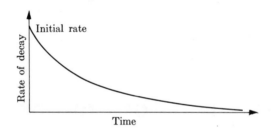

Fɪɢ. 24–2. The rate at which a sample of a radioactive substance decays is not constant, but varies with time in the manner shown in the curve. Some isotopes decay faster than others, but all obey curves having the same shape.

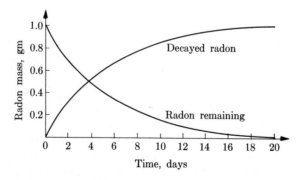

Fɪɢ. 24–3. The decay of radon, whose half-life is 3.8 days.

24–1 Radioactivity

In the previous chapter we discussed only stable nuclei. At the beginning of the 20th century it became known, as the result of research by Becquerel, the Curies, and others, that some nuclei exist which spontaneously transform themselves into other nuclear species with the emission of radiation. The observed radiation is of three types: *alpha particles*, which are the nuclei of $_2He^4$ atoms, *beta particles*, which are electrons, and *gamma rays*, which are composed of short wavelength (and hence energetic) photons (Fig. 24–1).

One of the characteristics of all types of *radioactivity* is that the rate at which the nuclei in a given sample decay always follows a curve whose shape is like that shown in Fig. 24–2. Some isotopes decay faster than others, but in each case a certain definite time is required for half of an original sample to decay. This time is called the *half-life* of the isotope. Radon has a half-life of 3.8 days, for instance: if we start with 1 gm of radon, $\frac{1}{2}$ gm will remain after 3.8 days, $\frac{1}{4}$ gm after 7.6 days, $\frac{1}{8}$ gm after 11.4 days, and so on (Fig. 24–3). Half-lives range from billionths of a second to billions of years. Samples of radioactive isotopes decay in the manner illustrated in Fig. 24–2 because a great many individual nuclei are involved, each having a certain probability of decaying. Radioactive decay involves individual events taking place within independent nuclei, rather than collective processes which involve more than one nucleus in interaction. This idea is confirmed by experiments which show that the half-life of a particular isotope is absolutely invariant under changes of pressure, temperature, electric and magnetic fields, and so on, which might, if sufficiently strong, influence internuclear phenomena.

To understand why alpha, beta, and gamma decays take place, let us examine Fig. 24–4, which is a plot showing the number of neutrons versus the number of protons in stable nuclei. For light, stable nuclei the numbers of neutrons and protons are approximately the same, while for heavier nuclei slightly more neutrons than protons are required for stability. (This is a consequence of the strong electrostatic repulsion exerted by protons on one another, which must be balanced by the attractive nuclear forces present between nucleons; to counterbalance the electrostatic repulsion of many protons, a comparatively greater number of neutrons, which produce only attractive forces owing to their electrical neutrality, is necessary for stability.) From Fig. 24–4 it is evident that an element of a given atomic number has only a very narrow range of possible numbers of neutrons if it is to be stable.

Suppose now that a nucleus exists which has either too many or too few neutrons relative to the number of protons present for stability. We

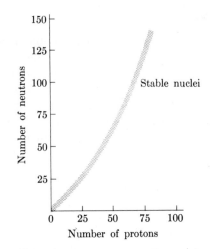

Fɪɢ. 24–4. The number of neutrons versus the number of protons in stable nuclei. The larger the nucleus, the greater the proportion of neutrons.

shall consider the former case first. If one of the excess neutrons transforms itself into a proton, this will simultaneously reduce the number of neutrons while increasing the number of protons. To conserve electric charge, such a transformation requires the emission of a negative electron, and we may write it in equation form as

$$n^0 \rightarrow p^+ + e^-. \tag{24–1}$$

The electron leaves the nucleus, and is detectable as a "beta particle." The residual nucleus may be left with some extra energy as a consequence of its shifted binding energy, and this energy is given off in the form of gamma rays. Sometimes more than one *beta decay* is required for a particular unstable nucleus to reach a stable configuration.

Should the nucleus have too few neutrons, the inverse reaction

$$p^+ \rightarrow n^0 + e^+, \tag{24–2}$$

in which a proton becomes a neutron with the emission of a *positive* electron (known as a *positron*), may take place. This is also called beta decay, since it resembles the emission of negative electrons from an unstable nucleus in every way save for the difference in charge.

Another way of altering its structure to achieve stability may involve a nucleus in *alpha decay*, in which an alpha particle consisting of two neutrons and two protons is emitted. Thus negative beta decay increases the number of protons by one and decreases the number of neutrons by

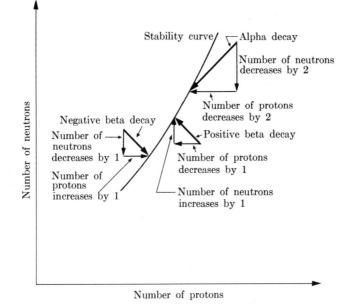

FIG. 24–5. How alpha and beta decays tend to bring an unstable nucleus to a stable configuration.

one; positive beta decay decreases the number of protons by one and increases the number of neutrons by one; and alpha decay decreases both the number of protons and the number of neutrons by two. These processes are shown schematically in Fig. 24–5. Very often a succession of alpha and beta decays, with accompanying gamma decays to carry off excess energy, is required before a nucleus reaches stability.

24–2 Alpha decay

Nuclei that contain more than about 210 nucleons are so large that the short-range forces holding them together are barely able to counterbalance the long-range electrostatic repulsive forces of their protons. Such a nucleus can reduce its bulk and thereby achieve greater stability by emitting an alpha particle, which decreases its mass number A by 4. It is appropriate to ask why it is that only alpha particles are given off by excessively heavy nuclei, and not, for example, individual protons or $_2\text{He}^3$ nuclei. The reason is a consequence of the high binding energy of the alpha particle, which means that it has significantly less mass than four individual nucleons. Because of this small mass, an alpha particle can be ejected by a heavy nucleus with energy to spare. Thus the alpha

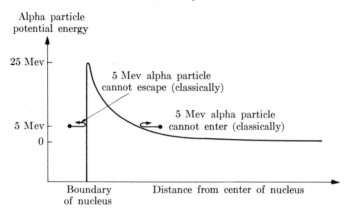

FIG. 24–6. The variation in alpha particle potential energy near a typical heavy nucleus. The alpha particle in this nucleus has 5 Mev available for its escape.

particle released in the decay of $_{92}U^{232}$ has a kinetic energy of 5.4 Mev, while 6.1 Mev would have to be supplied from the outside to this nucleus if it is to release a proton, and 9.6 Mev supplied if it is to release a $_2He^3$ nucleus.

Even though alpha decay may be energetically possible in a particular nucleus, it is not obvious just how the alpha particle is able to break away from the nuclear forces that bind it to the rest of the nucleus. Typically an alpha particle has available about 5 Mev of energy with which to escape. However, an alpha particle located at a point near the nucleus but just outside the range of its nuclear forces has an electrostatic potential energy of perhaps 25 Mev; that is, if released from this position it will have a kinetic energy of 25 Mev when it is an infinite distance away as a result of electrostatic repulsion. An alpha particle inside the nucleus therefore should require a minimum of 25 Mev in energy, five times more than is available, in order to break loose (Fig. 24–6). The alpha particle, then, is located in a box whose walls are of such a height that an energy of 25 Mev is needed to surmount them, while the particle itself has only 5 Mev for the purpose.

Quantum mechanics provides the answer to the paradox of alpha decay. Two assumptions are needed: (1) an alpha particle can exist as an individual entity within a nucleus, and (2) it is in constant motion there. According to quantum theory, a moving particle has a wave character, so that the proper classical analog of an alpha particle in a nucleus is a light wave trapped between mirrors and not a particle bouncing back and forth between solid walls. Now in order for a light wave to be reflected from a mirror, it must actually penetrate the reflect-

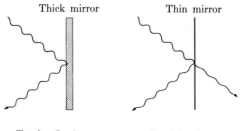

Total reflection   Partial reflection

FIG. 24–7. In the process of reflection, a wave penetrates the reflecting surface for a short distance and may pass through it if the mirror is sufficiently thin.

ing surface for a short distance (Fig. 24–7). The intensity of the wave drops off quite rapidly inside the reflecting surface, to be sure, but it *must* penetrate to some extent. If the mirror is thick, all of the incident light is reflected. However, if the mirror is very thin, some of the incident light can pass right through the mirror, as shown. The formal theory of this partial transmission, with only minor changes, is able to account quantitatively for alpha decay. The very existence of alpha decay, in fact, is further confirmation of the validity of quantum ideas, since the principles of physics that follow from Newton's laws of motion prohibit such decay.

Of course, a 25 Mev energy barrier is not very "transparent" to a 5 Mev alpha particle. A typical heavy nucleus might be 2×10^{-14} meters in diameter, and an alpha particle within it might oscillate back and forth with a speed of 2×10^7 m/sec. Hence the alpha particle strikes the confining nuclear wall 10^{21} times per second, but may nevertheless have to wait as much as 10^{10} years to escape from certain nuclei! The quantum-mechanical phenomenon of barrier penetration is sometimes called the *tunnel effect* because the particle escapes *through* the barrier and not over it.

24–3 Nuclear reactions

When two nuclei approach close enough together, it is possible for a rearrangement of their constituent nucleons to occur with one or more new nuclei formed. These are called *nuclear reactions*, by analogy with chemical reactions in which two or more compounds may combine to form new ones.

An immense number of nuclear reactions is known, all of which obey certain conservation laws. The three most important of these laws are:

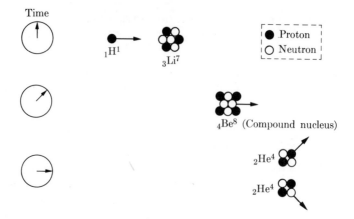

FIG. 24–8. The fusion reaction in which a lithium nucleus struck by a proton absorbs the proton and thereupon splits into two alpha particles. The liberated energy of 17.3 Mev is carried away by the alpha particles as kinetic energy.

(1) conservation of electric charge; (2) conservation of number of nucleons; and (3) conservation of mass energy. The reaction

$$_3\text{Li}^7 + {}_1\text{H}^1 \rightarrow 2\,{}_2\text{He}^4 \qquad\qquad (24–3)$$

involves the fusion of a lithium nucleus of atomic number 3 and mass number 7 with a proton of atomic number 1 and mass number 1 (Fig. 24–8). This reaction results in two alpha particles of atomic number 2 and mass number 4. A total of four protons is present on each side of the arrow, satisfying condition (1). Further, eight nucleons are present on each side of the arrow, taking care of condition (2). To see whether (3) is satisfied, we note that the mass of $_3\text{Li}^7$ is 7.01822, the mass of $_1\text{H}^1$ is 1.00812, and the mass of $_2\text{He}^4$ is 4.00387, all in atomic mass units. Thus the mass excess Q of the left-hand side of Eq. (24–3) is

$$Q = 7.01822\text{ amu} + 1.00814\text{ amu} - 2 \times 4.00387\text{ amu}$$

$$= 0.01862\text{ amu}$$

$$= 17.3\text{ Mev.}$$

This energy is divided between the two alpha particles.

It is important to note that the $_3\text{Li}^7$ and proton must have some initial relative velocity, hence initial energy, in order that they approach close enough for the reaction to take place. This initial energy must be distinguished from the initial energy required by a process whose products have *more* mass than the reactants. Let us examine the reaction

$$_7N^{14} + {}_2He^4 \rightarrow {}_8O^{17} + {}_1H^1. \tag{24-4}$$

Here, since the masses of the nuclei involved are respectively 14.00752, 4.00387, 17.00453, and 1.00814 amu, there is a mass *deficiency* of 0.00128 amu, which is 1.2 Mev. This energy must be supplied by the kinetic energy of either the alpha particle or the nitrogen nucleus or both if the reaction is to occur at all.

Often nuclear reactions occur in two separate phases. The first phase occurs when a particle (a nucleon or another nucleus) collides with a nucleus and joins with it to form a composite body called a *compound nucleus* (Fig. 24-8). In the second phase the compound nucleus decays into the products of the reaction. The notion of the compound nucleus is a useful one because a particular compound nucleus may be formed in a variety of ways, but it will decay in a characteristic manner that is independent of the way in which it came into being. The amount of excitation energy possessed by the compound nucleus is what determines how it will decay. The excitation energy is the energy equivalent of the mass difference between the initial particles and the compound nucleus plus whatever kinetic energy was brought in by the incident particle. For example, when $_7N^{14}$ is formed by the collision of a 6 Mev proton with a $_6C^{13}$ nucleus, its excitation energy is the sum of 6 Mev and the 7.5 Mev mass energy difference between $_7N^{14}$ and $_6C^{13} + {}_1H^1$. The compound nucleus $_7N^{14}$ whose excitation energy is 13.5 Mev usually decays into $_7N^{13}$ and a neutron:

$$_7N^{14} + 13.5 \text{ Mev} \rightarrow {}_7N^{13} + {}_1n^0 + 3 \text{ Mev}. \tag{24-5}$$

The 3 Mev that is liberated as kinetic energy is the difference between the 13.5 Mev excitation energy and the mass energy difference between $_7N^{14}$ and $_7N^{13} + {}_1n^0$. The essential point here is that the decay reaction (24-5) occurs regardless of how the $_7N^{14}$ compound nucleus acquired its 13.5 Mev of excitation; a 2 Mev alpha particle can react with a $_5B^{10}$ nucleus to give the identical compound nucleus, as can a 3.2 Mev deuteron with a $_6C^{12}$ nucleus.

The probability that a given pair of nuclear particles will react to form a compound nucleus depends both upon the identities of the particles and upon the energy brought into their interaction, even though the various conservation laws permit the reaction to take place. For example, $_{48}Cd^{113}$ is almost 10,000 times more likely to capture a 0.1 ev neutron than a 10 ev neutron, while it is highly improbable that $_{48}Cd^{112}$ will capture a neutron of either energy at all.

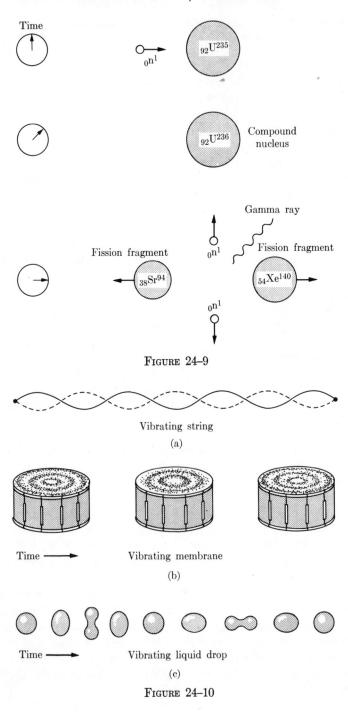

Time

$_{0}n^{1}$

$_{92}U^{235}$

$_{92}U^{236}$ Compound
nucleus

Gamma ray

$_{0}n^{1}$

Fission fragment   Fission fragment

$_{38}Sr^{94}$ $_{54}Xe^{140}$

$_{0}n^{1}$

$_{0}n^{1}$

FIGURE 24–9

Vibrating string
(a)

Time ⟶   Vibrating membrane
(b)

Time ⟶   Vibrating liquid drop
(c)

FIGURE 24–10

24-4 Nuclear fission

A particularly significant nuclear reaction that requires a neutron to initiate it is *fission*. In this process, which can take place only in certain very heavy nuclei such as $_{92}U^{235}$, the absorption of an incoming neutron causes the target nucleus to split into two smaller nuclei called *fission fragments* (Fig. 24–9). Because stable light nuclei have proportionately fewer neutrons than do heavy nuclei (Fig. 24–4), the fragments are unbalanced when they are formed and at once release one or two neutrons each. Usually the fragments are still somewhat unstable, and may undergo beta decays (accompanied by gamma decays) to achieve appropriate neutron:proton ratios. The products of fission, such as the fallout from a nuclear bomb burst, are accordingly highly radioactive. Although a variety of nuclear species may appear as fission fragments, we might cite as a typical fission reaction

$$_{92}U^{235} + _0n^1 \rightarrow _{92}U^{236} \rightarrow _{54}Xe^{140} + _{38}Sr^{94}$$

$$+ _0n^1 + _0n^1 + \gamma + 200 \text{ Mev}. \qquad (24\text{–}6)$$

The $_{92}U^{236}$ which is first formed is a compound nucleus, and it is this nucleus that splits in two. The fission fragments $_{54}Xe^{140}$ and $_{38}Sr^{94}$ are both beta radioactive; the former decays four successive times until it becomes the stable isotope $_{58}Ce^{140}$, and the latter decays twice in becoming the stable isotope $_{40}Zr^{94}$. About 84% of the total energy liberated during fission appears as kinetic energy of the fission fragments, about 2.5% as kinetic energy of the neutrons, and about 2.5% in the form of instantaneously emitted gamma rays, with the remaining 11% being given off in the decay of the fission fragments.

The liquid drop model of the nucleus that was mentioned in the previous chapter provides a plausible mechanism for the fission process. We are all familiar with the characteristic oscillations of a stretched string, Fig. 24–10(a), and of a taut membrane, Fig. 24–10(b). Less familiar perhaps are the characteristic oscillations of a liquid drop, shown in Fig. 24–10(c). If we accept the picture of a nucleus as a liquid drop, we can suppose that the absorption of a neutron by a heavy nucleus is enough to set it vibrating. The difference between an ordinary liquid drop and a nucleus is that, when the latter is distorted from a spherical shape, the short-range nuclear forces holding it together lose much of their effectiveness owing to the larger nuclear surface where nucleons have fewer bonds. Though the distortion may seriously weaken the attractive forces in the nucleus, the repulsive electrostatic forces of the protons are only slightly affected. If the attractive forces still predominate, the excitation energy of the nucleus eventually is lost through gamma decay: a neutron

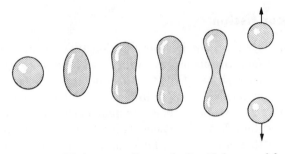

FIG. 24–11. Fission according to the liquid drop model.

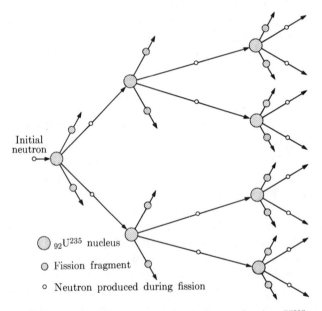

Initial
neutron

$_{92}U^{235}$ nucleus

Fission fragment

Neutron produced during fission

FIG. 24–12. Simplified sketch of a chain reaction in $_{92}U^{235}$.

is captured and a gamma ray is emitted. If the repulsive forces pre-dominate, however, the distortion grows larger and larger until the nucleus splits in two (Fig. 24–11), which is observed as fission.

Because each fission event liberates two or three neutrons while only one fission is required to initiate it, a rapidly multiplying sequence of fissions can occur in a lump of suitable material (Fig. 24–12). When uncontrolled, such a *chain reaction* evolves an immense amount of energy in a short time. If we assume that two neutrons emitted in each fission are able to induce further fissions (the average figure is lower in practice) and that 10^{-8} sec elapses between the emission of a neutron and its sub-sequent absorption, a chain reaction starting with a single fission will release 2×10^{13} joules of energy in less than 10^{-6} sec! An uncontrolled chain reaction evidently can cause an explosion of exceptional magnitude. When properly controlled so as to assure that exactly one neutron per fission causes another fission, a chain reaction occurs at a constant level of power output. A reaction of this kind makes a very efficient source of power: an output of about 1000 kilowatts is produced by the fission of 1 gram of a suitable isotope per day, as compared with the consumption of over 3 tons of coal per day per 1000 kilowatts in a conventional power plant. A device in which a chain reaction can be initiated and controlled is called a *nuclear reactor*.

24–5 Nuclear fusion

Virtually all of the energy in the universe originates in the fusion of hydrogen nuclei into helium nuclei in stellar interiors, where hydrogen is the most abundant element. Two different reaction sequences are possible, with the likelihood of each depending upon the properties of the star involved. The *proton-proton cycle* is the predominant energy source of stars whose interiors are cooler than that of the sun, perhaps 2×10^6 °K. The proton-proton cycle proceeds by means of the following reactions:

$$_1H^1 + {}_1H^1 \rightarrow {}_1H^2 + e^+ + 0.4 \, Mev,$$

$$_1H^1 + {}_1H^2 \rightarrow {}_2He^3 + 5.5 \, Mev, \qquad (24\text{–}7)$$

$$_2He^3 + {}_2He^3 \rightarrow {}_2He^4 + 2 \, {}_1H^1 + 12.9 \, Mev.$$

The first two of these reactions must each occur twice for every synthesis of $_2He^4$, so that the total energy produced per cycle is 24.7 Mev.

Stars hotter than the sun obtain their energy from the *carbon cycle*. This cycle requires a $_6C^{12}$ nucleus for its first step and in its last step regenerates a $_6C^{12}$ nucleus, so that this isotope may be thought of as a catalyst for the process. The carbon cycle proceeds as follows:

$$_1H^1 + {}_6C^{12} \rightarrow {}_7N^{13} + 2.0 \, Mev,$$

$$_7N^{13} \rightarrow {}_6C^{13} + e^+ + 1.2 \, Mev,$$

$$_6C^{13} + {}_1H^1 \rightarrow {}_7N^{14} + 7.6 \, Mev,$$

$$_7N^{14} + {}_1H^1 \rightarrow {}_8O^{15} + 7.3 \, Mev, \qquad (24\text{–}8)$$

$$_8O^{15} \rightarrow {}_7N^{15} + e^+ + 1.7 \, Mev,$$

$$_7N^{15} + {}_1H^1 \rightarrow {}_6C^{12} + {}_2He^4 + 4.9 \, Mev.$$

Here again the net result is the formation of an alpha particle and two positrons from four protons with a total of 24.7 Mev of energy evolved. In the sun both the proton-proton and carbon cycles take place with comparable probabilities.

The energy liberated by nuclear fusion is often called *thermonuclear energy*. High temperatures and densities are necessary for fusion reactions to occur in such quantity that a substantial amount of thermonuclear energy is produced: the high temperature assures that the initial light nuclei have enough thermal energy to overcome their mutual electrostatic repulsion and come close enough together to react, while the high density assures that such collisions are frequent. A further condition for the proton-proton and carbon cycles is a large reacting mass, such as that of a star, since a number of separate steps is involved in each cycle and much time may elapse between the initial fusion of a particular proton and its ultimate incorporation in an alpha particle. On the earth, where any reacting mass must be very limited in size, an efficient thermonuclear process cannot involve more than a single step. The reactions that appear most promising as sources of commercial power involve the combination of two deuterons to form a triton and a proton,

$$_1H^2 + {}_1H^2 \rightarrow {}_1H^3 + {}_1H^1 + 4.0 \text{ Mev,} \qquad (24\text{–}9)$$

or their combination to form a $_2He^3$ nucleus and a neutron,

$$_1H^2 + {}_1H^2 \rightarrow {}_2He^3 + {}_0n^1 + 3.3 \text{ Mev.} \qquad (24\text{–}10)$$

Both reactions have about equal probabilities. A major advantage of reactions (24–9) and (24–10) is that deuterium (the variety of hydrogen whose atomic nuclei are deuterons, sometimes called "heavy hydrogen") is relatively abundant on the earth, so that there should be no fuel problems in power plants operating on deuteron fusion. While there are many difficulties to surmount in achieving practical thermonuclear power, it will almost certainly become an eventual reality.

IMPORTANT TERMS

Radioactive nuclei spontaneously transform themselves into other nuclear species with the emission of charged particle radiation. The radiation may consist of *alpha particles,* which are the nuclei of helium atoms, or *beta particles,* which are positive or negative electrons. The emission of *gamma rays,* which are energetic photons, enables an excited nucleus to lose its excess energy. Positive electrons are known as *positrons.*

The time required for half of a given sample of a radioactive substance to decay is called its *half-life*.

The *tunnel effect* refers to the penetration of a potential barrier by a particle with insufficient energy to surmount the barrier. The tunnel effect occurs as a consequence of the wave nature of moving particles, and is responsible for alpha decay.

The first step in a *nuclear reaction* between two colliding nuclear particles is their joining together to form a *compound nucleus*. The compound nucleus then decays into the products of the reaction.

In nuclear fission, the absorption of neutrons by certain heavy nuclei causes them to split into smaller *fission fragments*. Because each fission also liberates several neutrons, a rapidly multiplying sequence of fissions called a *chain reaction* can occur if a sufficient amount of the proper material is assembled. A *nuclear reactor* is a device in which a chain reaction can be initiated and controlled.

The energy liberated by nuclear fusion is called *thermonuclear energy*.

PROBLEMS

1. The half-life of radium is 1600 years. How long will it take for 15/16 of a given sample of radium to decay?

2. Sixty hours after a sample of the beta emitter $_{11}Na^{24}$ has been prepared, only 6.25% of it remains undecayed. What is the half-life of this isotope?

3. The hydrogen isotope tritium, $_1H^3$, is radioactive and emits an electron with a half-life of 12.5 years. (a) What does $_1H^3$ become after beta decay? (b) What percentage of an original sample of tritium will remain 25 years after its preparation?

4. The isotope $_{92}U^{239}$ undergoes two negative beta decays in becoming an isotope of plutonium. State the atomic number and mass number of this isotope.

5. The nuclei $_8O^{14}$ and $_8O^{19}$ both undergo beta decay in order to become stable nuclei. Which would you expect to emit a positron and which an electron?

6. The nucleus $_6C^{11}$ undergoes positive beta decay. What is the atomic number, mass number, and chemical name of the resulting nucleus?

7. The isotope $_{90}Th^{233}$ undergoes two negative beta decays in becoming an isotope of uranium. State the atomic number and mass number of this isotope.

8. The nucleus $_{92}U^{238}$ decays into a lead isotope through the successive emissions of eight alpha particles and six electrons. State the atomic number and mass number of the lead isotope.

9. Radium undergoes spontaneous decay into helium and radon. Why is radium regarded as an element rather than as a chemical compound of helium and radon?

10. How much energy is evolved in the nuclear reaction

$$_1H^2 + {}_1H^3 \rightarrow {}_2He^4 + {}_0n^1 ?$$

11. The nuclear reaction

$$_3Li^6 + {}_1H^2 \rightarrow 2 \, {}_2He^4$$

evolves 22.4 Mev. Using the mass values of the deuteron and alpha particle given in Table 23–1, calculate the mass of $_3Li^6$ in amu.

12. In certain stars three alpha particles join in a single reaction to form a $_6C^{12}$ nucleus. The mass of $_6C^{12}$ is 12.0038 amu. How much energy is evolved in this reaction?

13. What are the difference and similarities between nuclear fusion and nuclear fission?

14. A nucleus of $_7N^{15}$ is struck by a proton. A nuclear reaction takes place with the emission of (a) a neutron, or (b) an alpha particle. Give the atomic number, mass number, and chemical name of the remaining nucleus in each of the above cases.

15. A nucleus of $_4Be^9$ is struck by an alpha particle. A nuclear reaction takes place with the emission of a neutron. Give the atomic number, mass number, and chemical name of the resulting nucleus.

16. A reaction often used to detect neutrons occurs when a neutron strikes a $_5B^{10}$ nucleus, with the subsequent emission of an alpha particle. What is the atomic number, mass number, and chemical name of the remaining nucleus?

17. If each fission in $_{92}U^{235}$ releases 200 Mev, how many fissions must occur per second to produce a power of 1 kw?

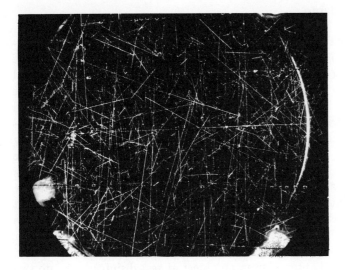

Elementary Particles | 25

The notion that the physical universe consists of myriad particles of a few basic kinds and that its evolution can be traced to the interactions of these particles originated long ago. Today, though convinced of the soundness of this notion, we are still far from a full understanding of these particles and of the reasons for their behavior. The existence of elementary particles was proposed in an effort to find order in the diversity of nature, and their generic name suggests their role as the ultimate, indivisible entities that constitute reality. Unfortunately for the quest for order and simplicity, a total of 30 different elementary particles have been discovered, and, though relatively few of them are stable when alone in space, all contribute in some way to the structure and behavior of matter. In this chapter we shall look into the evidence for these particles and into some ideas on their possible functions.

25–1 The neutrino

In radioactive decay, as in all other natural processes, energy (including mass energy) is conserved. For this reason the total mass of the products of a particular decay must be less than the mass of the initial nucleus, with the missing mass appearing as photon energy in the case of gamma decay and as kinetic energy in the cases of alpha and beta decay. In gamma and alpha decay the liberated energy is indeed precisely equal to the energy equivalent of the lost mass, but in beta decay a strange effect occurs: instead of all having the same energy, the emitted electrons from a particular isotope exhibit a variety of energies. These energies range from zero up to a maximum figure equal to the energy equivalent of the missing mass in the transformation. This effect is illustrated in Fig. 25–1, which shows the spread in electron energy in the decay of $_{83}\mathrm{Bi}^{210}$.

Momentum as well as energy is apparently not conserved in beta decay. As we learned in Chapter 5, when an object at rest disintegrates into two parts, they must move apart in opposite directions in order that the total momentum of the system remain zero. Experiments show, however, that the emitted electron and the residual nucleus do *not* in general travel in opposite directions after beta decay occurs, so that their momenta cannot cancel out to equal the initial momentum of zero.

A third difficulty concerns angular momentum. The spin quantum number of the neutron, the proton, the electron, and the positron is in every case one-half. The conversion of a neutron into a proton and an

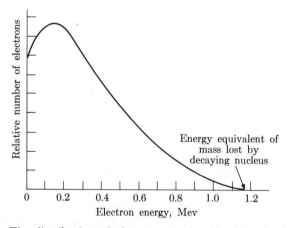

FIG. 25–1. The distribution of electron energies found in the beta decay of $_{83}\mathrm{Bi}^{210}$. The maximum electron energy is equal to the energy equivalent of the mass lost by the decaying nucleus.

electron or of a protron into a neutron and a positron therefore leaves an angular momentum discrepancy of $\frac{1}{2}\, h/2\pi$.

The process of beta decay is the first one we have encountered in which the conservation laws of energy, momentum, and angular momentum do not seem to hold. To account for the above discrepancies without abandoning three of the most fundamental and otherwise well-established physical principles, the existence of a new particle was postulated, the *neutrino*, symbol ν (Greek letter nu). The neutrino has no electric charge and no mass, but is able to possess both energy and momentum and has an intrinsic spin of one-half. (Lest this seem unlikely, we might reflect that the photon, also massless, has energy, momentum, and angular momentum. The neutrino is *not* a photon, however, but is an entirely different entity.) According to the neutrino theory, an electron and neutrino are simultaneously emitted in beta decay, as in Fig. 25–2, which permits energy and momentum to be conserved.

For a quarter of a century the existence of the neutrino was accepted despite the absence of any direct empirical evidence in support of it. Finally, in 1956, an experiment was performed in which a nuclear reaction that, in theory, could only be caused by a neutrino, was actually found to take place. The difficulty to be overcome in detecting neutrinos is their exceedingly feeble interaction with other particles. As we learned in the previous chapter, solar energy originates in nuclear reactions that take place in its interior. Beta decays occur at several stages in the sequences of reactions that participate in the conversion of hydrogen to helium, and a vast number of neutrinos is accordingly produced within the sun. Because neutrinos traverse matter freely, almost all of these neutrinos escape into space and take with them six to eight percent of

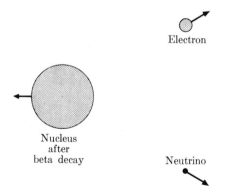

FIG. 25–2. An electron and a neutrino are simultaneously emitted in the beta decay of a nucleus, which makes possible the conservation of both energy and momentum in the process.

the total power generated by the sun. The flux of neutrinos from the sun is such that every cubic inch on the earth contains perhaps 100 neutrinos at any instant! In its operation a nuclear reactor also creates a great many neutrinos, and more than 10^{16} neutrinos per second may emerge from each square meter of the shielding around a reactor. A neutrino striking a proton has a small probability of inducing the reaction

$$\nu + p \rightarrow n + e^+ \qquad (25\text{-}1)$$

in which a neutron and positron are created. By placing a sensitive detecting chamber containing hydrogen near a nuclear reactor, the simultaneous appearance of a neutron and a positron could be registered each time the above reaction occurred. Calculations were made initially of how many such reactions per second should occur based on the known properties of the detecting apparatus and on the theoretical properties of the neutrino. When this reaction rate was actually found, there was no doubt that neutrinos indeed exist.

25–2 Antiparticles

Another recent experimental discovery of a particle whose existence had been predicted theoretically decades ago is that of the negative proton, or *antiproton*, whose symbol is $\bar{p}$. This is a particle with the same properties as the proton except that it has a negative electric charge. The existence of antiprotons was predicted largely on the basis of symmetry arguments: since the electron has a positive counterpart in the positron, why should the proton not have a negative counterpart as well? Actually, as sophisticated theories show, this is an excellent argument, and few physicists were surprised when the antiproton was actually found.

The reason positrons and antiprotons are so difficult to find is that they are readily *annihilated* upon contact with ordinary matter. When a positron is in the vicinity of an electron, they attract one another electrostatically, come together, and then both vanish simultaneously, with their missing mass appearing in the form of two gamma-ray photons (Fig. 25–3):

$$e^+ + e^- \rightarrow \gamma + \gamma. \qquad (25\text{-}2)$$

The total mass of the two particles is the equivalent of 1.02 Mev, and so each photon has an energy of 0.51 Mev. (Their energies must be equal and they must be emitted in opposite directions in order that momentum be conserved.) While the similar reaction

$$p + \bar{p} \rightarrow \gamma + \gamma \qquad (25\text{-}3)$$

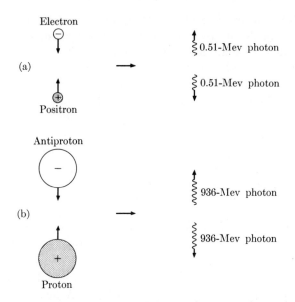

FIG. 25-3. The mutual annihilation of (a) an electron and a positron, and (b) a proton and an antiproton.

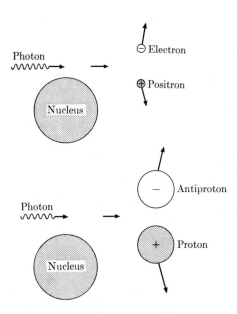

FIG. 25-4. The production of (a) an electron-positron pair and (b) a proton-antiproton pair by the materialization of sufficiently energetic photons. Pair production can occur only in the presence of a nucleus.

can occur when a proton and antiproton undergo annihilation, it is more usual for the vanished mass to reappear in the form of several π-mesons, particles which we shall consider in Section 25–3.

The reverse of annihilation can also take place, with the electromagnetic energy of a photon materializing into a positron and an electron or, if it is energetic enough, into a proton and an antiproton (Fig. 25–4). This phenomenon is known as *pair production*, and requires the presence of a nucleus in order to happen. Any photon energy in excess of the amount required to provide the mass of the created particles (1.02 Mev for a positron-electron pair, 1872 Mev for a proton-antiproton pair) appears as kinetic energy.

Antineutrons (symbol $\bar{n}$) and antineutrinos (symbol $\bar{\nu}$) have also been identified. Antineutrons can be detected through their mutual annihilation with neutrons, while more indirect, though equally definite, evidence supports the existence of antineutrinos. The antineutrino differs from the neutrino in that, while the spin axes of both are parallel to their directions of motion, the spin of the former is clockwise and that of the latter is counterclockwise when viewed from behind. A moving neutrino may be thought of as resembling a left-handed screw, and a moving antineutrino as resembling a right-handed screw. An antineutrino is released during a beta decay in which an electron is emitted, and a neutrino is released during a beta decay in which a positron is emitted. Thus the fundamental equations of beta decay are

$$\mathrm{p} \rightarrow \mathrm{n} + \mathrm{e}^+ + \nu, \qquad (25\text{–}4)$$

$$\mathrm{n} \rightarrow \mathrm{p} + \mathrm{e}^- + \bar{\nu}. \qquad (25\text{–}5)$$

Annihilation and pair production are consequences of the facts that matter is a form of energy and that conversions from matter to energy and from energy to matter are no more improbable than conversions from, say, gravitational potential energy to kinetic energy. Relativity and quantum theory underlie the reality of our experience as much as more familiar notions such as the kinetic theory of heat, and the full realization of their significance has within the past fifteen or twenty years profoundly affected the lives of all of us.

25–3 Mesons

If nuclear forces indeed arise from the exchange of mesons between nearby nuclei, as we mentioned in Chapter 23, it should be possible to detect their existence by imparting enough energy to a nucleus for conversion into meson mass without affecting the nuclear mass itself. We can estimate the meson mass from the uncertainty principle,

$$\Delta E \Delta t \approx \frac{h}{2\pi}. \qquad (25\text{--}6)$$

We assume that the temporary energy discrepancy ΔE is of the same magnitude as the rest energy mc^2 of the meson, and that the meson speed is of the same magnitude as the speed of light c as it travels from one nucleon to another. The time Δt the meson spends between creation and absorption cannot be greater than R/c, where R is the maximum distance that can separate interacting nucleons. We therefore have

$$\Delta E \Delta t \approx \frac{h}{2\pi},$$

$$mc^2 \times \frac{R}{c} \approx \frac{h}{2\pi},$$

$$m \approx \frac{h}{2\pi R c}. \qquad (25\text{--}7)$$

Nuclear forces have a range of about 1.4×10^{-15} meter. When we substitute $R = 1.4 \times 10^{-15}$ meter into Eq. (25–7) we find a predicted meson mass of 2.5×10^{-28} kg, 275 times the mass of the electron, which we write as 275 m_e. Of course, the above calculation is hardly a precise one, but, if Yukawa's theory has any validity, the mesons he postulated should have masses of several hundred times that of the electron.

Not long after Yukawa's work, charged particles of about the right mass were experimentally discovered in the cosmic radiation. Their discovery was not unexpected because a sufficiently energetic nuclear collision should be able to liberate mesons by providing enough energy to create them without violating conservation of energy, and nuclear collisions between fast cosmic-ray protons from space and oxygen and nitrogen nuclei occur constantly in the atmosphere. However, these particles did not behave at all in the way they were expected to behave. Far from strongly interacting with nuclei, as Yukawa's mesons were supposed to do, they barely interacted at all. Instead of being absorbed in at most a meter of earth, they penetrated thousands of meters into the ground. Finally, in 1947, the explanation for their peculiar behavior was found. The weakly interacting mesons, known as μ-mesons (μ is the Greek letter mu; μ-mesons are often called *muons*) and found so profusely in cosmic rays at sea level, are not the direct products of nuclear collisions, but are secondary particles that result from the decay of the Yukawa mesons. The meson postulated by Yukawa is known as the π-meson (or *pion*), and when outside a nucleus it usually decays in

2.54×10^{-8} sec into a μ-meson and a neutrino plus kinetic energy. That is,

$$\pi^+ \rightarrow \mu^+ + \nu, \tag{25-8}$$

$$\pi^- \rightarrow \mu^- + \bar{\nu}. \tag{25-9}$$

(A small percentage of charged π-mesons decay directly into an electron and a neutrino.) The π-meson is observed to have the properties Yukawa attributed to it, and may be loosely thought of as the "glue" that holds nuclei together so strongly. Charged π-mesons have a mass of $273\, m_e$. Neutral π-mesons also exist; their mass is $264\, m_e$, a little less than that of their charged counterparts, and they decay in about 10^{-16} sec into two gamma-ray photons:

$$\pi^0 \rightarrow \gamma + \gamma. \tag{25-10}$$

The brief lifetimes of π-mesons are responsible for the delay in their discovery. In the past decade, high-energy accelerators have been built whose particles are able to violently disrupt nuclei with the creation of many π-mesons; considerable progress has been made in our understanding of these mesons and their behavior. The π-meson is the antiparticle of the π^+-meson, and the neutral π^0-meson is its own antiparticle.

The μ-meson, whose mass is $207\, m_e$, is less well understood than the π-meson. The relatively long μ-meson lifetime of 2.22×10^{-6} sec and their feeble interaction with matter accounts for the penetrating ability of these mesons. In contrast to the π-meson, which fits in well with the theory of nuclear forces, the μ-meson seems to have no particular function in the scheme of things; why such a particle should exist is one of the current problems of theoretical physics. The μ^+- and μ^--mesons decay into positrons and electrons together with a neutrino and an antineutrino in each case:

$$\mu^+ \rightarrow e^+ + \nu + \bar{\nu}, \tag{25-11}$$

$$\mu^- \rightarrow e^- + \nu + \bar{\nu}. \tag{25-12}$$

The μ^+-meson is the antiparticle of the μ^--meson. There is no neutral μ-meson.

25–4 Strange particles

Not long after the discovery of the π-meson a succession of other, quite unpredicted elementary particles came to light. Even though a good deal is now known about these new particles, they may still be called by their original name of *strange particles* with some justice. It is convenient to divide strange particles into two classes, *K-mesons*, whose masses are all almost exactly 967 m_e, and *hyperons*, whose masses exceed the mass of the proton. Both classes of strange particles, like π-mesons, are found as products of energetic nuclear collisions, and are unstable.

Charged K-mesons may decay in a variety of ways after a mean lifetime of 1.22×10^{-8} sec. The K^+-meson has been observed to have the following six decay schemes, listed in the order of probability:

$$
\begin{aligned}
K^+ &\rightarrow \mu^+ + \nu \\
&\rightarrow \pi^+ + \pi^0 \\
&\rightarrow \pi^+ + \pi^+ + \pi^- \\
&\rightarrow \pi^+ + \pi^0 + \pi^0 \\
&\rightarrow e^+ + \pi^0 + \nu \\
&\rightarrow \mu^+ + \pi^0 + \nu.
\end{aligned}
\tag{25-13}
$$

The K^--meson, the antiparticle of the K^+-meson, has similar decay possibilities. Two different versions of the neutral K-meson are known. The K_1^0-meson decays in 10^{-10} sec in either of these ways:

$$
\begin{aligned}
K_1^0 &\rightarrow \pi^+ + \pi^- \\
&\rightarrow \pi^0 + \pi^0.
\end{aligned}
\tag{25-14}
$$

The K_2^0-meson has the much longer mean lifetime of 6×10^{-8} sec and six possible decay schemes:

$$
\begin{aligned}
K_2^0 &\rightarrow \pi^+ + \pi^- + \pi^0 \\
&\rightarrow \pi^0 + \pi^0 + \pi^0 \\
&\rightarrow \pi^- + \mu^+ + \nu \\
&\rightarrow \pi^+ + \mu^- + \bar{\nu} \\
&\rightarrow \pi^- + e^+ + \nu \\
&\rightarrow \pi^+ + e^- + \bar{\nu}.
\end{aligned}
\tag{25-15}
$$

Unlike the π^0-meson, which is its own antiparticle, an anti-K^0-meson, the $\bar{K}^0$-meson, is believed to exist.

There are three known families of hyperon, all of whose members are heavier than nucleons and all of which have a nucleon as one of their decay products. There is only a single kind of Λ-hyperon (Λ is the Greek capital letter lambda), a neutral particle whose mass is 2182 m_e that decays in an average of 2.5×10^{-10} sec into either a proton and

π^--meson or into a neutron and a π^0-meson, the latter alternative being the less likely:

$$\Lambda^0 \rightarrow p^+ + \pi^-$$
$$\rightarrow n^0 + \pi^0. \tag{25-16}$$

The $\overline{\Lambda}^0$-hyperon is the antiparticle of the Λ^0-hyperon, and decays into antiparticle counterparts of the above products.

The Σ-hyperons may be neutral or may have either charge (Σ is the Greek capital letter sigma). The Σ^+-hyperon has a mass of 2327 m_e, a mean lifetime of 0.8×10^{-10} sec, and an equal probability of decaying into a proton and a π^0-meson or into a neutron and a π^+-meson:

$$\Sigma^+ \rightarrow p^+ + \pi^0$$
$$\rightarrow n^0 + \pi^+. \tag{25-17}$$

Only one decay mode is known for the Σ^--hyperon, whose mass is 2341 m_e and whose mean lifetime is 1.6×10^{-10} sec:

$$\Sigma^- \rightarrow n^0 + \pi^-. \tag{25-18}$$

The Σ^0-hyperon has a mass of 2329 m_e and a mean lifetime of less than 10^{-12} sec. The Σ^0-hyperon decays into a Λ^0-hyperon and a gamma ray:

$$\Sigma^0 \rightarrow \Lambda^0 + \gamma. \tag{25-19}$$

The $\overline{\Sigma}^+$-, $\overline{\Sigma}^-$-, and $\overline{\Sigma}^0$-hyperons are antiparticles of the Σ^+-, Σ^--, and Σ^0-hyperons.

The third family of hyperons has two members, the Ξ^--hyperon and the Ξ^0-hyperon (Ξ is the Greek capital letter xi). Their masses are about 2583 m_e, and their lifetimes are in the range 10^{-9} to 10^{-10} sec. Both decay into a Λ^0-hyperon and a π-meson,

$$\Xi^- \rightarrow \Lambda^0 + \pi^-, \tag{25-20}$$

$$\Xi^0 \rightarrow \Lambda^0 + \pi^0, \tag{25-21}$$

and both have appropriate antiparticles.

25–5 Categories of elementary particles

In all, 30 elementary particles have been discovered, of which we have been introduced to no less than 25 in this chapter. Such abundance where scarcity had been expected inspired considerable interest among physicists in the past decade, with the result that partial success has been attained in understanding elementary particles. But there is still a long way to go before this fundamental puzzle is solved.

Elementary particles fall naturally into four categories, namely, *photons, leptons, mesons,* and *baryons.* The members of each category share certain basic properties despite differences among themselves in other properties. Photons, as we know, are quanta of electromagnetic energy; they are stable and have zero rest mass. The category of leptons includes neutrinos, electrons, and μ-mesons. Leptons exhibit very weak interactions with nuclei and have a spin of one-half. (We note that in this classification, μ-mesons are considered leptons rather than mesons.) The meson category includes both π- and K-mesons. These particles interact quite strongly with nuclei and all have a spin of zero. Baryons include nucleons and hyperons; they also interact strongly with nuclei, but have spins of one-half. Table 25–1 lists the 30 elementary particles together with some of their characteristic properties.

While it is obvious from Table 25–1 that the various elementary particles fit into the indicated categories on the basis of their masses and spins, experimental evidence of an unusual kind further substantiates this division. Let us assign all leptons the special quantum number $L = +1$, all antileptons $L = -1$, and all other particles $L = 0$. It is found that in every known process involving leptons, including their production, decay, or interaction with other particles, the total value of L remains constant. Also, let us assign all baryons the special quantum number $B = +1$, all antibaryons $B = -1$, and all other particles $B = 0$. It is found that in every known process involving baryons, the total value of B remains constant. A simple example is the beta decay of the neutron:

$$n^0 \rightarrow p^+ + e^- + \bar{\nu} \qquad (25\text{–}22)$$

$$L = \quad 0 \qquad 0 \qquad +1 \qquad -1$$

$$B = +1 \qquad +1 \qquad 0 \qquad 0$$

The usual conservation laws of energy, momentum, angular momentum, and charge plus the new (and still rather mysterious) conservation laws of lepton and baryon numbers help in understanding why certain processes involving elementary particles occur and why others do not occur. For instance, the free neutron can decay according to Eq. (25–22) and still obey the above conservation laws, but in no other way. The free proton, which has less mass than the neutron, cannot decay at all: energy conservation prohibits it from decaying with any other baryon as a product, and baryon number conservation prohibits it from decaying into any lighter particles. The other stable particles are also prohibited from decaying by one or more of the above laws when they are in free space.

TABLE 25-1

Elementary particles*

Name	Particle	Antiparticle	Mass m_e	Spin	L	B	S	Mean lifetime (sec)	Category
Photon	γ	(γ)	0	1	0	0	0	Stable	Photon
Neutrino	ν	$\bar{\nu}$	0	$\frac{1}{2}$	$+1$	0	0	Stable	Leptons
Electron	e^-	e^+	1	$\frac{1}{2}$	$+1$	0	0	Stable	
Mu-meson	μ^-	μ^+	207	$\frac{1}{2}$	$+1$	0	-1	2.22×10^{-6}	
Pi-meson	π^+	π^-	273	0	0	0	0	2.54×10^{-8}	Mesons
	π^0	(π^0)	264	0	0	0	0	10^{-16}	
K-meson	K^+	K^-	967	0	0	0	$+1$	1.22×10^{-8}	
	K^0	$\bar{K}^0$	967	0	0	0	$+1$	$10^{-10}; 6 \times 10^{-8}$	
Proton	p	$\bar{p}$	1836	$\frac{1}{2}$	0	$+1$	0	Stable	Baryons
Neutron	n^0	$\bar{n}^0$	1839	$\frac{1}{2}$	0	$+1$	0	1.11×10^3	
Lambda-hyperon	Λ^0	$\bar{\Lambda}^0$	2182	$\frac{1}{2}$	0	$+1$	-1	2.5×10^{-10}	
Sigma-hyperon	Σ^+	$\bar{\Sigma}^+$	2327	$\frac{1}{2}$	0	$+1$	-1	0.8×10^{-10}	
	Σ^-	$\bar{\Sigma}^-$	2341	$\frac{1}{2}$	0	$+1$	-1	1.6×10^{-10}	
	Σ^0	$\bar{\Sigma}^0$	2329	$\frac{1}{2}$	0	$+1$	-1	10^{-12}	
Xi-hyperon	Ξ^-	$\bar{\Xi}^-$	2583	$\frac{1}{2}$	0	$+1$	-2	$10^{-9} - 10^{-10}$	
	Ξ^0	$\bar{\Xi}^0$	2583	$\frac{1}{2}$	0	$+1$	-2	$10^{-9} - 10^{-10}$	

* Decay schemes given in text.

Introducing the quantum numbers L and B did not clarify all aspects of elementary particle behavior. Three of the still-outstanding problems were the following:

(1) Some heavy particles decay into lighter ones plus gamma rays while others do not. For example, the Σ^0-baryon decays into a Λ^0-baryon and a gamma ray,

$$\Sigma^0 \rightarrow \Lambda^0 + \gamma,$$

while the Σ^+-baryon never decays in the analogous manner

$$\Sigma^+ \nrightarrow p^+ + \gamma.$$

(2) Physical processes that release large amounts of energy almost always occur faster than processes involving lesser amounts. However, relatively long-lived particles are known whose decays evolve tens or hundreds of Mev's.

(3) Strange particles never come into being singly, but always two or more at a time.

Observations such as these led to the introduction of a further quantum number, the *strangeness number S*. Table 25–1 shows the values of S that are assigned to the various elementary particles. Only the photon and π^0-meson have $L = B = S = 0$. Since these particles also have no charge, it is impossible to distinguish between them and their antiparticles, and the photon and π^0-meson are accordingly regarded as their own antiparticles. Before we discuss how the strangeness number may be interpreted, it will be necessary to look into another aspect of elementary particle physics.

25–6 Fundamental interactions

Elementary particles can interact with one another in four different ways, and these interactions are responsible for all of the physical processes in the universe. Feeblest is the gravitational interaction, which follows the inverse-square law discovered by Newton. Next is the *weak interaction*, characteristic of leptons, that acts between leptons and either mesons, baryons, or other leptons (besides any electromagnetic forces that may also be present). The weak interaction is responsible for reactions involving neutrinos, such as beta decays. Of greater strength than gravitational and weak interactions are the electromagnetic interactions between all charged particles and also between those with electric or magnetic moments. Most powerful of the four is the *strong interaction* between mesons, between baryons, and between mesons and baryons that gives rise to nuclear forces.

The relative magnitudes of the strong, electromagnetic, weak, and gravitational interactions are in the ratios $1:5 \times 10^{-4}:3 \times 10^{-15}:10^{-40}$. To be sure, the ranges of the forces produced by these interactions are quite different: the nuclear force between two nucleons vastly exceeds the gravitational force between them when they are 10^{-15} m apart, but when they are a meter apart the situation is reversed. Nuclear structure is a consequence of the properties of the strong interaction, while atomic structure is a consequence of those of the electromagnetic interaction. Because bulk matter is electrically neutral and the strong and weak forces are severely limited in their range, gravitational forces, negligible on a small scale, become all-important on a large scale. The role of the weak force in the structure of matter seems to be restricted to causing beta decays that adjust the neutron:proton ratios of nuclei to stable values.

In all processes involving strong and electromagnetic interactions, the strangeness number S is conserved. Thus the decay

$$\Sigma^0 \quad \rightarrow \quad \Lambda^0 + \gamma$$
$$S = -1 \qquad -1 \qquad 0$$

that we mentioned earlier conserves S and actually occurs, and the apparently similar decay

$$\Sigma^+ \quad \rightarrow p^+ + \gamma$$
$$S = -1 \qquad 0 \qquad 0$$

does not conserve S and has never been found to occur. The multiple production of strange particles in high-energy nuclear collisions is necessary in order to conserve S, since such collisions involve strong interactions. The relatively long lifetimes of all unstable particles except the π^0-meson and Λ^0-hyperon follows from the assumption that weak interactions are also characteristic of mesons and baryons as well as leptons, though normally dominated by strong or electromagnetic interactions in the former cases. With strong or electromagnetic decay processes impossible except for the π^0-meson and Λ^0-hyperon because S must be conserved in such processes, only the weak interaction is available for particle decay. Events governed by weak interactions are very slow, in agreement with observation. However, even in weak interactions S does not change by more than $+1$ or -1. Thus the Ξ^--hyperon does not decay directly into a neutron in the single step

$$\Xi^- \quad \rightarrow n^0 + \pi^-$$
$$S = -2 \qquad 0 \qquad 0$$

but instead requires two separate decays:

$$\Xi^- \;\rightarrow\; \Lambda^0 + \pi^- \qquad\qquad \Lambda^0 \rightarrow n^0 + \pi^0$$
$$\text{and}$$
$$S = -2 \qquad\quad -1 \qquad 0 \qquad\qquad S = -1 \qquad 0 \qquad 0$$

Despite the elegance with which the introduction of the quantum numbers L, B, and S permits us to organize our knowledge of elementary particle phenomena, much remains to be understood about the origins of these numbers and their relation to the four fundamental interactions. The study of elementary particles is accordingly one of the most active branches of current scientific research.

IMPORTANT TERMS

Elementary particles are particles found in nature that do not consist of combinations of other particles. Thirty elementary particles are known, of which all but the photon, neutrino, antineutrino, positron, electron, proton, and antiproton are unstable outside a nucleus and decay into other particles.

The *neutrino* is a massless, uncharged particle that is emitted together with an electron during beta decay. It can possess energy, momentum, and angular momentum.

The *antiproton* has properties similar to the proton except that it is negatively charged. When a proton and an antiproton come together they *annihilate* each other, and the vanished mass reappears in the form of two gamma-ray photons. Positrons and electrons can also undergo annihilation. In the presence of a nucleus, a photon can materialize into an electron-positron pair or a proton-antiproton pair if it has the required energy. All elementary particles except the photon and neutral pi-meson have antiparticles.

It is convenient to divide elementary particles into four categories, namely *photons, leptons* (neutrinos, electrons, mu-mesons), *mesons* (pi- and K-mesons), and *baryons* (protons, neutrons, lambda-, sigma-, and xi-hyperons). The production and decay of elementary particles can be partially understood in terms of the conservation of three quantities: lepton number L, baryon number B, and strangeness number S.

The four fundamental types of interaction between elementary particles are, in order of increasing strength, *gravitational, weak, electromagnetic,* and *nuclear* (or *strong*). The strong and weak interactions have very short ranges.

PROBLEMS

1. State which of the following are spontaneously emitted by radioactive nuclei, and which are not: electrons, positrons, neutrinos, protons, antiprotons, neutrons, mesons, alpha particles, electromagnetic waves.

2. How much energy must a gamma-ray photon have if it is to materialize into a proton-antiproton pair with each particle having a kinetic energy of 10 mev?

3. How much energy (in ev) must a gamma-ray photon have if it is to materialize into a neutron-antineutron pair? Is this more or less than that required to form a proton-antiproton pair?

4. A 1 Mev positron collides head on with a 1 Mev electron, and the two are annihilated. What is the energy and wavelength of each of the resulting gamma-ray photons?

5. One proton strikes another, and the reaction

$$p + p \to n + p + \pi^+$$

takes place. What is the minimum energy the incident proton must have had?

6. Find the energy of each of the gamma-ray photons produced in the decay of the neutral pi-meson. Why must their energies be the same?

7. Why does the neutron not decay into an electron, a positron, and a neutrino?

8. Why does the Λ-hyperon not decay into a π^+- and a π^--meson?

9. A proton can interact with a μ^--meson to form a neutron and another particle. What must this particle be?

10. According to the theory of the continuous creation of matter, the evolution of the universe can be traced to the spontaneous appearance of neutrons in free space. Which conservation laws would this process violate?

MATHEMATICAL REVIEW

A knowledge of elementary algebra and trigonometry is necessary for understanding physics. A brief review of these subjects is given here for the benefit of those readers whose acquaintance with them needs refreshing.

Algebra

Algebra is arithmetic with symbols used in place of specified numbers. The advantage of algebra is that with its help we can perform calculations without knowing in advance the numerical values of all the quantities involved. Sometimes algebra is no more than a help, perhaps by showing us how to simplify complex calculations, and sometimes it is the only way in which we can solve a problem.

The symbols of algebra are normally letters of the alphabet. If we have two quantities A and B and add them together to give the sum C we would write

$$A + B = C.$$

If we subtract B from A to give the difference D, we would write

$$A - B = D.$$

Multiplying A and B together to give E may be written

$$A \times B = E$$

or, more simply, as just

$$AB = E.$$

Whenever two algebraic quantities are written together with nothing between them, it is understood that they are to be multiplied together.

Dividing A by B to give the quotient F is usually written

$$\frac{A}{B} = F,$$

but it may sometimes be more convenient to write

$$A/B = F,$$

which has the same meaning.

Parentheses and brackets are used to show the order in which various operations are to be performed. Thus

$$\frac{(A + B)C}{D} - E = F$$

means that we are first to add A and B together, multiply their sum by C, then divide by D, and finally subtract E.

Exponents

It is often necessary to multiply a quantity by itself a number of times. This process is indicated by a superscript number called the exponent, according to the following scheme:

$$A = A^1,$$
$$A \times A = A^2,$$
$$A \times A \times A = A^3,$$
$$A \times A \times A \times A = A^4,$$
$$A \times A \times A \times A \times A = A^5.$$

We read A^2 as "A squared" because it is the area of a square of length A on a side; similarly A^3 is called "A cubed" because it is the volume of a cube each of whose sides is A long. More generally we speak of A^n as "A to the nth power." Thus A^5 is read as "A to the fifth power."

When we multiply a quantity raised to some particular power (say A^n) by the same quantity raised to another power (say A^m), the result is that quantity raised to a power equal to the sum of the original exponents. That is,

$$A^n A^m = A^{(n+m)}.$$

For example,

$$A^2 A^5 = A^7,$$

which we can verify directly by writing out the terms:

$$(A \times A) \times (A \times A \times A \times A \times A) = A \times A \times A \times A \times A \times A \times A.$$
$$A^2 \quad \times \quad A^5 \quad\quad = \quad\quad A^7$$

From the above result we see that when a quantity raised to a particular power (say A^n) is to be multiplied by itself a total of m times, we have

$$(A^n)^m = A^{nm}.$$

For example,

$$(A^2)^3 = A^6,$$

since

$$(A^2)^3 = A^2 \times A^2 \times A^2 = A^{(2+2+2)} = A^6.$$

Reciprocal quantities are expressed in a similar way with the addition of a minus sign in the exponent, as follows:

$$\frac{1}{A} = A^{-1},$$

$$\frac{1}{A^2} = A^{-2},$$

$$\frac{1}{A^3} = A^{-3},$$

$$\frac{1}{A^4} = A^{-4}.$$

Exactly the same rules as before are used in combining quantities raised to negative powers with one another and with some quantity raised to a positive power. Thus

$$A^5 A^{-2} = A^{(5-2)} = A^3,$$
$$(A^{-1})^{-2} = A^{-1(-2)} = A^2,$$
$$(A^3)^{-4} = A^{-4 \times 3} = A^{-12},$$
$$A A^{-7} = A^{(1-7)} = A^{-6}.$$

It is important to remember that any quantity raised to the zeroth power, say A^0, is equal to 1. Hence

$$A^2 A^{-2} = A^{(2-2)} = A^0 = 1.$$

This is more easily seen if we write A^{-2} as $1/A^2$:

$$A^2 A^{-2} = A^2 \times \frac{1}{A^2} = \frac{A^2}{A^2} = 1.$$

Fractional exponents are frequently useful. The simplest case is that of the *square root* of a quantity A, commonly written $\sqrt{A}$, which when multiplied by itself, equals the quantity:

$$\sqrt{A} \times \sqrt{A} = A.$$

Using exponentials we see that, because

$$(A^{1/2})^2 = A^{2 \times (1/2)} = A^1 = A,$$

we can express square roots by the exponent $\frac{1}{2}$:

$$\sqrt{A} = A^{1/2}.$$

Other roots may be expressed similarly. The *cube root* of a quantity A, written $\sqrt[3]{A}$, when multiplied by itself twice equals A. That is,

$$\sqrt[3]{A} \times \sqrt[3]{A} \times \sqrt[3]{A} = A,$$

which may be more conveniently written

$$(A^{1/3})^3 = A,$$

where $\sqrt[3]{A} = A^{1/3}$.

In general the nth root of a quantity, $\sqrt[n]{A}$, may be written $A^{1/n}$, which is a more convenient form for most purposes. Some examples may be helpful:

$$(A^4)^{1/2} = A^{(1/2) \times 4} = A^2,$$
$$(A^{1/4})^{-7} = A^{-7 \times (1/4)} + A^{-7/4},$$
$$(A^3)^{-1/3} = A^{-(1/3) \times 3} = A^{-1},$$
$$(A^{1/2})^{1/2} = A^{(1/2) \times (1/2)} = A^{1/4}.$$

Fractional exponents may also be expressed as decimals:

$$A^{1/2} = A^{0 \cdot 5},$$
$$A^{1/3} = A^{0 \cdot 333},$$
$$A^{1/4} = A^{0 \cdot 25},$$
$$A^{7/4} = A^{1 \cdot 75}.$$

Quantities raised to decimal exponents are manipulated in the same manner we have become accustomed to:

$$(A^{1 \cdot 8})^{-4} = A^{-4 \times 1.8} = A^{-7 \cdot 2},$$
$$(A^{0 \cdot 6})^{0 \cdot 5} = A^{0.5 \times 0.6} = A^{0 \cdot 3}.$$

Proportionality

The notion of proportionality is frequently encountered in physics. This is natural, because physics has its roots in experiment and a frequent result of an experiment is a proportional relationship. Let us look into just what is implied in relationships of this kind.

When the value of a quantity A depends upon the value of another quantity B in such a way that doubling B causes A to double, tripling B causes A to triple, and so on, A is said to be *directly proportional* to B. If A is directly proportional to B and

$$A = a \quad \text{when} \quad B = b,$$

where a and b are some specific numbers, then it is also true that

$$A = 2a \quad \text{when} \quad B = 2b,$$
$$A = 3a \quad \text{when} \quad B = 3b,$$
$$A = 4a \quad \text{when} \quad B = 4b,$$
$$A = \tfrac{1}{2}a \quad \text{when} \quad B = \tfrac{1}{2}b,$$

and so on. No matter what the value of B, the *ratio* between A and B remains constant:

$$\frac{A}{B} = \frac{a}{b} = \text{constant} = c.$$

Hence we may express the fact that A is proportional to B by the equation

$$A = cB,$$

where c is called the constant of proportionality.

TABLE I-1

The altitude of an airplane at
1-min intervals after its take-off

Time, min	Altitude, ft
0	0
1	900
2	1800
3	2700
4	3600
5	4500

A simple example of a direct proportionality is provided in Table I–1, which lists the altitude of an airplane at various times after it has taken off from the ground. From the table we see that the airplane's altitude is directly proportional to the time, with the constant of proportionality given by

$$c = \frac{\text{altitude}}{\text{time}} = \frac{h}{t} = 900 \, \frac{\text{ft}}{\text{min}} \, .$$

In other words, the airplane is climbing at the rate of 900 ft/min. To find the altitude of the airplane at any time t after its take-off, we need only substitute for t in the equation

$$h = ct = 900 \, \text{ft/min} \times t.$$

If t is in minutes, we will then have h in feet. Thus, after 2 min and 20 sec, which is $2\frac{1}{3}$ min,

$$h = 900 \, \text{ft/min} \times 2\tfrac{1}{3} \, \text{min} = 900 \times \tfrac{7}{3} \, \text{ft} = 2100 \, \text{ft}.$$

Often one quantity which depends upon another is not directly proportional to it. We might find, for example, a situation in which doubling some quantity B causes the quantity A to quadruple, tripling B causes A to increase ninefold, quadrupling B causes A to increase sixteenfold, and so on. In this case, if

$$A = a \quad \text{when} \quad B = b,$$

then

$$A = 4a \quad \text{when} \quad B = 2b,$$
$$A = 9a \quad \text{when} \quad B = 3b,$$
$$A = 16a \quad \text{when} \quad B = 4b,$$

and so on. The ratio between A and B is *not* constant here, but we observe that the ratio between A and B^2 *is* constant, no matter what the value of B:

$$\frac{A}{B^2} = \frac{a}{b^2} = \text{constant} = c.$$

(In this example, of course, the value of the constant of proportionality c is different from what it was in the previous example.) Since this equation may be rewritten

$$A = cB^2,$$

we say that A is directly proportional to B^2.

There are situations in physics and elsewhere in science in which a direct proportionality may exist between some quantity A and a power of B which is neither 1 nor 2, in general in the form

$$A = cB^n.$$

Here we would say that A is directly proportional to the nth power of B.

A different type of proportionality occurs when A decreases as B is made larger. For instance, doubling B might halve A, tripling B might reduce A to one-third its original value, quadrupling B might reduce A to one-fourth its original value, and so on. If

$$A = a \quad \text{when} \quad B = b,$$

in this case

$$A = \frac{a}{2} \quad \text{when} \quad B = 2b,$$

$$A = \frac{a}{3} \quad \text{when} \quad B = 3b,$$

$$A = \frac{a}{4} \quad \text{when} \quad B = 4b,$$

and so on. We observe that the *product* of A and B is always constant, no matter what the value of B:

$$AB = ab = \text{constant} = c.$$

Hence we may write

$$A = \frac{c}{B},$$

and describe the relationship between A and B by saying that A is *inversely proportional* to B. In general, we may write inverse relationships in the form

$$A = \frac{c}{B^n},$$

or, equivalently, as

$$A = cB^{-n}.$$

Trigonometry

In physics it is often necessary to know the relationships among the various sides and angles of a *right triangle*, which is a triangle two of

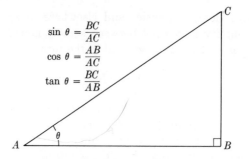

$$\sin \theta = \frac{BC}{AC}$$

$$\cos \theta = \frac{AB}{AC}$$

$$\tan \theta = \frac{BC}{AB}$$

FIG. I–1. The trigonometric functions.

whose sides are perpendicular. The three basic trigonometric functions are defined in terms of the triangle shown in Fig. I–1 as follows:

We shall label θ the angle at A included between the side AB of the triangle and its hypotenuse AC. The *sine* of this angle, which is abbreviated sin θ, is the ratio between the side BC of the triangle *opposite* to θ and the hypotenuse AC. Hence

$$\sin \theta = \frac{BC}{AC} = \frac{\text{opposite side}}{\text{hypotenuse}}.$$

The cosine of the angle θ, abbreviated cos θ, is the ratio between the side AB of the triangle adjacent to θ and the hypotenuse AC. Hence

$$\cos \theta = \frac{AB}{AC} = \frac{\text{adjacent side}}{\text{hypotenuse}}.$$

The *tangent* of the angle θ, abbreviated tan θ, is the ratio between the side BC opposite to θ and the side AB adjacent to θ. Hence

$$\tan \theta = \frac{BC}{AB} = \frac{\text{opposite side}}{\text{adjacent side}}.$$

From these definitions we can obtain a useful result:

$$\frac{\sin \theta}{\cos \theta} = \frac{\text{opposite side/hypotenuse}}{\text{adjacent side/hypotenuse}}$$

$$= \frac{\text{opposite side}}{\text{adjacent side}}$$

$$= \tan \theta.$$

The tangent of an angle is equal to its sine divided by its cosine.

Numerical tables of sin θ, cos θ, and tan θ for angles from 0° to 90° are given in Appendix V. These tables may be used for angles from 90°

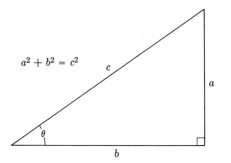

Fɪɢ. I–2. The Pythagorean theorem.

to 180° with the help of the following formulas:

$$\sin (90° + \theta) = \cos \theta,$$
$$\cos (90° + \theta) = -\sin \theta,$$
$$\tan (90° + \theta) = -\frac{1}{\tan \theta}.$$

Trigonometric functions can be treated algebraically, just as any other quantity. Suppose that we know the length of the side b and the angle θ in the right triangle of Fig. I–2, and wish to find the lengths of the sides a and c. From the definitions of sine and tangent we see that

$$\tan \theta = \frac{a}{b}, \qquad \sin \theta = \frac{a}{c},$$

and so

$$a = b \tan \theta, \qquad c = \frac{a}{\sin \theta}.$$

Another useful relationship in a right triangle is the Pythagorean theorem, which states that the sum of the squares of the sides of such a triangle adjacent to the right angle is equal to the square of its hypotenuse. For the triangle of Fig. I–2,

$$a^2 + b^2 = c^2.$$

Hence we can always express the length of any of the sides of a right triangle in terms of the other sides:

$$a = \sqrt{c^2 - b^2},$$
$$b = \sqrt{c^2 - a^2},$$
$$c = \sqrt{a^2 + b^2}.$$

Appendix II

USEFUL PHYSICAL CONSTANTS

Quantity	Symbol	Value
Gravitational constant	G	6.67×10^{-11} n·m^2/kg^2
Acceleration of gravity at earth's surface	g	9.81 m/sec^2 = 32.2 ft/sec^2
Atmospheric pressure at sea level		14.7 lb/in^2 = 1.01×10^5 n/m^2
Absolute zero	$0°$K	$-273°$C
Boltzmann's constant	k	1.38×10^{-23} j/$°$K
Electrostatic constant	ϵ	8.85×10^{-12} coul/n·m^2
Electromagnetic constant	μ	1.26×10^{-6} weber/amp·m
Charge of electron	e	1.60×10^{-19} coul
Electron rest mass	m_e	9.11×10^{-31} kg
Proton rest mass	m_p	1.67×10^{-27} kg
Neutron rest mass	m_n	1.67×10^{-27} kg
Speed of light	c	3.00×10^8 m/sec
Planck's constant	h	6.63×10^{-34} j·sec

Appendix III

CONVERSION FACTORS

---

1 meter (m) $=$ 100 cm $=$ 39.4 in. $=$ 3.28 ft

1 centimeter (cm) $=$ 10 millimeters (mm) $=$ 0.394 in.

1 kilometer (km) $=$ 1000 m $=$ 0.621 mi

1 foot (ft) $=$ 12 in. $=$ 0.305 m

1 inch (in.) $=$ 0.0833 ft $=$ 2.54 cm

1 mile (mi) $=$ 5280 ft $=$ 1.61 km

1 m/sec $=$ 3.28 ft/sec $=$ 2.24 mi/hr $=$ 3.60 km/hr

1 ft/sec $=$ 0.305 m/sec $=$ 0.682 mi/hr $=$ 1.10 km/hr
  (*Note:* it is often convenient to remember that 88 ft/sec $\doteq$ 60 mi/hr.)

1 mi/hr $=$ 1.47 ft/sec $=$ 0.447 m/sec $=$ 1.61 km/hr

1 radian (rad) $=$ 57.30° $=$ 57°18′

1° $=$ 0.01745 rad

1 revolution/minute (rev/min) $=$ 0.1047 rad/sec

1 kilogram (kg) $=$ 1000 grams (gm) $=$ 0.0685 slug
  (*Note:* 1 kg corresponds to 2.21 lb in the sense that the *weight* of 1 kg is
  2.21 lb.)

1 slug $=$ 14.6 kg
  (*Note:* 1 slug corresponds to 32.2 lb in the sense that the *weight* of 1 slug is
  32.2 lb.)

1 atomic mass unit (amu) $=$ 1.66×10^{-27} kg $=$ 1.49×10^{-10} j $=$ 931 Mev

1 newton (n) $=$ 0.225 lb

1 pound (lb) $=$ 4.45 n

1 joule (j) $=$ 0.738 ft·lb $=$ 2.39×10^{-4} kcal $=$ 6.24×10^{18} ev

1 kilocalorie (kcal) $=$ 4185 j

1 foot·pound (ft·lb) $=$ 1.36 j

1 electron volt (ev) $=$ 10^{-6} Mev $=$ 1.60×10^{-19} j $=$ 1.18×10^{-19} ft·lb
  $=$ 3.83×10^{-23} kcal

1 watt $=$ 1 j/sec $=$ 0.738 ft·lb/sec

Appendix IV

THE ELEMENTS

Atomic number	Element	Symbol	Atomic weight*	Atomic number	Element	Symbol	Atomic weight*
1	Hydrogen	H	1.008	52	Tellurium	Te	127.6
2	Helium	He	4.003	53	Iodine	I	126.9
3	Lithium	Li	6.940	54	Xenon	Xe	131.3
4	Beryllium	Be	9.013	55	Cesium	Cs	132.9
5	Boron	B	10.82	56	Barium	Ba	137.4
6	Carbon	C	12.01	57	Lanthanum	La	138.9
7	Nitrogen	N	14.01	58	Cerium	Ce	140.1
8	Oxygen	O	16.00	59	Praseodymium	Pr	140.9
9	Fluorine	F	19.00	60	Neodymium	Nd	144.3
10	Neon	Ne	20.18	61	Promethium	Pm	(147)
11	Sodium	Na	22.99	62	Samarium	Sm	150.4
12	Magnesium	Mg	24.32	63	Europium	Eu	152.0
13	Aluminum	Al	26.98	64	Gadolinium	Gd	157.3
14	Silicon	Si	28.09	65	Terbium	Tb	158.9
15	Phosphorus	P	30.98	66	Dysprosium	Dy	162.5
16	Sulfur	S	32.07	67	Holmium	Ho	164.9
17	Chlorine	Cl	35.46	68	Erbium	Er	167.3
18	Argon	A	39.94	69	Thulium	Tm	168.9
19	Potassium	K	39.10	70	Ytterbium	Yb	173.0
20	Calcium	Ca	40.08	71	Lutetium	Lu	175.0
21	Scandium	Sc	44.96	72	Hafnium	Hf	178.5
22	Titanium	Ti	47.90	73	Tantalum	Ta	181.0

Number	Name	Symbol	Weight	Number	Name	Symbol	Weight
23	Vanadium	V	50.95	74	Tungsten	W	183.9
24	Chromium	Cr	52.01	75	Rhenium	Re	186.2
25	Manganese	Mn	54.94	76	Osmium	Os	190.2
26	Iron	Fe	55.85	77	Iridium	Ir	192.2
27	Cobalt	Co	58.94	78	Platinum	Pt	195.1
28	Nickel	Ni	58.71	79	Gold	Au	197.0
29	Copper	Cu	63.54	80	Mercury	Hg	200.6
30	Zinc	Zn	65.38	81	Thallium	Tl	204.4
31	Gallium	Ga	69.72	82	Lead	Pb	207.2
32	Germanium	Ge	72.60	83	Bismuth	Bi	209.0
33	Arsenic	As	74.91	84	Polonium	Po	210
34	Selenium	Se	78.96	85	Astatine	At	(210)
35	Bromine	Br	79.92	86	Radon	Rn	222
36	Krypton	Kr	83.80	87	Francium	Fr	(223)
37	Rubidium	Rb	85.48	88	Radium	Ra	226.1
38	Strontium	Sr	87.63	89	Actinium	Ac	227
39	Yttrium	Y	88.92	90	Thorium	Th	232.1
40	Zirconium	Zr	91.22	91	Protactinium	Pa	231
41	Niobium	Nb	92.91	92	Uranium	U	238.1
42	Molybdenum	Mo	95.95	93	Neptunium	Np	(237)
43	Technetium	Tc	(99)	94	Plutonium	Pu	(242)
44	Ruthenium	Ru	101.1	95	Americium	Am	(243)
45	Rhodium	Rh	102.9	96	Curium	Cm	(245)
46	Palladium	Pd	106.4	97	Berkelium	Bk	(249)
47	Silver	Ag	107.9	98	Californium	Cf	(249)
48	Cadmium	Cd	112.4	99	Einsteinium	Es	(253)
49	Indium	In	114.8	100	Fermium	Fm	(255)
50	Tin	Sn	118.7	101	Mendelevium	Md	(256)
51	Antimony	Sb	121.8	102	Nobelium	No	(253)

*Those in parentheses refer to the most stable isotope of the element.

Appendix V

NATURAL TRIGONOMETRIC FUNCTIONS

Angle					Angle				
De-gree	Ra-dian	Sine	Co-sine	Tan-gent	De-gree	Ra-dian	Sine	Co-sine	Tan-gent
0°	.000	0.000	1.000	0.000					
1°	.017	.018	1.000	.018	46°	0.803	0.719	0.695	1.036
2°	.035	.035	0.999	.035	47°	.820	.731	.682	1.072
3°	.052	.052	.999	.052	48°	.838	.743	.669	1.111
4°	.070	.070	.998	.070	49°	.855	.755	.656	1.150
5°	.087	.087	.996	.088	50°	.873	.766	.643	1.192
6°	.105	.105	.995	.105	51°	.890	.777	.629	1.235
7°	.122	.122	.993	.123	52°	.908	.788	.616	1.280
8°	.140	.139	.990	.141	53°	.925	.799	.602	1.327
9°	.157	.156	.988	.158	54°	.942	.809	.588	1.376
10°	.175	.174	.985	.176	55°	.960	.819	.574	1.428
11°	.192	.191	.982	.194	56°	.977	.829	.559	1.483
12°	.209	.208	.978	.213	57°	.995	.839	.545	1.540
13°	.227	.225	.974	.231	58°	1.012	.848	.530	1.600
14°	.244	.242	.970	.249	59°	1.030	.857	.515	1.664
15°	.262	.259	.966	.268	60°	1.047	.866	.500	1.732
16°	.279	.276	.961	.287	61°	1.065	.875	.485	1.804
17°	.297	.292	.956	.306	62°	1.082	.883	.470	1.881
18°	.314	.309	.951	.325	63°	1.100	.891	.454	1.963
19°	.332	.326	.946	.344	64°	1.117	.899	.438	2.050
20°	.349	.342	.940	.364	65°	1.134	.906	.423	2.145
21°	.367	.358	.934	.384	66°	1.152	.914	.407	2.246
22°	.384	.375	.927	.404	67°	1.169	.921	.391	2.356
23°	.401	.391	.921	.425	68°	1.187	.927	.375	2.475
24°	.419	.407	.914	.445	69°	1.204	.934	.358	2.605
25°	.436	.423	.906	.466	70°	1.222	.940	.342	2.747
26°	.454	.438	.899	.488	71°	1.239	.946	.326	2.904
27°	.471	.454	.891	.510	72°	1.257	.951	.309	3.078
28°	.489	.470	.883	.532	73°	1.274	.956	.292	3.271
29°	.506	.485	.875	.554	74°	1.292	.961	.276	3.487
30°	.524	.500	.866	.577	75°	1.309	.966	.259	3.732
31°	.541	.515	.857	.601	76°	1.326	.970	.242	4.011
32°	.559	.530	.848	.625	77°	1.344	.974	.225	4.331
33°	.576	.545	.839	.649	78°	1.361	.978	.208	4.705
34°	.593	.559	.829	.675	79°	1.379	.982	.191	5.145
35°	.611	.574	.819	.700	80°	1.396	.985	.174	5.671
36°	.628	.588	.809	.727	81°	1.414	.988	.156	6.314
37°	.646	.602	.799	.754	82°	1.431	.990	.139	7.115
38°	.663	.616	.788	.781	83°	1.449	.993	.122	8.144
39°	.681	.629	.777	.810	84°	1.466	.995	.105	9.514
40°	.698	.643	.766	.839	85°	1.484	.996	.087	11.43
41°	.716	.658	.755	.869	86°	1.501	.998	.070	14.30
42°	.733	.669	.743	.900	87°	1.518	.999	.052	19.08
43°	.751	.682	.731	.933	88°	1.536	.999	.035	28.64
44°	.768	.695	.719	.966	89°	1.553	1.000	.018	57.29
45°	.785	.707	.707	1.000	90°	1.571	1.000	.000	∞

ANSWERS TO ODD-NUMBERED PROBLEMS

CHAPTER 1

1. 6.46×10^9; 1×10^6; 3.516×10^5; 8.4×10^3; 7×10^1; 3.81×10^0; 1.4×10^{-1}; 7.890×10^{-3}; 1.3×10^{-5}; 7.819×10^{-9}.
3. 451,000,000; 200,000; 781.9; 5.1; 0.08; 0.000,095,6; 0.000,000,010,03; 0.000,000,000,1.
5. 14.8 mi/hr.
7. 11.2 hr.
9. 345 m/sec; 0.214 mi/sec; 770 mi/hr.
11. 9.4 sec.
13. 48 mi/hr.
15. 144 mi/hr; 720 mi.
17. 8.33 min.
19. 12,5 (mi/hr)/sec; 18.3 ft/sec².
21. 6.4 sec.
23. 180 ft.
25. 3.53 sec.
27. 56.6 ft/sec.
29. 1.41 sec; 65.1 ft/sec.
31. Vector quantities: displacement, velocity, acceleration, force. Scalar quantities: time, distance, speed, volume.

CHAPTER 2

1. 39.2 n; 5 m/sec².
3. 0.816 kg; 24.5 m/sec².
5. From $F = ma$, which is the second law of motion, we see that, when $a = 0$, $F = 0$, which is the first law of motion. The third law of motion cannot be derived from the second.
7. 13,200 n.
9. 5.33 lb.
11. $F = ma$ and $W = mg$, and so $m = F/a = W/g$. Hence $a/g = F/W$.
13. 72 lb.
15. 4.9 m/sec²; 3.92 m/sec².
17. 4.33 sec; 9.22 m/sec.
19. Downward and away from the wall; upward and toward the wall; downward and against the wall; upward and away from the wall.

CHAPTER 3

1. 94 mi.
3. 8.25 mi/hr at an angle of 76° relative to the river bank; 0.19 hr; 0.38 mi.
5. 16.7 m.
7. 27.5 n at 44° clockwise from reference direction.
9. 64.3 lb vertical, 76.6 lb horizontal.
11. 93 mi/hr; 141 mi/hr.
13. 118 ft/sec at 32° below the horizontal.
15. 600 ft; 139 ft/sec; 139 ft/sec.
17. 7950 m; 9.4 sec.
19. 23.4 m/sec; 33°; 51°.
21. 20.6 lb at 29° above the horizontal.
25. 2450 n.
27. 156 n.
29. 2.6 ft from the 30-lb child.
31. 120 lb.

CHAPTER 4

1. 1.74 ft/sec; 6.1 ft/sec^2.   3. 384 n.   5. 41.3 ft/sec; 1.64 rev/sec.
7. 8 ft/sec.
9. A greater frictional force is required to keep the car to the road at the higher speed.
11. 6.3 m/sec^2.
13. No, because the earth's orbit is an ellipse and the distance between the earth and the sun varies through the year.
15. The rotation of the earth causes an outward centrifugal reaction force to act on particles at the equator; more than, owing to greater centrifugal reaction force.
17. 5.27 ft/sec^2; 16.5 lb.
19. 6.67×10^{-10} n; 6.67×10^{-10} n; 2-kg mass, 3.33×10^{-10} m/sec^2; 5-kg mass, 1.33×10^{-10} m/sec^2.

CHAPTER 5

1. 1040 j; 1040 j.   5. 0.   9. 1.72×10^5 ft·lb.
3. 13.6 n.   7. 7.52×10^4 j.   11. 7.07 m/sec.
13. 900 ft·lb; 800 ft·lb; friction in pulleys.
15. 2000 ft·lb; 1600 ft·lb; 400 ft·lb.
17. At lowest point; at highest point.
19. 38,000 lb.   21. 5870 slugs·ft/sec.
23. The larger boy rolls in opposite direction with speed of $3\frac{1}{3}$ mi/hr.
25. 12.5 mi/hr; 4.9×10^5 ft·lb.   27. 8 bullets.
29. 0.99 m/sec at angle of 63° west of north; 3.75 j.
31. Neutron speed is 3.33×10^4 m/sec; deuteron speed is 6.67×10^4 m/sec (in opposite direction).
33. 5×10^5 n.

CHAPTER 6

1. 5329°; 0.349 rad.   13. 10 rad/sec^2; 100 rad/sec.
3. 0698 rad.   15. The solid cylinder.
5. 181 rad/sec; 18.1 m/sec.   17. 15.75 rad/sec; 2.48 j; 2.48 j.
7. 10 ft/sec; 5 ft/sec; 0.   19. 0.091 ft·lb; 0.018 slug·ft/sec.
9. 9450 rad.   21. 113 ft·lb.
11. 31.5 rad/sec^2; 1000 rev.   23. See argument on p. 118.
25. The length of the day will increase since the earth's moment of inertia will be increased, and hence its angular velocity will be reduced when the water from the icecaps becomes uniformly distributed.

CHAPTER 7

1. 6×10^4 lb/ft.   7. 1250 lb; 11,250 lb.   13. 377 atm.
3. 7.84×10^{-2} m.   9. 3.26×10^{-4} ft.   15. $777.
5. 0.294 mm.   11. 72 lb/ft^2.
17. The density will be greater by 9.06×10^{-5} kg/m^3 than it is at the top of the column.

19. 0.26 ft.   23. 882 n.   27. 25.9 lb/in.^{2}
21. 90%.   25. 0.21 lb.   29. 4.4 m/sec; 9.85×10^3 n/m^2.
31. 3.59×10^3 n/m^2; 2.15×10^4 n/m^2.

CHAPTER 8

1. 5.17 ft/sec.   9. 247 m/sec^2; 1.23 n; 123 m/sec^2; 0.617 n.
3. 0.365 m.   11. 9.66 m/sec^2.
5. 0.556 sec; 1.80 sec^{-1}.   13. 0.9 sec^{-1}; 1.11 sec.
7. 24.7 ft · lb; 494 ft/sec^2.   15. 1.33×10^{-4} m.

CHAPTER 9

1. 626°F; 2138°F.   3. −40°F.   5. −80°C.
7. Yes, because the specific heat of alcohol is much less than that of water.
9. 20.7°C.   15. 0.742 kg.   21. 4.5 lb/in^2.
11. 620 kcal.   17. 327 m/sec.   23. 10.3 m.
13. 0.0263 kg.   19. 0.854.
25. As the steam condenses, the pressure inside the can drops below atmospheric pressure, and the can collapses.
27. The boiling point of water increases with pressure.
29. Liquid.

CHAPTER 10

1. Homogeneous: salt, diamond, iron, solid carbon dioxide, gaseous carbon dioxide, helium, rust. Heterogeneous: leather, stone, blood.
3. 27.095 m.   5. 7.8×10^{-6} 1°C.
7. Glass expands less than copper with a rise in temperature, so the bulb would crack as it heats up. Lead-in wires must have the same coefficient of thermal expansion as glass if they are to be used successfully.
9. 6.84 cm^3.   11. 4×10^5 n/m^2; 2.67 m^3; 2.67 m^3.
13. 118 cm^3; 45.5 cm^3; 53.7 cm^3.
15. −460°F; °F = °R − 460°, °R = °F + 460°; 492°F, 672°F.
17. The thermal energy of a solid resides in oscillations of molecules about their equilibrium positions.
19. Intermolecular attraction tends to decrease gas pressures.
21. H$_2$; UF$_6$.   23. 3.93×10^2 m/sec; 1092°K.
25. 3.33×10^{-9} m; 16.7 diameters.
27. Doubling the number of molecules doubles the number of molecular collisions per second with the walls, which doubles the pressure.

CHAPTER 11

1. Because steel is a better conductor of heat.
3. When heat is added to a gas at constant pressure, the gas expands and does work in the expansion.
5. 41.9%.   7. 62%.
9. There is no suitable low-temperature reservoir in space.
11. No.   13. 5.0×10^4 j.

CHAPTER 12

1. Static electricity; Millikan's experiments.
3. 0.025 n; 0.001 n; 2.5 n.
5. There are many reasons, but the most general is that all gravitational forces are observed to be attractive, while if they were electrostatic in origin some would have to be repulsive.
7. 2.74×10^3 n toward the -1×10^{-4} coul charge.
9. 5.08 m.   11. 180 n, attractive; 22.5 n, repulsive.
13. Between the charges 0.0828 cm from the smaller one.
15. 3.6×10^4 n/coul.   19. 2.84 ev.
17. 5×10^{-4} n; 5×10^{-4} j.   21. 4.2×10^6 m/sec.

CHAPTER 13

1. 2.88×10^5 coul; 3.46×10^6 j.
3. 180 coul/min.   7. Energy.
5. No, because electrons have little inertia.   9. 2.52 hr.
13. 100 m; the wire has a resistance 10^4 that of the rod.
15. 2.5×10^{-2} coul.   21. 6.25.
17. 0.375 j.   23. 2.
19. 44 $\mu\mu$f; 2.0×10^{-9} coul; 4.5×10^{-8} j.   25. 0.025 sec; 0.2 amp; 0.025 sec.
27. 40 j/m^3.

CHAPTER 14

1. See text.   5. 7.94×10^5 amp.
3. 0.1 n.   7. 14.1 webers/m^2.
11. The magnetic force is over 10^{13} times greater than the gravitational force.
13. 3.2×10^{-16} n; 3.2×10^{-16} n.
15. $B = E/v$ perpendicular to both E and v.
17. 3.37×10^{-2} m.
19. 5.6×10^6 sec^{-1}; 20 Mev; 3.1×10^7 m/sec.

CHAPTER 15

1. 5×10^{-6} weber; 1.67×10^{-6} volt.   3. 10^3 webers/m$^2 \cdot$ sec.
5. Because the flux through the loop undergoes both an increase and a decrease in each rotation.
7. This behavior follows from Lenz's law.
9. Downward.   13. 4.8×10^{-3} v.   17. 0.634 mh.
11. 6 v.   15. 1.4 h.   19. 3.97×10^6 j; 3×10^{10} v/m.

CHAPTER 16

1. 341 m.
3. 3.25×10^5 km.
5. 3.3×10^{-13} w/m^2; 1.3×10^{-11} watts/m^2; 4.4×10^{-20} n/m^2.

CHAPTER 17

1. Sound waves, vibrational (i.e., transverse) waves, electromagnetic waves; sound waves, electromagnetic waves; sound waves, electromagnetic waves; electromagnetic waves.
3. 2.999×10^8 m/sec; 1.24×10^8 m/sec; 1.97×10^8 m/sec; 2.24×10^8 m/sec.
5. This is due to refraction.
7. Omitted.
9. Since both are transparent and have the same index of refraction, light passes through the bottle unaffected by the presence of the rod.
11. The flashes of color are the result of dispersion. In red light the flashes would be red only.
13. If the other fork had a frequency of 246 cycles/sec, 10 beats per second would also be heard.
15. See text.   17. 5×10^{-7} m.
19. 8.74×10^{-3} m; 1.31×10^{-2} m; 1.09×10^{-2} m.
21. 4.06×10^{-7} m.   23. This is a diffraction effect.

CHAPTER 18

1. No; yes; yes.   5. 2.6×10^8 m/sec.   9. 3.7×10^{-12} kg.
3. 1:40.3; 1:40.3.   7. 4.2×10^7 m/sec.   11. 6.05×10^{-8} %.
13. 2.6×10^8 m/sec.   15. 4.2×10^7 m/sec; 9.3×10^6 ev.

CHAPTER 19

1. Even a faint light involves a great many photons.
3. 1.71×10^{31} photons/sec.   5. 4.64×10^4 sec^{-1}.
7. Diffraction, interference, and polarization can be exhibited much more simply than the photoelectric or other quantum effect.
9. 4.14×10^4 volts.   11. 2.41×10^{18} sec^{-1}; x-rays.
13. 9.95×10^{15} j; 3.32×10^{-23} kg·m/sec.
15. 9.8×10^{18} sec^{-1}; 1.5×10^{-16} j.
17. 1.34×10^{19} sec^{-1}; 1.10×10^{19} sec^{-1}.
19. 2.73×10^{-39} m.   21. 2.87×10^{-14} m.
23. Apply electric or magnetic field and look for deflection, or try Compton-type experiment.
25. 7.3×10^5 m/sec; 4.0×10^2 m/sec.

CHAPTER 20

1. They agree that the positive matter in the atom has much greater mass than the negative matter, but in the Thomson model the positive matter occupies the entire atomic volume, while it occupies only a tiny region at the center of the atom in the Rutherford model.
3. 2.2×10^6 m/sec.
5. A hydrogen sample contains a great many atoms, each of which has a variety of possible transitions.
7. 12 ev.

9. 4.05×10^{-6} m.   11. 8.3×10^{-20} j; 0.52 ev.   13. Lyman series.
15. 9.73×10^{-8} m; 6.81×10^{-27} kg·m/sec; 4.08 m/sec.
17. Emission.

CHAPTER 21

3. 2.
7. The outer shells of these atoms all have either 1 or 2 electrons, since the increases in atomic number in this sequence go into filling up the inner $n = 3$, $l = 2$ subshell.

CHAPTER 22

1. Both Li and Na atoms have one electron outside a closed shell.
3. Lithium atoms must *lose* an electron each in forming a molecule, and sharing electrons does not permit this to occur.
5. 0.5 ev; 2.9×10^{-9} m.
7. Li atoms have one electron outside a closed shell, which is more easily removed than any of the electrons of F atoms whose outer shells lack an electron of completion.
9. Na^+ ions have closed shells, while Na atoms have a single, easily detached outer electron.
11. All rare gas atoms contain only closed electron shells.
13. Both are nonconductors at very low temperatures; germanium is the better conductor at room temperature, because its forbidden band is narrower.

CHAPTER 23

1. 2500 volts.   3. 6.1 mm; 81% B^{11}, 19% B^{10}.
5. 5 p, 5 n; 10 p, 12 n; 16 p, 20 n; 38 p, 50 n; 72 p, 108 n.
7. 289 Mev.   9. 18.8 Mev; 19.4 Mev.
11. 0.85 Mev; difference in binding energy is 0.62 Mev, so nuclear forces cannot be strongly charge-dependent.

CHAPTER 24

1. 6400 years.   3. $_2He^3$; 25%.
5. $_8O^{14}$ emits a positron, and $_8O^{19}$ emits an electron.   7. 92, 233.
9. Helium and radon cannot be combined chemically to form radium nor can radium be broken down into helium and radon by chemical means.
11. 6.017 amu.
13. In fission a large nucleus splits into smaller ones, while in fusion small nuclei join together to form a larger one. In both, the product nucleus or nuclei have less mass than the initial one or ones, with the missing mass being evolved as energy.
15. 6, 12, carbon.   17. 3.1×10^{13}.

CHAPTER 25

1. Electrons, positrons, neutrinos, alpha particles, and electromagnetic waves.
3. 1879 Mev; more.   5. 280 Mev.
7. Such a decay would not conserve lepton number or baryon number.
9. A neutrino.

INDEX